Competitive Te

Mike,

Best wishes,

Other McGraw-Hill Communications Book of Interest

BENNER • *Fibre Channel*
BEST • *Phase-Locked Loops, Second Edition*
BUSH • *Private Branch Exchange System and Applications*
COOPER • *Computer and Communications Security*
DAYTON • *Integrating Digital Services*
FORTIER • *Handbook of LAN Technology*
GOLDBERG • *Digital Techniques in Frequency Synthesis*
GORALSKI • *Introduction to ATM Networking*
GRINBERG • *Seamless Networks*
HEBRAWI • *OSI: Upper Layer Standards and Practices*
HELD, SARCH • *Data Communications*
HELDMAN • *Information Telecommunications*
HELDMAN • *Information Telecommunications Millenium*
HELDMAN • *Future Telecommunications*
HUGHES • *Data Communications*
INGLIS • *Electronic Communications Handbook*
KESSLER • *ISDN, Second Edition*
KESSLER, TRAIN • *Metropolitan Area Networks*
LEE • *Mobile Cellular Telecommunications, Second Edition*
LINDBERG • *Digital Broadband Networks and Services*
LOGSDON • *Mobile Communication Satellites*
MACARIO • *Cellular Radio*
RHEE • *Error Correction Coding Theory*
RODDY • *Satellite Communications, Second Edition*
ROHDE ET AL. • *Communication Receivers, Second Edition*
SIMON ET AL. • *Spread Spectrum Communications Handbook*
SMITH • *Cellular System Design and Optimization*
WINCH • *Telecommunication Transmission Systems*

Competitive Telecommunications

How to Thrive Under the Telecommunications Act

Peter K. Heldman

With contributions by

R. K. Heldman

T. A. Bystrzycki

McGraw-Hill

New York San Francisco Washington, D.C. Auckland Bogotá
Caracas Lisbon London Madrid Mexico City Milan
Montreal New Delhi San Juan Singapore
Sydney Tokyo Toronto

Library of Congress Cataloging-in-Publication Data

Heldman, Peter K.
Competitive Telecommunications : how to thrive under the Telecommunications Act / Peter K. Heldman, with contributions by R.K. Heldman, T.A. Bystrzycki.
p. cm.
Includes bibliographical references and index.
ISBN 0-07-028113-0 (hardcover)
1. Telecommunication policy—United States. 2. Telecommunication—Deregulation—United States. 3. United States. Telecommunication Act of 1996. 4. Telecommunication—Law and legislation—United States. 5. Competition—United States. I. Heldman, Robert K. II. Bystrzycki, Thomas A. III. Title.
HE7781.H448 1996
384'.0973—dc20 96-36805
CIP

McGraw-Hill

A Division of The ***McGraw-Hill*** *Companies*

1 2 3 4 5 6 7 8 9 0 DOC/DOC 9 0 2 1 0 9 8 7

ISBN 0-07-028113-0

Portions of this book originally appeared in *Global Telecommunications* by Robert K. Heldman, published by McGraw-Hill.

The sponsoring editor of this book was Steve Chapman. The editing supervisor was Scott Amerman, and the production supervisor was Pamela Pelton. This book was set in New Century Schoolbook. It was composed by the Desktop Composition Unit in Hightstown, NJ.

Printed and bound by R. R. Donnelley & Sons Company.

To Dad

He paused along his journey to build a bridge for others. . .
that we might travel further during our watch.

Contents

Foreword

Never before in the history of the 100-year-old telecommunications industry has it been positioned to experience such dramatic and rapid change. During the close of the twentieth century, the Telecommunications Act of 1996 has opened up the voice-based communication networks. Intent on fostering change, it will substantially restructure the communications industry to facilitate the transport of fully integrated voice, data, and video information. Highly complex communication networks will merge with highly sophisticated computer applications to formulate and establish the new, exciting information marketplace.

At this point in time, it is essential to pause and formulate a feasible, realistic vision for the future, as well as a reasonable, acceptable plan to achieve this vision. It is equally mandatory to have a well-conceived, balanced plan of action to establish the proper information networks, with their many services, for both today's world as well as tomorrow's. For society to grow and flourish over the next millennium, it is essential to have an economic, universally available, switched public data/video network, enabling "any-to-any" communications over shared facilities, providing access to distributed databases, promoting full interoperability among diverse systems, and facilitating the internetworking of private networks.

With these needs in mind, this analysis will help minimize the current atmosphere of crisis and confusion that has recently engulfed many providers, suppliers, and users during this intense period of voluminous, complex change. With an identified direction, as well as an appropriate, orderly program to achieve it, we can ensure that each of our endeavors becomes a building block for establishing a competitive telecommunications infrastructure that will hopefully improve the quality of life for ourselves, our children, and our children's children.

Tom Bystrzycki

Introduction: The Missing Infrastructure

We should all be interested in the future because we will have to spend the rest of our lives there.

CHARLES F. KETTERING

In 1960, the U.S. Air Force established its exciting new data networks to interconnect its air command and missile control centers with its airplanes and launching pads, as it constructed "L" systems, networking everything together. Then in the late 1960s and early '70s, the Autovon/Autodin networks were constructed for facilitating full-featured voice and data networking across the globe. During this period, various data networks were overlaid over voice facilities, transporting low-speed-type (guard-start-message-stop) formatted packets at 60 words per minute and then at faster rates of 600, 1,200, and 2,400 bits per second.

In the mid-1970s, the Bell system tested its networks and data error rates, determining that the 2,400-bits-per-second systems would achieve error rates of 1 error in 10^4 bits of information. A bit is a binary number of 1 or 0, indicated by voltage-level shifts or differing frequencies. A character such as the letter A or Z was represented by 8 bits of 1s and 0s so that 2^8, or 256, alphanumeric characters could be successfully described using the 8-bit code. In this manner, computer designers sent characters of information between large mainframes and remote terminals. Gradually, transport facilities changed technologies from electromechanical to stored program control analog systems, enabling transport rates to successfully increase to 4,800, 9,600, 14,400, 19,200, and 28,800 bits per second, depending on coding schemes and error detection/correction algorithms. So the 1960s and

'70s blissfully passed, but no public switched data-handling infrastructure was constructed, since private networks simply leased raw transport on which to overlay their data internetworking capabilities as they attempted to interconnect their internal private networks from customer premise to customer premise.

Then the late 1970s and '80s saw the deployment of digital networks in which voice conversations were packaged in 64,000-bits-per-second channels. Voice calls were sampled 8,000 times a second and quantized into 8-bit codes denoting frequencies and power (amplitude) levels, so $8 \times 8{,}000 = 64{,}000$ bits per second of information represented a voice conversation. These voice calls were integrated with other calls so that 24 conversations were simultaneously packaged together over single lines at rates of 1.544 million bits per second.

Now you may say, "So what—these digital networks are simply a cost-savings tool for the telco again," but, in truth, what it really means is that not only voice but also data and video calls can be digitized and integrated together in number streams with error rates of 1 in 10^7 and, in some cases, 1 in 10^{11}. Thus is born the digital revolution!

These integrated transport facilities were described as highways to transport different forms of communications traffic at varying speeds. These future pathways have been embraced by the International Telecommunications Union (ITU) in its CCITT working groups, which noted the opportunity to build new fully switched, fully interconnected data-handling networks using integrated services digital network (ISDN) interconnect standards, thereby enabling users to interconnect over digital facilities at increasing rates of 64,000, 128,000, 1.544 million, 6.3 million, 45 million, 50 million, 155 million, and 620 million bits per second as they advance from twisted-pair copper wire to fiber facilities.

It is interesting to note that these facilities transport fully switched information from addressable originations to addressable destinations. Yes, just like a telephone call!

But where are they today? What happened to early public data highways that would enable different types of information vehicles to travel here and there? Where are the fully integrated digital highways for voice, data, and video conferencing? Where are the superhighways for faster and faster speeds? Some have compared the needs for the new supporting infrastructure of the information revolution to those required for the industrial revolution, when it became a major objective to build extensive roads throughout the local communities together with state multilane parkways and the turnpikes to form the needed highway infrastructure, which enabled the Model As and Ts to travel here and there at rates of 30, 40, and sometimes 50 miles per hour. Later, the national goal became one of establishing high-speed cross-country four-lane interstate facilities. (These were originally sponsored by federal funding under the national defense umbrella for transporting missiles across the country.) Now, 50 to 70 years later, it is obvious that there were indeed proven reasons for building this supporting infrastructure that enabled the fantastic revolutionary changes in society due to automobiles. Today, the new "Information Society" needs a correspondingly new, revolutionary information-handling infrastructure to serve as a supporting catalyst for change for a substantial but different form of transformation and growth. But where is it? What is it?

It certainly is not the Internet. Even though the mid-1990s have seen various forms of high-speed access servers, the Internet is basically a store-and-forward, router-type system constructed from evolutionary additions and changes to the earlier ARPAnet models, which enabled research professors to send data files and messages to colleagues. It is also not the specialized networks for privately linking together small internal buses for local area messaging with similar (and now dissimilar) computer systems. This new revolutionary infrastructure must be much more than a specialized wide area network (WAN) for internetworking dispersed local area networks (LANs) using slow-speed, complex, error-prone protocols. So what is it?

We must step back and ask what support infrastructure is needed to meet the requirements of new data, video conferencing, and multimedia applications. Relating to the analogy of the highway model, where are the fast and faster lanes? Where are the on ramps, the off ramps, the rules of the road, the traffic cops, the tickets, the courts? As we compare the similarities of cruising on the information highway and the dream of driving the American car anywhere, anytime on the automobile highway, all is not allowed. We are not allowed to drive dangerously, weave through traffic, shoot at another driver who made a mistake, make lewd gestures or comments, dress indecently, nor present phallic images on or in the vehicle. We must act in a responsible manner, or we will inevitably be arrested or fined by the law. Where are similar features and services in a public information highway? Where are the security checks, the survivability requirements, the privacy protectors, the audit trails, the password verifications? What happens if a hospital drops its point-to-point leased facilities in favor of using a public switched network? Where are the closed user group protectors, the backups, the safeguards to ensure that radiologists can indeed perform remote consultations and not have the network go down in the middle of their conversations? What about the priority override that allows a doctor to consult with another specialist or have protection to block unwanted parties from interrupting or monitoring a call (for example, a local TV station news announcer who wants to know the status of a hit-and-run injury)?

If we are going to have an information revolution, we need a real information infrastructure. This infrastructure is currently missing in the United States, as well as in many of the industrial countries around the world. Looking at the benefits of the forthcoming information age in terms of jobs, jobs, and jobs, we must first have infrastructure, infrastructure, and infrastructure.

Peter Heldman

Part

1

The Competitive Information Arena

They ride to and fro upon tigers, which they dare not dismount, and the tigers are getting hungry.

SIR WINSTON SPENCER CHURCHILL

Chapter

1

Competing in an Open Competitive Marketplace

You can't achieve anything without getting in someone's way. You can't be detached and effective.

ABBA EBAN

It began just like any other day, but for those who knew and understood what was happening, the cold February winds of 1996 were winds of change. Just 12 years had passed since the 1984 divestiture ruling that precipitated the separation of local and long distance telephone services, breaking up the Bell monopoly. Under the Telecommunications Act of 1996, the local monopoly becomes a fully open, competitive arena for all to participate and formulate the forthcoming open information marketplace. The 1996 act is analogous to the dropping of the second shoe, the first being the 1984 divestiture.

Before assessing the issues and ramifications of this congressional decision—the first rewriting of the Communication Act since 1934—let's review what happened since the dropping of the first shoe in 1984. It is interesting to note how quickly the playing field changed—not only the games changed, but also the terrain—as players moved from telephone services to real estate services and then on to remote regions of the world, offering cable services in England, packet switching in Czechoslovakia, and high-speed transport in Russia, Hong Kong, and China; providing telephone services in New Zealand; purchasing shares of Mexico's telephone company; and offering wireless cellular services in the southwestern United States. So time progressed from the wild 1980s to the somber 1990s, first expanding, then contracting, with not millions but billions of dollars at stake in each merger, acquisition, and partnership. In fact, the 1990s could be called the time of "the game of games" for the traditional telephone companies, when a multitude of games were simultaneously played in different arenas. These games included

- Splitting firms into separate parts to participate in an expanding array of new ventures so as to be anywhere, doing anything, from financial services to credit cards to cable companies to cellular franchises.
- Overlaying on the traditional voice network a variety of ad hoc services, initially to leverage off the existing structure by offering such items as second voice line, CLASS, extended custom calling, etc., and then to burden the voice facilities with data calls that have different traffic-loading impacts that affect holding time, attempts, and concentration parameters.
- Fixing whatever could be fixed of traditional telephony capabilities as work forces were reduced and reduced again by upgrading and automating support capabilities.
- Attempting to change network capabilities into a more vibrant and growing infrastructure offering a proliferation of new services.
- Filling the voids in service offerings by constructing new value-added networks for selected markets while catering to the internetworking needs of private networks.
- Setting of new telecommunication charters and directions for states by governmental agencies and councils.
- Offering "on a shoe string" overlay networks, transporting data over voice-grade facilities to interconnect personal computer users to databases, previously supported by free governmental funding.
- Broadcasting entertainment over cable, satellites, and microwave towers directly to residents.
- Adapting mobile phones for marine, automobile, and personal applications.
- Extending local, regional, and national multimedia transporting capabilities to the global networking arena not only for industrial countries, but the world.
- Using fiber to relieve growth capacity needs in local interoffice and long distance networks, as well as constructing various forms of integrated wireline, fiber hookups, and hybrid fiber-coax transports.
- Employing application service centers in attempts to create a service-based marketplace for selected applications, such as legal searches, missing children files, shared stolen vehicle files, and stock exchange information, etc., as well as Internet access menus.

We saw the Department of Justice working with and without the Federal Communications Commission (FCC) and public utilities commissions (PUCs), making policy and regulatory decisions that established the changing rules and boundaries of the evolving game. In parallel, the industry played an economically deadly version of Simon Says in following the changing decisions of the Justice Department as it slowly allowed companies to advance to more and more forms of information handling and transport. Without strong industry leadership pursuing a clear vision and strategic plan, the future direction was left to one's imagination and countless consulting reports. These reports suggested various controversial and thought-provoking views, such as the death of the Regional Bell Operating Companies (RBOCs) or the demise of three of the seven RBOCs by the turn of the century.

The competitive arena changed and changed again, as judges decided and the players pushed here, there, and everywhere. But by February 1996, as the representatives and senators looked across the Potomac and surveyed the terrain, they could see little progress in actual results—namely in the expansion of the local communications marketplace. Long distance costs had been substantially reduced, but end customers still had few new information-handling services as the local bottlenecks—the "limited last mile" issues—were raised again and again.

Congress saw that the Germans had established plans for providing businesses and residences with high-speed data and video service, the Japanese had a full broadband fiber 20/20 vision, and the French had data networks and Mintel data terminals, but Americans had their same voice-grade telephone networks overlaid with low-speed dial-up analog modems to the Internet. Urban American cities were plagued with crime and violence, and rural communities were drying up as businesses shifted to more global communities offering higher-quality telecommunications capabilities. The RBOCs and their counterpart cable companies continued to merge and partner in arenas outside the RBOC regions, such as in Europe and Asia. So, where was America in the global telecommunications game? Where was rural America in the American telecommunications game? Many began to ask questions: What is happening to our telecommunications capabilities and the supporting infrastructure for American business, for homes, schools, hospitals, and for our society? Where are we going? Where do we want to go? How can we get there?

Clearly, America still lacked the robust communications infrastructure needed to support not only data but also video and multimedia applications in the exciting array of information services and applications that would become the new products of the information marketplace. Thus, America was confronted with a two-part problem:

1. How to establish the new communications infrastructure.
2. How to ensure that once the infrastructure was established, a host of exciting information services for new users would be provided at competitive prices.

Unfortunately, as these questions were debated and solutions were proposed, it became apparent that Congress was consolidating issues to treat everything as a singular problem to be resolved by a singular solution—a free, competitive marketplace that would, over time, eventually establish the right infrastructure. This resulting conclusion was one that any student of history or elementary economics could see was not necessarily true. For example, the significant growth of railroads occurred only when large tracts of land and tax incentives were granted as incentives, and the interstate highway system was deployed only when built by federal government financing. To date both railroads and highways are still supported by federal money, as well as other subsidies and tax incentives.

Congress focused on one and only one message, based on its desire for more and more competition. Simply stated, it said that what we need is an open competitive information marketplace, one in which anyone can participate in the local arena. We need to foster competition in the intrastate long distance market as well. We need to remove barriers from RBOCs competing with new information services in their local markets instead of driving them to remote regions of the world. If content manipulation and processing are their desire, let them do it. If they desire long distance end-to-end service, let them do it. But also let the current long distance service providers provide local end-to-end customer service. If price wars brought down the cost of long distance service, then let price wars bring down the price of local telephone and cable services. Let the cable companies offer telephone service, and let the telephone companies offer cable services. If the RBOCs have had to go to foreign manufactures to get their products, then let them manufacture their own products, as long as this arena is competitive. If product suppliers have no buyers for their products in the local arena, then let them merge or partner with alternative transport providers (ATPs) or competitive access providers (CAPs), or let them become alternative exchange carriers (AECs) or value-added resellers (VARs) in the local arena. So on and on the talk went, and the ideas flowed.

Unfortunately, as discussions intensified and heightened, the possibilities for the future grew and grew in the eyes of the beholders. Few, if any, of the Washington gang actually defined these new possibilities. The reality of this "new competitive arena" was left for the players to decide. Congress believed that the competitive arena would bring down the price of local telephone and cable service, even though it acknowledged that these prices would likely rise in the short term.

The lawmakers believed they must not weaken in their resolve and must let the free market decide. This process was similar to academics going to Russia and preaching the advantages of a free marketplace, letting the best win and the losers fall out. They missed the other part of the answer in comparing this approach to post-World War II progress—namely the Marshall Plan—in which Americans spent billions in Europe to ensure success, nurturing fledgling economies until they were up and on their feet and establishing legitimate governments that were willing to lay out the proper game for the future.

Unfortunately, if we do not learn the lessons of history, we are destined to repeat them, and such was the case on that cold day in February. There were no government billions to be spent to ensure success, to achieve a larger vision like the interstate highway system. They simply followed the academic community approach of promoting freedom of the marketplace. Whatever results, so be it! But where was the vision? Where were the resources and incentives? Where were the plans, the commitment to construct a solid robust information infrastructure, to successfully support this free-blossoming open competitive information marketplace?

With these thoughts in mind, let us look at this new, open, competitive information marketplace and review the Telecommunications Act of 1996 (Fig. 1-1). Let's consider the new game with its new rules, its new players, its new roles to play. Let's specifically determine: What is this competitive arena? What is required to really establish its competitiveness? What is its supporting infrastructure? What is this information marketplace? What does it mean to whom? How is it open? competitive? sustainable? supportable? survivable? What are its key

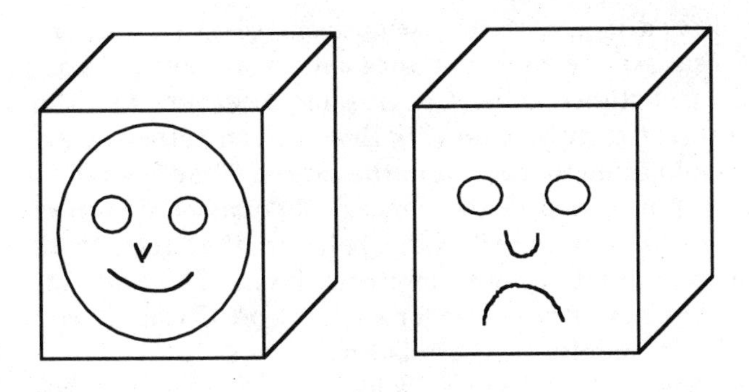

Figure 1-1. A new game . . . new rules . . .

provisions? Where are its opportunities? dangers? threats? limitations? It is time to identify and decide: How does it affect me? How are others reacting? What does it open up? Where does it open up? What can I do with it? What issues need to be solved? What challenges need to be overcome? What are the new boundaries? new regulations and nonregulations? new legislative goals? What about wireless/spectrum usage? What are the building blocks for the future? (See Appendix A for a survey of the key provisions and requirements of the 1996 Telecom Act, including the checklist that local exchange carriers—the LECs and RBOCs—must meet to offer long distance services.)

Let's step backward and take a fresh look at the competitive information arena and its open information marketplace and take time to reconsider what's happening. Let us consider the full range and scope of the issues, the visions, and the views of the players, their perspectives, potentials, possibilities, and probabilities, as well as the key concerns, complexities, and considerations, the major aspects and facets, their extent and degree, their resulting impacts and ramifications, and, of course, final conclusions and recommendations.

Competitive Arena

In summarizing, the purpose of the Telecommunications Act was to

- Eliminate the telco/cable cross-ownership ban.
- Take back policy from the Department of Justice and return it to Congress.
- Transfer from a monopoly model to a free-competition model.

Congress had watched long distance costs decrease by 50 percent while the costs of local service short distance calls were, in some cases, higher than long distance calls. Congressional leaders felt the need to change corporate culture, believing that the competitive marketplace benefits would come to fruition in the future. They desired to address the "law of unintended consequences" of "monopoly powers abusing their monopolistic powers." Many believed that regulators should avoid trying to regulate by customer complaints. The lawmakers believed that as the industry moves forward, the FCC would need to manage and the PUC monitor this transition period.

In requiring the opening of the local network for all types of competitive uses, lawmakers believed that the cost of basic voice service would go down eventually, while the initial cost might go up as key business customers were lost to competitors, throwing traditional rate-balancing parameters off kilter. They also believed that the initial cost of the feature-rich packages of new services might go up until competition arose to bring the cost of these services down as the players jockeyed for positions with joint ventures. They assumed that, in the face of competition, traditional network providers would be motivated to upgrade their local plants with feature-rich packages of new network services and capabilities to promote growth. They promoted the view that faith in the marketplace is needed as players are provided an equal opportunity to compete with better services, different services, and quality services. In fairness, we should also point out that a few knowledgeable lawmakers noticed that a considerable amount of local plants would need to be upgraded accordingly to promote growth and competition.

Spectrum

These issues of growth of new services and trust in the providers to do the "right thing" are captured in the issues and concerns noted in the release of government spectrum. What is the public policy perspective as broadcasters' media go digital? Considerable spectrum is released in the shift from analog to digital. So what will be needed for high-definition, high-resolution television, and how much for PC-television, since there is some overlap of the entertainment and information industries? Several issues need to be addressed concerning the use of spectrum for broadcast and other purposes, such as exactly how the digital revolution will take place as the programming shifts to digital, requiring users to buy digital TVs. How much time will be allowed for this transition? Similarly, will free television diminish as broadcasters go digital? How will the price charged for the new digital spectrum affect this decision? What about the public-interest requirement to provide programming for the common good as broadcasters gain access to spectrum?

In a congressional forum, only two months after the signing the Telecommunications Act into law, congresspersons offered the following questions as they attempted to understand the complexities of opening up spectrum, which is just one piece of the open marketplace: Is this just another way of taxing Americans, since the government makes billions on the sale of spectrum and users must then go out and buy new TVs and lose previously free programming? Can't we have it both ways? How do we finance public television? To survive in the future, broadcasters must go to digital for a better-quality, high-definition picture, but how will they pay for the digital? Will free TV be eliminated? Is the role of Congress simply a stop sign or even a cheerleader?

Many believe that they must inject themselves into the issue to protect and encourage the full availability of competing technologies—where public airwaves belong to the people—especially noting the possibility of dozens of derelict TVs in closets, along with Betamax video systems and 8-track players.

Where are the standards? The intensity and complexity in the standards arena is considerable, as seen in the issue of high-definition television (HDTV) standards. How will networking standards be established and who will do it as various players decide among the conflicting issues of wireline broadband fiber to the home versus wireless satellite digital broadcast systems (DBS) versus microwave multichannel-multipoint distribution systems (MMDS) versus upgrades to the CATV systems alternatives?

Some believe that broadband fiber will take one or two lifetimes to deploy, while wireless systems will take only one or two years. For them the issue is not fiber to the home, but digital to the home, any way, the fastest way, with whatever types of services can be achieved. Alternatively, numerous countries have realized that fiber provides a robust, secure, interactive service like no other. They see the need for fiber in the local arena—for everyone on the neighborhood block—without blockage or concentration. So the "any deployment" strategy is not a cure-all, because it usually offers finite capacity to the home. Views of "any deployment, though limited" are not the answer, although America loves to jump quickly into the action, even if doing so means throwing it all away after a short time and doing it all over.

Unfortunately, an all-encompassing vision must be established with a plan to go forward. Networking standards must first be set before application development begins in earnest; otherwise, the open competitive arena becomes something on the order of the old Roman circus or a traveling three-ring circus or a cheap sideshow (Fig. 1-2).

As the focus shifts to the telecommunications revolution, what role does digital wireless play? What transport will interactive computer services require as consumers demand better and better quality, service, price, and capacity? In this regard, it is evident that legislation has a long way to go to fully understand the differences between wireless and wireline offerings to adequately chart the future of telecommunications. Congress simply left it to the prevailing forces and factors to shape and position an open and competitive arena. Did it create an environment in which success (or anarchy) is viable?

Let's take a harder look at this open and competitive information marketplace. Let's remove the hype and hoopla and try to grasp a crisper and clearer understanding of it so we can appreciate the differences between the various forms of wireless and wireline service offerings that indeed really make the big differences for the customers who use them. In so doing, we will better appreciate the purpose and need to set a national goal for fiber to the home by 2020, not in multiple lifetimes; the ramifications of an "anything goes" evolutionary marketplace; the power of the current copper plant, if groomed properly, to phase offerings in a sequenced pattern of phased

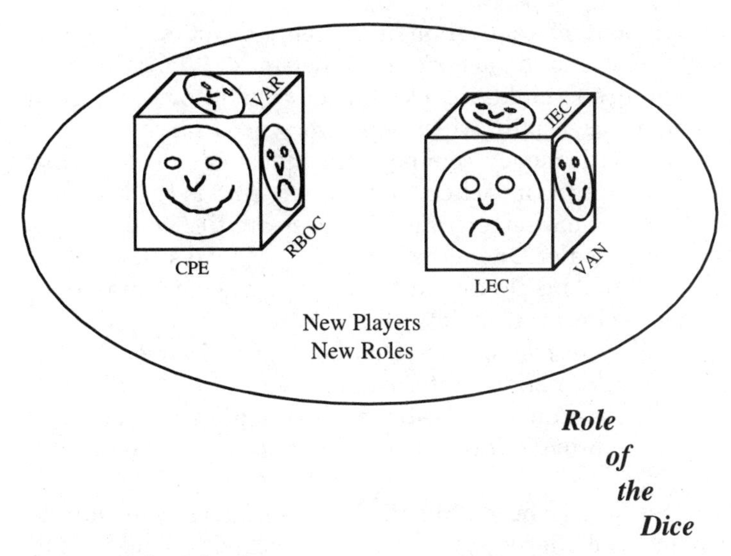

Figure 1-2. Role of the dice.

parallel overlays, achieving greater and greater transport capabilities and capacities; and the appropriate and timely use of the spectrum in sync with wireline services.

Competition

It is important to remember one particular human trait: Few people take the time to do extensive planning.[1] Given a choice of performing an easy short-term task or a long-term complex task, most people will roll up their sleeves and address the more easily defined short-term tasks first. They will address the easier opportunities, especially if the financial rewards can be soon forthcoming. So, in a very competitive environment—especially one with few rules or goals for establishing long-term sustainable structures that support future offerings—the choice is to pursue the easy short-term tasks and not waste one's time and energy. As one executive put it, "Why should I build something to make my successor a success? Wall Street and its stockholders want immediate reward for their bucks and that translates into quicker offerings to the marketplace and immediate revenues, with less costs—all of which means more for me in bonuses and stock options."

---

[1](One can't help but recall the story of two boys starting out on a day of adventure who didn't take the time to plan, but instead elected to flip a coin at every intersection and see where they ended up at the end of the day. They ended up two blocks from where they started.)

In this environment of rollover management, buyouts, shootouts, and poison pills, where is the keeper of the network? Who cares what is built for our children's children, let alone our children? What is the role of government, especially when legislators need large funds for media campaigns, with money coming from thousands of PACs that want their firms to get in the action first? Again, who is the keeper of the network? Is it the marketeer whose bonuses are tied to the yearly doubling of revenue goals? Or the technologist, who says "It's not my decision. All I need to do is figure out how to squeeze more out of this already squeezed medium! If society wants to walk away from fiber, which has the ability to transport not millions, not billions, but trillions of pieces of information—interactively both ways—and go to cheap, quick mediums that have faster penetration but are substantially less capable to provide future feature-rich services, then that is their decision, not mine."

The technologist simply needs to satisfy the marketeer, usually by overlays of compressed offerings, for example, overlaying voice calls on the telephone network to order downstream movies on the cable network and then overlaying quick-selection messaging upstream transport on the cable system. Of course, there are substantial limits to these types of facilities for long-holding, large-bandwidth conversations. In time, marketeers will push the envelope in subsequent periods of hype and promotions as they introduce more complex and extensive requirements, like a high-quality interactive videophone for grandma. Grandma can call her grandson over the upstream bandwidth of a cable network, but there is only one problem; if Grandma is using the medium for high-quality, high-resolution terminals, she is probably the only one in the neighborhood capable of calling until she frees up the system. So the other grandmas on the block will most likely only be able to listen in on her conversation with the appropriate box code overrides.

The issue of competition needs to be understood, especially as the competitive marketplace providers offer numerous parallel networks. How many times can we plow up the street and jackhammer the concrete sidewalks, inserting blacktop patches that become potholes as the information highway destroys the existing automobile and pedestrian highways? What happens after billions of dollars are spent to make quick millions, as everyone gets a "piece of the action," but the pieces get smaller as the pie remains the same (or only slightly bigger) since we are using dead-end technologies? What happens after limited technologies are pursued that offer only one or two services and crumble under their own success? The technologist's statement might later become "I didn't know that you wanted to do that with this. It is not designed to do that. If you wanted to do that, you should have used a different technology in the first place!"

It is sort of like looking at the old mines that are still dumping raw minerals into the streams in Colorado—who cleans up the mess? On the other hand, different technologies can play useful roles in different locations; for example, wireless can play a major role in the remote regions of a bankrupted country. But it is important to know just where the Dick Tracy watch leaves off and the sophisticated, secure, interactive, high-quality video conversation takes over. Trying to make one into the other later is nearly impossible. Each technology has its inherent strengths and limits; these must be understood and the technologies applied accordingly.

Similarly, tradeoffs must be understood when considering selling off digital TV spectrum to commercial players who have little, if any, concern for the public trust, the public good. They are simply in it as an investment for the short-term money. It has been said that if the telephone and television were to be done over again, the spectrum should be saved for interactive mobile conversations and the wireline for passive television. Prices, security, robustness, and availability objectives must be established and maintained while attempting to achieve a reasonable balance between short-term competitive ventures and establishing the building blocks for long-term sustaining structures. This is not like deregulating long distance voice service providers, who have already collected users for defined services. Once billions are spent in the local arena, constructing structures for limited or one-sided feature growth, it will take substantially many more billions of inflated money to construct entirely new structures to support the new host of voice, data, video, and multimedia applications and features that the future market requires. It can become an economical nightmare to pursue a costly patchwork approach of attempting to build new capabilities into limited networks. Unfortunately, other long-lasting repercussions and changes can also occur. For example, during the race to capture the lucrative urban markets, the rural areas may have dried up and died, large companies may have been forced to cluster in urban areas to be competitive, and the small businesses may be slowly phasing out to more competitive shared consortiums that have private network syndicates, enabling them to compete globally.

Open

Open is another term being championed as the key to a free market. This notion of openness translates into access anywhere, at any point, to the traditional network infrastructure, with little concern for the resulting destruction of its operational ability to support not three or four but hundreds of thousands of new providers of different services with different characteristics and attributes. This destructiveness is even more profound when using structures that were not originally designed to be opened at all, let alone opened anywhere, anyway. *Order* and *timing* are the words that are missing: "an open, orderly, and timely manner." This openness comes at a price—who will pay for all this openness?

The issues of openness need to be considered in light of new offerings by traditional RBOCs or LECs. Does openness mean that every new technology they use and deploy must be designed for openness? Do their competitors have the same requirements? For example, as a cable firm begins to offer telephony, the traditional phone company may request similar interfaces to the cable provider's networks anywhere, requiring hookups to pass their phone messages.

A specialized service provider may obtain access to a traditional provider's database and have calls routed to its platform, but some other service provider sees the offerings, establishes similar offerings, and requests that a choice menu be established by the local telco provider so that customers will know that it is also available to provide competing services. Soon, both specialized service providers go under, because this small "electronically shared" market is unable to support them. Firms in this new electronic marketplace have much greater risk as many new vendors rush to the newly discovered electronic services goldstrike and deplete it. But who pays to clean up the residue? Who pays for the traditional providers to establish the menus and for the network providers to lay in new facilities to route the calls to the now-defunct service provider? Who pays the cost for removing or maintaining the dormant network or the cost of leaving idle facilities in now-empty buildings? Who pays when the network goes down due to abuse in excess forms of traffic?

In the past, these facilities were protected under the "rules of the road" of the traditional five-level traffic-sensitive hierarchy. Now bursty or extremely long-holding-time uncontrolled traffic can enter at any point. How far can the Erlang and Poisson traffic-engineering tables for the existing structure be stretched to address the new applications' uses and abuses of the existing infrastructure? Who pays for the new network that is better constructed and suited to meet this new requirement of "competitive openness"? What induces the RBOCs and LECs to build this "openness," which they themselves might not be using, when issues of resale and unbundling become economically illogical for the LECs as they try to comply with FCC and PUC rules and regulations, as inconsistencies and disagreements are presented to the appeals court. Indeed a complex game! (See Appendix A.)

Information

Just when we think we have an understanding of the complexity, we begin to find out what the last two words (*information* and *marketplace*) mean! Let us begin with information. No longer are we simply discussing the current voice network, which is already in place—fully visible and available for cutting up and opening up—we must consider the new forthcoming multimedia voice, data, and video world. In fact, just the data world will be an exciting challenge, as players wrestle through the various technologies, enabling circuit, packet, frame, and channel movement of data in bursty variable bit rate and continuous bit rate forms. Connection-oriented or connectionless formats will be used for datagram-type traffic over permanent virtual paths or switched virtual paths as each of the technologies employs various call setup, routing translations, and takedown protocols, as well as complex error detection/correction mechanisms.

As they say, “too many cooks spoil the broth”; so it is in the standards arena. This is especially true if we are in a “bull in the china shop” situation, when political bulls have invaded areas in which they have little understanding of the damage they may do, especially when they bring in their wide-eyed views of immediate openness to any and all information markets. On the other hand, traditional carriers have successfully resisted real movement into new information services. They clutched to the basic telephone services, typically offering only second line, caller ID, custom voice calling, and voice mail as enhanced services while basically ignoring the data and video world. So like it or not, following the passage of the Telecommunications Act of 1996, the tired old voice structure and the tree-like cable structure will be inundated with requests to handle new requirements for all forms and types of data and video offerings integrated together in multimedia offerings.

Marketplace

So what is the telecommunications information marketplace of the future? Well, for one thing, as the Internet has demonstrated, it is online databases, enabling users to access, search, page, browse, retrieve, delete, store, and present information electronically. Such information is in the form of voice, voice and data, voice and video, graphics, text, and fully integrated multimedia—voice/data/video—communications. This information needs to be transferred in the mode of asymmetrical broadcast (one-way), as well as in an interactive (two-way), symmetrical manner. It is an ongoing dynamic marketplace, continually overlaid on the existing infrastructure, which is predominantly in a switched-star configuration based on an analog/digital voice architecture. It is augmented by customer-premise star-type voice PBXs as well as shared-bandwidth facilities of private ring and bus structures, enabling higher speeds of data transport. Parallel to these endeavors is the tree-and-branch nonswitched structure using coaxial cable to broadcast 750 MHz or so of downstream frequencies of broadcast entertainment, with the potential to expand to 45 million or so frequencies to handle upstream asymmetrical information interexchange, thereby passing a customer request for a movie, shopping catalogue, or game to be selectively transported to the customer's box receiver.

With these various types of network infrastructures, marketeers play with their various types of application possibilities, attempting to deliver tailored services to their particular application. As the marketplace blossoms with more and more service providers and network providers and hopefully more and more users, their success swells the network traffic, providing more and more revenues for growth and expansion. At least this is the academic view of a free enterprise marketplace; specifically, it is the type preached by economic advisors from academia to Russia. Unfortunately, as we look back on what happened during the transition phase, we see that the objective was right, but much was lost in the implementation.

What should be the role of government to make all this happen in a legal and orderly manner? Or is it left for all to fend for themselves? In Russia we saw the black market swell as it bypassed the slow process of removing existing roadblocks in the legal open marketplace. Unfortunately, if the process of change is misunderstood or mismanaged and moves forward too slowly, while at the same time it opens up everything to anyone, the result is crisis and chaos with only the strong surviving over the weak. In the new free market, the Russian Mafia now controls major portions of the local economies, having made inroads into every major sector.

In the American marketplace, this process is called cream skimming. LECs and RBOCs can assume the position that since they are no longer monopolies, they are no longer responsible to deploy "universal, ubiquitous service." They may only cater to the needs of 20 percent of the customers, from whom they obtain 80 percent of their revenues (the old Marketing-101 rule). So much for rural towns and small businesses. There's a story about one marketeer who was brought in to streamline an RBOC to play in the new game, who said, when given the possibility of offering a new product to service the overall public, "I am only interested in disposable technologies that provide services for our greatest growth customers [the 20 percent], thereby guaranteeing a high and instant rate of return. I only cut big deals with big players for big opportunities." So it goes. So where does all this lead?

The FCC, per the Telecommunications Act, is left to monitor, assess, and legislate.[2] It has its work cut out for it, if indeed it also wishes to protect the basic telephone network services users who may not be offered new information services and may indeed lose their existing lifeline voice telephone services. Watching the endeavors of the media,

---

[2]See Appendix A for the FCC response to the Telecommunications Act.

one would think that the world revolves around that industry, as it informs the public that the information highway exists so that it can offer 500 channels of "questionable value" entertainment and as it gleefully discuss how the cable industry will merge with the old traditional telephone industry to create an expanding cable entertainment industry—taking over the world. Sad to say, the business community, the education community, the trade community, and the government agencies have been ignored as part of the future users of the information highway. They have watched what is happening, painfully noting that their means of communication for the most part remains a twisted pair of wires supporting the dial-up voice telephone, which now comes in numerous colors and shapes, or a local LAN offered by a computer networking support center, interconnected over point-to-point facilities to selected locations, or a crumbling Internet, burdened with its own success. As one businessman asked, "What is the Act and when will it make services available in my area to save our rural town from extinction? Ask the FCC or PUC! We are fed up with waiting and waiting and waiting."

So what holds up this competitive information marketplace? What is the supporting bridge on which traffic, both public and private, traverses? It is quite apparent that simply dissecting and dividing the existing voice-grade leased-line/trunk facilities does not really achieve much in the way of developing a growing and expanding universal marketplace. It simply ends up with a lot of "quick buck" players dissecting the remains of the dying tele-dinosaur. The old slogan that you have to "spend money to make money and you have to spend big money to make big money" has been interpreted as a Wall Street banner for pursuing mergers, acquisitions, and partnerships, with few, if any, new products or new infrastructures. Many firms are streamlining operations by cutting costs or removing research-and-development (R&D) endeavors (especially long-term ventures), downsizing the work force, rightsizing the work force, or using outsourcing providers, but there has been little talk or effort on new long-term endeavors.

One marketeer was heard to say, "Technology is moving so fast, it is hard to keep up with it, so each day simply pick off-the-shelf items, put them in, and then throw them away—build a 'throwaway network.'" Does this mean we should eliminate setting standards for the more complex, survivable, growable infrastructures and only concentrate on here today, gone tomorrow endeavors? This appears to be the game for some. What is the new sandbox, the new arena, the new information marketplace that enables all players to play their games, which does not disintegrate by the weight of its own success? Many ventures work for a few users, but when everyone becomes involved, they collapse, sort of like the Internet, which proved the case for access to databases—especially if that access is inexpensive or even "free" for some. The Internet did demonstrate how information can be helpful to a world community, as the World Wide Web made it easy to send e-mail across the world. But what will hold it together? It is simply an ad hoc router-type network for accessing distributed databases. As more and more users use it, it demonstrates traffic delays similar to the morning rush hour on L.A.'s freeways. Don't look to its initial design for robustness, security, survivability, safety, or privacy—all the key needs of a robust growing marketplace. So what's next; where do we go from here (Fig. 1-3)?

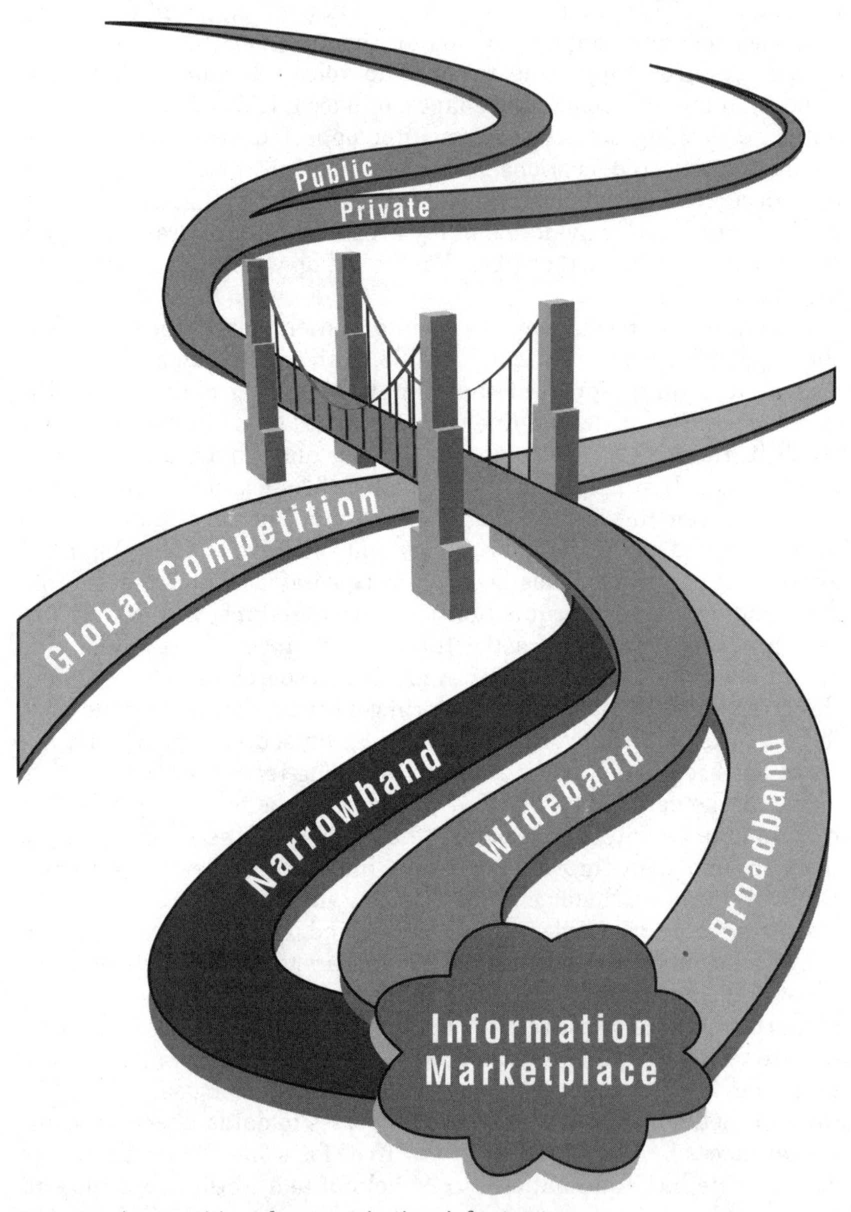

Figure 1-3. A competitive telecommunications infrastructure.

New Players, New Game, New Marketplace

The Telecommunications Act of 1996 has opened a door to a new society, a new world. It is indeed an opportunity for exciting things to come. As we look at the visions and views of the various players and their impacts and ramifications, it is quite apparent that what is needed now—before it is too late—is a new model, a blueprint for success. This blueprint must establish a supportive information infrastructure designed to foster openness and competitiveness and to facilitate the exchange of all forms and types of information, creating a universal, ubiquitous marketplace. It is not only a network for the select few, from which 80 percent of current revenues can be quickly obtained, but one that enables the growth and expansion of society—rural and urban, local and national, and international. Only then, as the new sandbox expands and changes to embrace all the new players in the new game, does it become the right supporting, long-lasting infrastructure. It should not be designed and established for short-term, limited, narrowly focused markets, but rather for an expanding and blossoming information marketplace, creating jobs, jobs, jobs, as we shift from a society based on heavy industries to one based on "light" wave information.

Chapter

2

Different Views, Different Perspectives, Different Paths, Different Steps

You take the high way,
I'll take the low way,
and together we'll meet in Tipperary.

In assessing the Telecommunications Act, there are different views and different visions, as different players play different roles, having different paths to take and steps to make. So let's pause here to take a deeper look at these differences to see how the different players play different games.

On January 1, 1996, the first wave of baby boomers in America (the large block of post-World War II babies) turned 50 years old. The working force turned grayer with a higher number of dependent workers to support, thereby raising the issue of how those working and competing within a global economy will be able to generate sufficient revenue to meet their increasing support needs. By 2050 the United States population, now 260 million, will grow to 400 million, as populations continue to increase throughout the world. Hence the following questions need to be addressed: How will future workers support future societies in an information age? What form of telecommunications will create what type of workplace?

Will all these 400 million-plus people be living in urban city highrises with no one in rural communities? Where is society going? Where will it end up as different players take different paths to where? The Telecommunications Act and its related bills, together with new FCC/PUC regulations, provide the industry players a singular opportunity to reassess their previous strategies and activities while refocusing their attention on defining future roles and endeavors. Viewing the situation from the different vantage points of the various players is analogous to looking through the lens of a camera, where one may wish to focus on the slow blooming of a flower or take a faraway view of a race horse rounding a bend. The manipulations of the aperture and f-stops are adjusted to the eye of the photographer; so it is for the local exchange carriers (LECs), interexchange carriers (IXCs), and value-added networks (VANs) as they focus on different facets of the new information arena through the lens of the new Telecommunications Act (Fig. 2-1).

Every arena must be reassessed. For example, as discussed in Chapter 1, new spectrum opportunities exist concurrent with the bill for digital TV broadcasting. The government released new frequencies, enabling owners to expand channel capacity as much as fivefold. This situation, being quite lucrative for those who owned the frequencies, prompted the "right and proper" questioning of how the networks would fulfill their public-interest obligations as they sought free access to these new frequencies. The broadcast industry was forced to reassess its relationship to the public in terms of its members' status as public trustees by clarifying how they would fulfill their license obligations, such as how they would deliver the benefits of free public TV programs and political advertising. In this continuum of change and opportunity, everything needs to be reviewed and reconsidered as the camera lens focuses and refocuses on the various different arenas (Fig. 2-2). As the LECs and RBOCs change, with some expanding and others fading away, many new players will join while others regroup in traditional and nontraditional roles.

Figure 2-1. Different views, different perspectives.

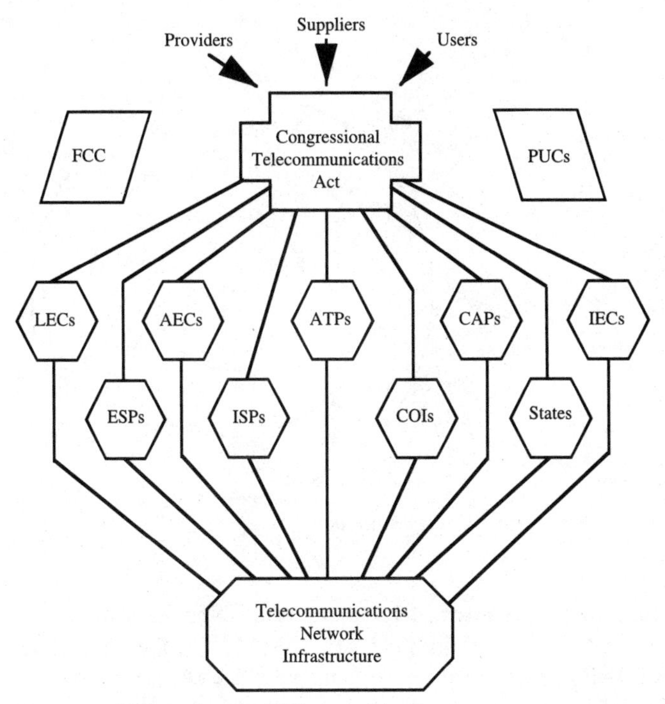

Figure 2-2. Different goals, different objectives.

Different Players, Different Perspectives

Some have the view (Fig. 2-3) from the perspective of moving forward from a TV network baseline. In this scenario, based on the broadcast entertainment model (Model E), the services of a telephone network are migrated to the broader asymmetric bandwidth of the cable industry, thereby expanding the cable endeavors. This TV-based approach is subsequently expanded by promoting multichannel-multipoint distribution systems (MMDS) using microwave broadcasts to spray a city with frequencies or using local multipoint distribution systems (LMDS) of cable/fiber feeders to deliver information to broadcast distribution nodes, covering 1-kilometer or 4-kilometer cells. Here the major objective is one of delivering more and more broadcast entertainment, as well as other selective entertainment (movies, games, etc.), using the landline telephone network for upstream requests.

Similarly, direct broadcast satellite (DBS) systems distribute entertainment programs to small-aperture antennas throughout the rural and urban countryside as an alternative to cable. This technology has less opportunity for local programming than MMDS, since DBS covers the entire nation while MMDS has a more local, 50-mile radius.

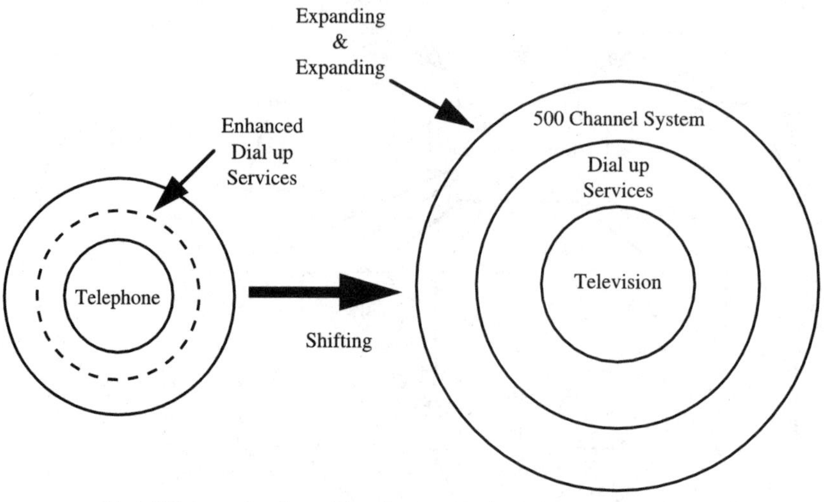

Figure 2-3. Model E: base for the cable television industry.

With a different market in mind, the traditional carriers, the LECs and IXCs and their alternative transport providers (ATPs), competitive access providers (CAPs), and value-added networks (VANs), focus not only on the use of the existing copper plant to provide traditional telephone services, but enhanced and expanded facilities, augmented with fiber, to support dataphone and videophone, as they offer symmetrical interactive services to businesses, hospitals, education groups, and industrial complexes, as well as work-at-home users. Hence, their perspective of evolving and revolutionizing the wireline infrastructure, supplemented with overlays of wireless telephone services, is noted in the telephone model, or Model T, displayed in Fig. 2-4.

Different Goals, Different Objectives

Let us begin with the congressional view of the Telecommunications Act by noting the issues and complexities, indicated in Fig. 2-5, in terms of the following observations, concerns, and ramifications.

Congressional view

The goal of legislation was to foster competition in every facet of the local arena. Its main provisions were to

- Open the local monopoly, including interconnection, resell, unbundling, and collocation.
- Require LEC compliance by meeting a special checklist before providing long distance interregion, intraregion, and interLATA (Local Access and Transportation Area) services.

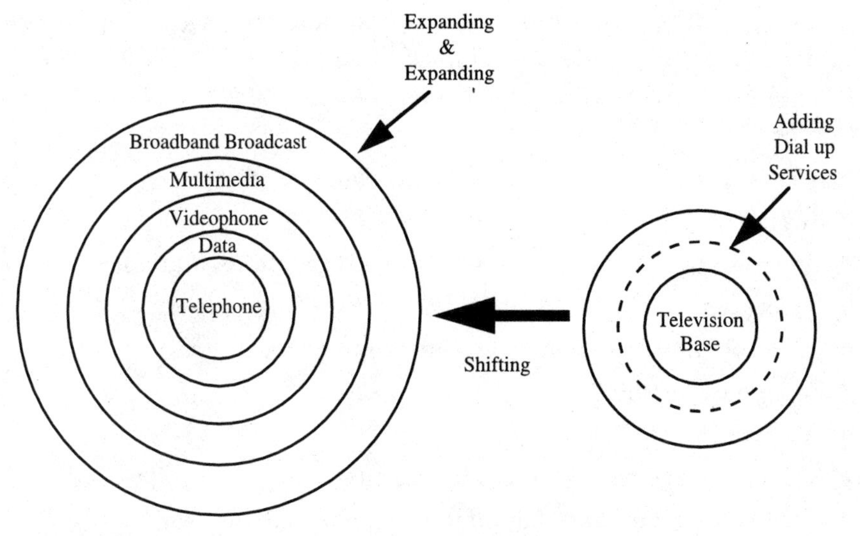

Figure 2-4. Model T: base for the wireline telephone industry.

- Enable phone companies to become cable companies in-region and vice versa.
- Obtain local universal information services from multiple providers.
- Restrict obscenity and violence.
- Allow LEC product/service manufacturing.
- Return IXCs to the local services arena, enabling "origination to termination" IXC transport from local customer to customer.

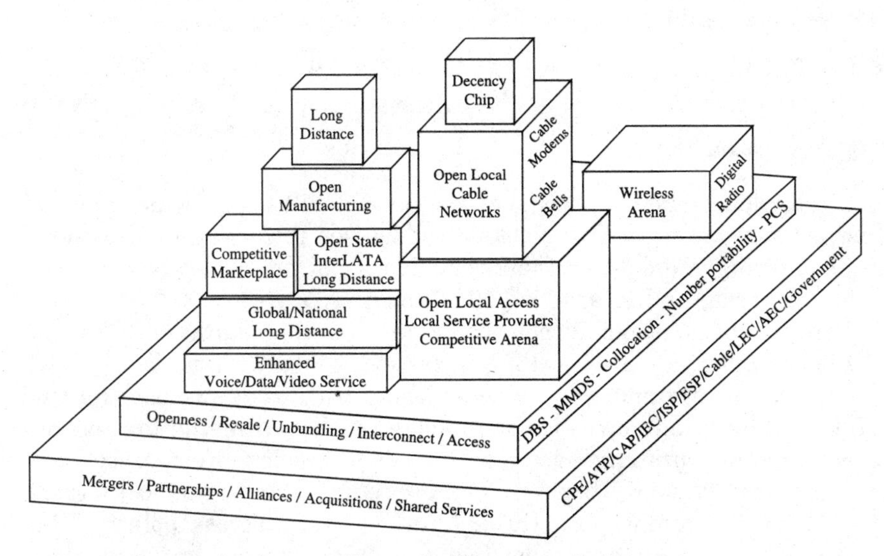

Figure 2-5. Congressional perspective on the Telecommunications Act.

In reviewing this legislation in terms of these objectives, it is quite evident that this is indeed a complex arena, as we watch the various players attempt to apply their particular twist and spin as they sift through the following issues:

- Access
- Resale
- Unbundling
- Collocation
- "Equivalent" collocation
- Foreign ownership
- Spectrum sales
- Appeals court's decisions and judgments
- Unregulated hardware manufacturing and software services
- Cable Bells
- Decency
- FCC rulings and regulations
- PUC monitoring
- Interconnection
- Open LANs

Therefore, in stepping back for a moment, it is interesting to observe how the marketplace is developing over these initial phases of the "information millennium." Players have found that in attempting to meet the goal of opening everything to anyone, anywhere, anytime in a random, uncontrolled manner, it is impossible, both economically and technically, to protect the network from

- Burnouts
- Brownouts
- Blowouts
- Nonrecoverable outages
- Overloads
- Severe delays
- Congestion
- Emergency crises/confusion
- Blockage

Concentration
Contention
Conflicting interactions
Clashes
Limitations

Protection problems become especially difficult over a wide range of issues such as providing blocks of addresses to numerous providers and offering number portability to blocks of these newly assigned country, area, and local office NNX codes (N = 2–9, X = 0–9; where quantities of unassigned codes cause considerable address scarcities). Other complex issues—reselling "what services" at "what lower prices" than the LECs? unbundling where? when? with what incentive for the LECs? collocating how?—cause problems such as bringing the central switching network office down with radio frequency interference from system to system, with power bus grounding shorts, and with overloads (both attempts and throughput) on weak access points. Who pays for constructing parallel facilities to the customers when they drop access to one provider over another as price wars arise? How are other difficulties resolved? Issues such as delays in interconnectivity and access cause disagreements and conflicts that result in long periods for FCC/PUC reviews and appeals court decisions as customers wait. What about the numerous network integrity, security, privacy, robustness, and survivability issues that will arise here and there and remain unsolved, unresolved, and so on and on?

It is quite evident that at the time the act was passed, there was no game plan to protect the local unprofitable areas from obsolescence as the initial game shifted to the more lucrative urban markets, leaving only voice in the rural areas and the new multimedia services in the urban areas. As initial players mainly concentrated on the voice world, cutting up this pie and overlaying more and more ad hoc services while believing that the pie was getting bigger, they were in fact simply pushing the existing voice network to its limits, straining switch and transport capabilities and capacities. In reality, the numerous opportunities of interactive data, videophone, and video conferencing were overlooked and missed. There were few incentives for offering fully switched fiber throughout urban communities as players attempted to squeeze more and more from local voice networks.

Soon the game turned to cherry-picking the Fortune 500 firms, as players carved up the local business communities with new alliances between IXCs and ATPs. (Soon after the act was signed, several IXCs began to cherry-pick the local access market; for example, AT&T bid to let business customers in 70 cities connect direcly with AT&T's long distance service through alliances with five ATPs: American Communications Services, Brooks Fiber Properties, Hyperion Telecommunications, IntelCom Group, and Time Warner.) Next, specialized data-handling multimedia network providers entered to address specifically targeted markets with specific client/server services. So the plot thickened as Congress continued to reassess the situation, stepping back into the arena with new laws, new rules, new legislation, and new roles to play.

LECs, RBOCs, and independents

LECs, RBOCs, and independents have different perspectives as they assess the advantages and disadvantages of the new arena in terms of short-term versus long-term goals. Unfortunately, or fortunately, depending on one's perspective, Pavlov's dog has been well trained, and local providers work under the proven technique of "connect a user to the network and get a coin for every call." It doesn't matter if the call is a telephone call or a television call. So count up the numbers of users, peg-count their call usage, monitor their calling patterns, construct a network server to handle their traffic, and take their money to the bank.

$$\text{Profit} = \text{Revenue} - \text{Cost}$$

This has been a simple formula for success as more and more players addressed POTs—be it "plain old telephone services" or "plain old television services." They counted the number of customers lined up for telephone or TV. So "talk or entertain" translated into the easiest and quickest, most economical pursuit, as some traditional LEC and IXC providers shifted focus from the unglamorous telephone to the more glamorous (and supposedly more lucrative) Hollywood ventures of dial-up movies and games, while others reconcentrated their efforts to provide selected offerings to selected communities of interest, such as the business community. Some saw entertainment as the major opportunity (Fig. 2-6A), with their new networks delivering "choice of selection" of TV programs and movies. These offerings were sometimes accompanied by Internet access services. Some of these players had little inclination to expand the voice "tele" network services, especially in the rural environments; rather they simply migrated basic telephony into upstream cable facilities.

Others view the world from the "tele" operational business perspective (Fig. 2-6B) of expanding the telephone network to include interactive dataphone and videophone opportunities, enabling the business and residential communities to access global and national carriers over fully integrated voice, data, and video multimedia information-handling facilities. They expand offerings over an ever-increasing range of technical breakthroughs while moving from narrowband to wideband to broadband fully switched transport mediums.

In reviewing Figs. 2-6A and B, we need to further assess these opportunities, perhaps with the following considerations: As LECs spread their resources to meet the competition while they attempt to play in many arenas, usually with limited sustaining resources, they have

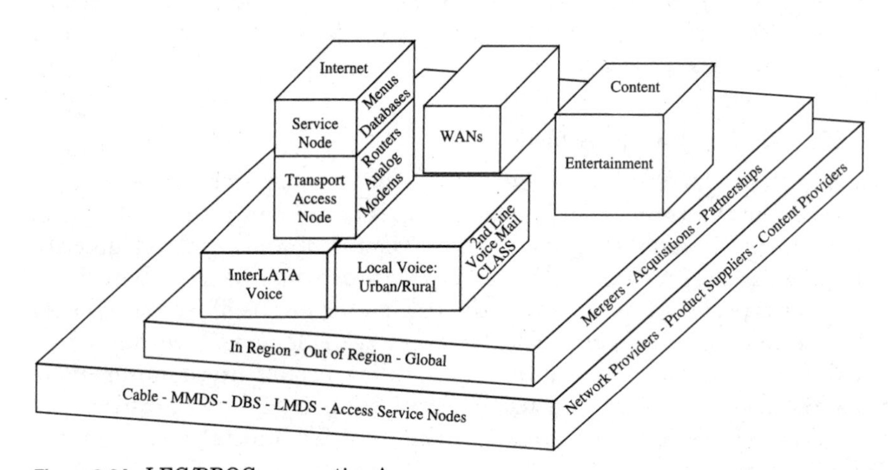

Figure 2-6A. LEC/RBOC perspective A.

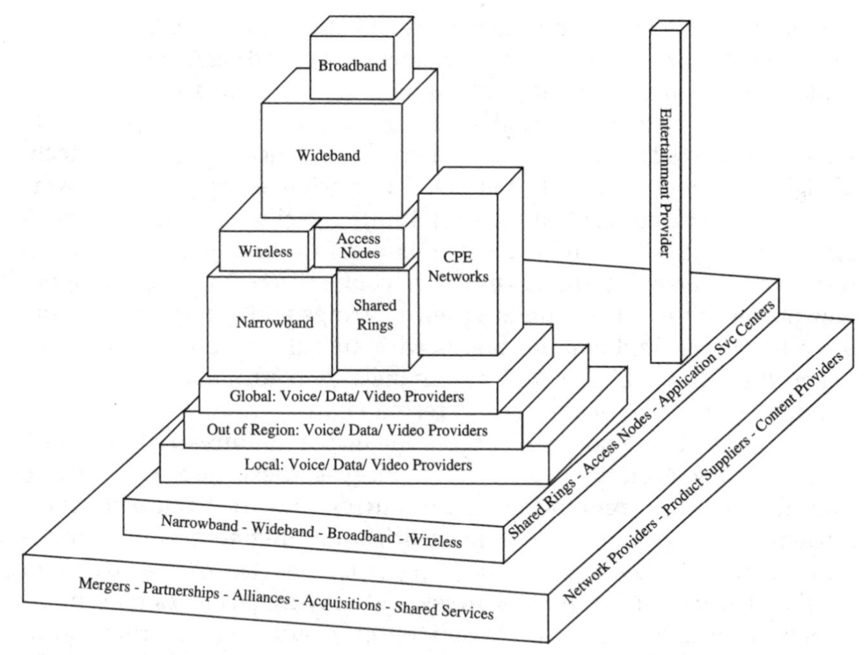

Figure 2-6B. LEC/RBOC perspective B.

made considerable expenditures for spectrum, old cable delivery systems, international buy-ins, and numerous capital-intense diverse ventures, with little interest in constructing new significant infrastructures that require a slower build-up curve for attracting new customers, educating them to new capabilities, and waiting for their usage to become profitable. Similarly, there have been few, if any, incentives to become the "keeper of the network" for this new open arena. With limited governmental regulations to foster the universal availability of the new services, many providers (old and new) have said that it is no longer their responsibility, thereby deploying limited network-management mechanisms to ensure survivability, security, and robustness as they demonstrate a willingness to overlay numerous options on the existing voice network. In this new arena, LECs have shown an inability to stop cream-skimming as the rate of growth of their voice-grade service revenues began to be seriously affected by the numerous new players in more lucrative areas of the local arena. A nonbiased observer might certainly note that it has been the viewpoint of short-term Wall Street analyses that the LECs should not make major investments to build new infrastructure. It is interesting to further note that, following this strategy, the telcos may in time go the way of gas lights and railroads, making way for the new ATPs and CAPs as they simply maintain existing voice-grade services, with a few data-handling overlays, while being bombarded for open access everywhere by anyone.

As some LECs shift their focus (attempting to play in both the telco and the cable arenas) to devise a means for expanding existing analog cable systems into hybrid digital carriers to include upstream tele services, this shift has put considerable pressure on product suppliers to refocus their R&D onto the more constrained, concentration-type technologies as they attempt to make do with what is out there by overlaying compressed applications on limited facilities. This approach also places the LECs and cable companies in a limited game, both in terms of technologies and in arenas of competition, as more and more new players attempt to emulate their offerings and cut up the existing pie into smaller and smaller pieces without baking new pies or growing new crops for bigger and better pies, as traditional services are unbundled, resold, packaged, and resold again.

Perhaps the LECs need to reassess the impact of unrestricted access. Perhaps they need to aggressively address these new competitive frontal attacks by redefining, reconfiguring, and restructuring their infrastructure, expanding its capabilities and augmenting its limitations, rather than simply traversing from one expensive arena to another without sufficient resources and staying power to play in all arenas, especially as new players enter every facet of their local domains. The impact of the Telecommunications Act can and will continue to be quite devastating, as noted in Table 2-1.

Telecommunications Act

Impact

- Open Local Loop
  - Open access to market over *existing* network
  - Focus on Services not Infrastructure; thereby missing the key to achieving enhanced services
- Threat to Existing Infrastructure
  - No *Keeper of the network* resulting in responsibility for the States to become the Keeper!
  - Burden & abuse - resulting in over-engineering, higher costs, & brown outs
  - Cherry picking / cannibalizing - resulting in *Telecom have's & have nots*
  - Lost incentive to upgrade & improve infrastructure
- Shift in Power to FCC & States from LECs
  - Large role for FCC & States - describing, defining: extent, degree, & boundaries of Act
- Interregnum
  - Time between kings: resulting in great excitement & uncertainty in the Industry & Marketplace
  - Opportunity for the States to establish leadership; fill the vacuum
- Heightened Expectations
  - ESPs & ISPs looking for robust Infrastructure - those with it will profit!
- Wide Open Game
  - 3 Games played in 3 Arenas: Larger Pie vs. Larger Slice
  - Myriads of confusing overlays

Table 2-1. The impact of the Telecommunications Act.

The ramifications of these considerations are quite evident as noted in the interesting observations contained in the front-page story of the *Wall Street Journal* about the demise of PacTel in its merger with Southwestern Bell:

> It is spending hugely to get into the cable-television business. It also is trying to rebuild its wireless business from scratch. PacTel has spent $700,000,000 for licenses to offer new wireless "Personal Communications Services" and must lay out several billion dollars more for construction. Even then, PacTel was to face a formidable rival: AirTouch, which dominates the California market.
>
> AirTouch enjoyed other advantages in the split. Management wanted the wireless unity to start out debt-free, so PacTel "forgave" nearly $1 billion in debt that the wireless unit had racked up during the years. Likewise, the cellular spin-off should not have to help fund PacTel's rich dividend payments. Nor would it have to pay PacTel for its cellular licenses, worth billions on the open market today.
>
> Likewise, [a senior executive] had [previously] played up Personal Communications Services as PacTel's path to rebuilding its wireless muscle. But most experts have come to view such services as a supplement to, rather than a replacement for, cellular service. SBC has one of the largest cellular operations in the nation and views PacTel's wireless PCS licenses as its complementary entry into the California market.
>
> Many of the expectations and underlying reasons for the spin-off have proved to be mistaken. For example, a primary goal was to free AirTouch from federal restrictions that bar Bell subsidiaries from entering the long-distance business or making equipment. Those hindrances were largely removed with the new telecom law that went into effect last month, granting the same freedoms to the other Bells that held onto their coveted cellular assets.
>
> The diverging fortunes of PacTel and its offspring have sparked criticism. Consumer groups complain that California customers—not PacTel shareholders—paid the cost of building the cellular network that PacTel simply gave away. Now the Bell is trying to dig deep into customers' pockets again as it struggles to get its own wireless service off the ground again (Cauley 1996).

Of course, in fairness, PacTel might offer a different perspective of its course of action, but this article does show the intensity of the game.

ATPs, CAPs, AECs, VANs, and VARs

Alternative transport providers (ATPs), competitive access providers (CAPs), alternative exchange carriers (AECs), value-added networks (VANs), and value-added resellers (VARs) all have something in common. They each want a "piece of the action," a segment of the network. They demand, not ask, traditional carriers to "open it up," enabling access, access, access, so their customers can "jump on and off" public transport to private transport to public transport, here, there, and everywhere. These providers are establishing parallel network offerings for their customers, using LEC capacity for resell, selective deployment of fiber, selected insertion of switching nodes, resale of shared Centrex-type services, and LECs' unbundled local loop at selected line-trunk-switch distribution points such as the mainframe, intermediate mainframe, remote switch units, remote line units, multiplexers, concentrators, TR303/008 interface points, and frame relay packet nodes, using "economic bypass" pricing strategies.

In this arena, ATPs, AECs, and VARs will insert selective service nodes to offer the business community the most cost-effective time-of-day routing capabilities to meet their changing transport capacity needs, adding such capabilities as advanced information networking (AIN) services, video conferencing, videophone, and data networking, as well as constructing switched LAN-type services locally and nationally for specific *communities of interest* (COIs). As noted in Fig. 2-7, these specialized carriers are quite involved in every aspect of this diverse marketplace, packaging and bundling their value-added offerings for the fully integrated multimedia world of interactive voice, data, and video PC communications. Virtual networks are constructed—both switched and permanent point-to-point—to support an exciting array of new services specialized to affect the business-type users, hospitals, and various selective communities of

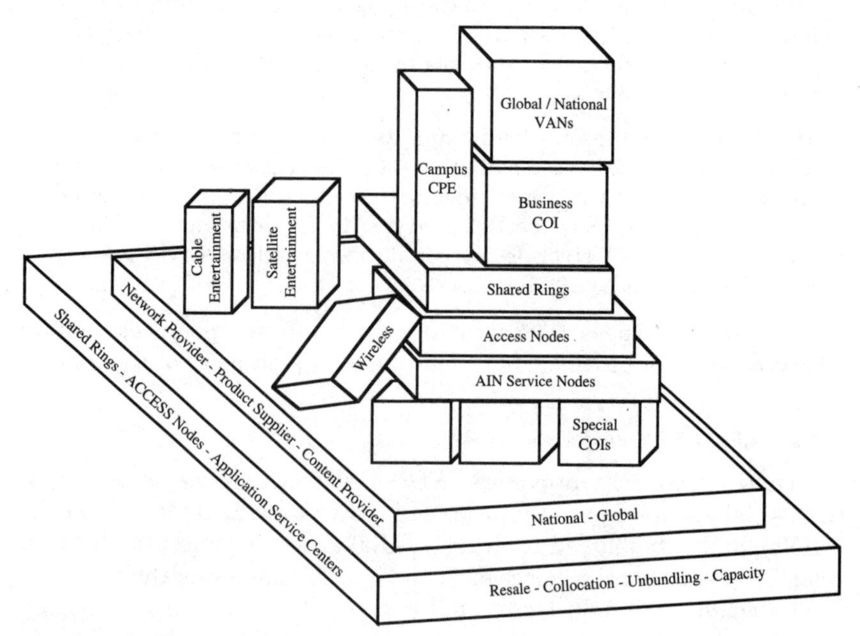

Figure 2-7. Perspective of the AECs, ATPs, CAPs, and VANs.

interest. This form of integrated, concentrated, specifically targeted offerings removes many of the traditional business users' revenue streams from the LECs. This competition can and will continue to cause quite a devastating impact across the universal services monopoly, requiring a reassessment of many of the underlying strategies, such as the extent and degree that business communities subsidize residential communities.

COIs

Communities of interest cover the full scope and range of interconnected networks specific to a particular area or community, such as health care, law enforcement, education, state government, or manufacturing. This marketplace is discussed in depth in *Future Telecommunications*, by Robert Heldman (New York: McGraw-Hill, 1993). Numerous diverse but interdependent users will be coalesced in COIs, using expanding wide-area internetworking facilities, both functionally and geographically (see Fig. 2-8). Over the initial phases of the Telecommunication Act's open competitive marketplace, numerous ATPs, CAPs, AECs, and VANs are rushing to serve these COIs, using parallel overlay networks to transport interactive variable bit rate bursts or continuous bit rate voice, data, and video conversations. These networks enable users to obtain access to specialized servers offered by numerous application centers, which provide their custom capabilities to search local and remote specialized databases for their users' needs.

These COI users will generate increasing amounts of traffic as they not only internetwork together but cross-internetwork to each other's databases; for example, a lawyer may wish to access an insurance company's databases or a law enforcer's database containing an accident report. As these community-of-interest users continue to overlay their increasing volume of data networking traffic on existing voice networks, they will indeed cause burnouts and degradation of voice-grade facilities, interfering with normal voice traffic. For these communities to be successful, they require networks that offer additional data-handling capabilities (see Chapter 6). Hence, using the analog data modem transport to access Internet-type open databases is not the answer! IXCs and VANs will be encouraged to bring their customers to their networks as quickly as possible, requiring direct access to many diverse points, distributed throughout the local network, causing irregular traffic-congestion patterns to permeate throughout the traditional local networks as these specialized carriers attempt to quickly address the needs of these expanding COIs.

ISPs, ESPs, ASCs, DBNs

Information service providers (ISPs), enhanced service providers (ESPs), applications service centers (ASCs), and database service

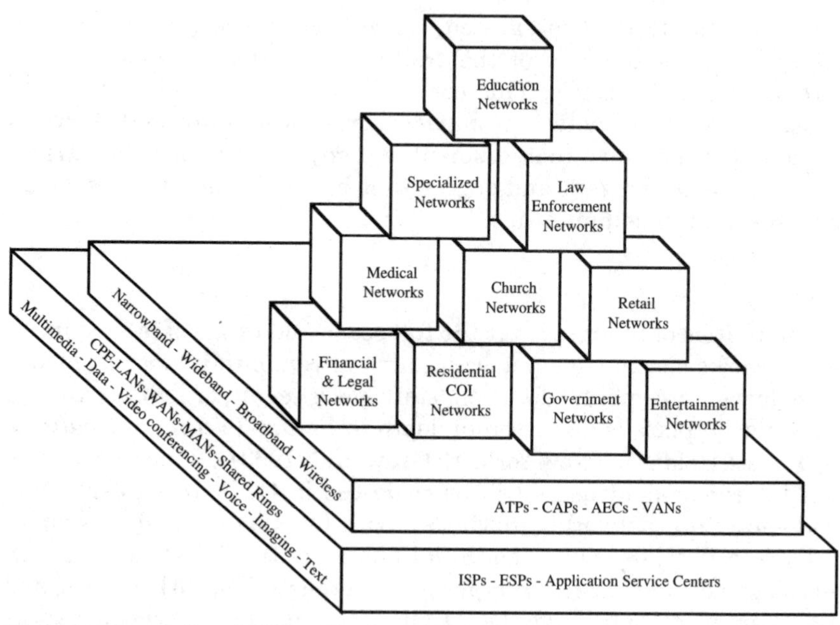

Figure 2-8. Perspective of COI applications networks.

nodes (DBNs) will expand from the 2000 to 4000 conglomeration of specialized servers of the 1990s to several orders of increasing magnitude, in tens of thousands, once the appropriate data transport infrastructure is constructed to ensure privacy, security, and robustness. Today's Internet's free wild ride has demonstrated the public interest, or should we say the world community interest, as the World Wide Web of servers are connected to the worldwide audience. It has been some 30 years since military networks first demonstrated the awesome power of accessing remote databases quickly, easily, and confidentially to help determine the equipment status, deployment availability, and troop readiness of their forces. (Note: The Internet doesn't support confidential access or easy access). Similarly, some 20 years ago, as inventory control and logistics mechanisms went online, manufacturers found the economical attractiveness of just-in-time delivery systems to reduce inventories. As process controls required extensive access to internal, as well as external, supplier databases, using extensive private data transport communication networks, it became quite evident that few data-networking capabilities have been offered to the universal public in terms of public data networks, other than leased lines/trunks or low-speed, dial-up, voice-grade data transport.

Today, as users advance to more sophisticated and specialized information presentation systems, they require greater and greater transport capabilities. One researcher noted, "Our computer capabilities are doubling every 18 to 24 months, so where are our communications capabilities doubling? When will they advance just a little—let alone double in data/video capabilities? How long must we wait? I cannot believe that anyone actually believes that the answer is Internet." In fact, most online service providers have requested higher-speed facilities for their dial-up data customers time and time again to enable them to shop or browse through databases, simulate scenarios, and obtain solutions in graphic image form. One computer database server noted the tremendous savings in computer-processing utilization (over a large number of users) if they did not have to wait for slow-speed input/output inquiry and response delivery systems. This savings was so great that it was willing to pay exorbitant costs to obtain narrowband ISDN-type offerings, hoping that transport prices would diminish over time. Unfortunately, access still usually consists of a maze of mixed transports and protocols, adding levels and levels of complexity, as various access devices attempted to mix and match, as users searched for an easy, straightforward public data network underlying infrastructure (see Fig. 2-9).

Government

As the states formulate their governance and policy statements and procedures, it is essential to note that their views must not be high-level, classroom-type, academic assessments, noting the need to formulate an open, free-information, enterprise republic. Their visions and views must be essential life-giving blueprints for survival of their

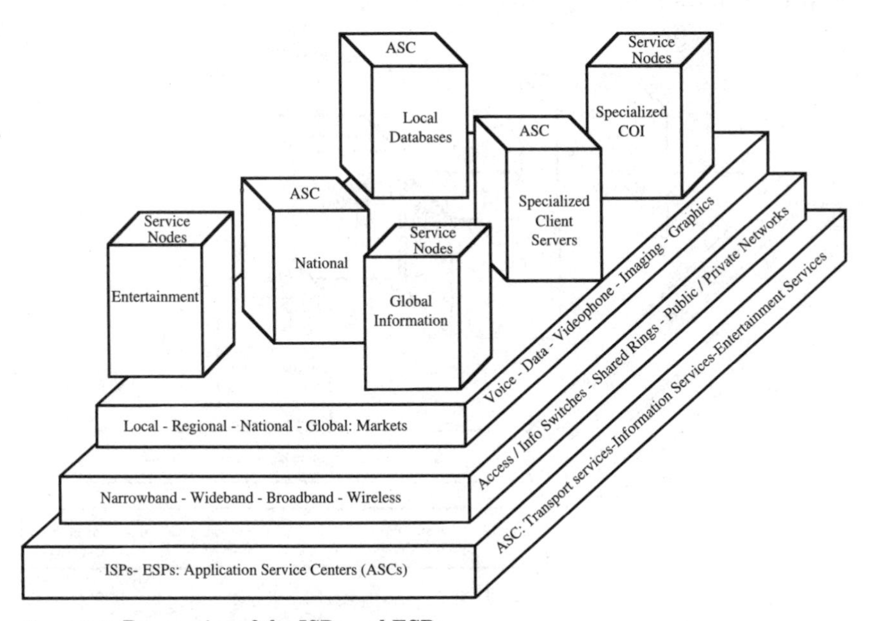

Figure 2-9. Perspective of the ISPs and ESPs.

populations. The future information age will see a change from "haves" and "have-nots" to those "with" or "without" access to information as populations learn to use information in every aspect of their daily lives. It is not only an opportunity for the rural communities in third-world countries to advance, but for those in industrial nations as well. As urban populations grew over shrinking rural communities during the industrial revolution, the information revolution provides an opportunity for a reversal. In fact, there will most likely be a definite shift as urban congestion and violence increase, thereby encouraging megametropolitan populations to leave to more rural habitats in pursuit of a better quality of life. Similarly, many of the former socialistic twentieth-century industrial nations may collapse under their own stagnation in the twenty-first century, as the more aggressive nonindustrial nations quickly embrace the new, less capital-intensive, information-based technologies to successfully compete in the blossoming global information marketplace. Here in this new arena, governments have several options on how they may plan, lead, organize, and control the fortunes and destinies of those within their domain of concern (Fig. 2-10).

It is interesting to note the direction-setting guidance given by the United States Advisory Council on the National Information Infrastructure amidst this highly complex and changing marketplace. One may question whether these are simply meaningless platitudes or if they contain meaningful, life-giving directions to enable the nation's participants to achieve a viable information structure supporting a new information society.

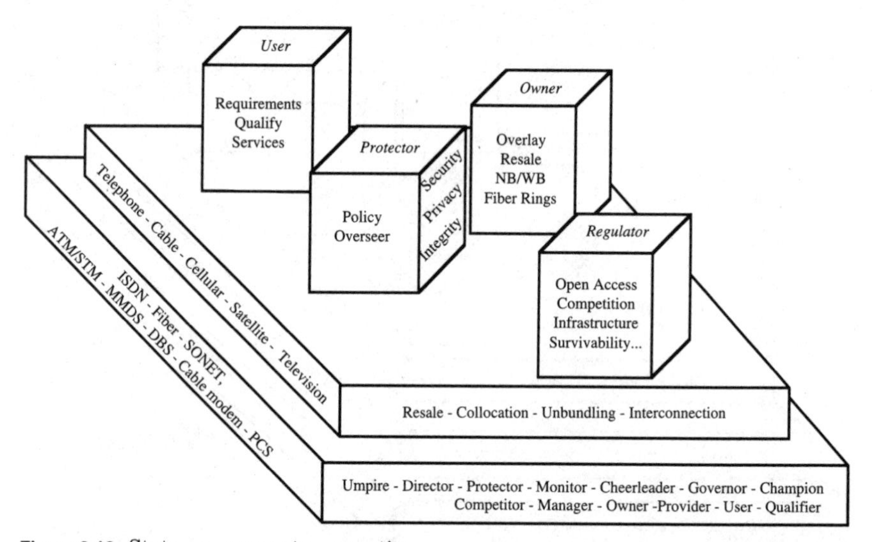

Figure 2-10. State government perspective.

NII's definition of information superhighway. The Information Superhighway is a term that . . . captures the vision of a nationwide, invisible, seamless, dynamic web of transmission mechanisms, information, appliances, content, and people.

The Council's vision. The United States stands today in the midst of one of the great revolutions in recorded history: the Information Age. The Information Superhighway provides the infrastructure that enables enormous benefits in education, economic well being, and quality of life.

The Council urges that the Nation adopt the following five fundamental goals.

First, let us find ways to make information technology work for us, the people of this country, by ensuring that these wondrous new resources advance American constitutional precepts, our diverse cultural values, and our sense of equity.

Second, let us ensure, too, that getting America on-line results in stronger communities, and a stronger sense of national community.

Third, let us extend to every person in every community the opportunity to participate in building the Information Superhighway. The Information Superhighway must be a tool that is available to all individuals—people of all ages, those from a wide range of economic, social, and cultural backgrounds, and those with a wide range of functional abilities and limitations—not just a select few. It must be affordable, easy to use, and accessible from even the most disadvantaged or remote neighborhood.

Fourth, let us ensure that we Americans take responsibility for the building of the Superhighway—private sector, government at all levels, and individuals.

And, fifth, let us maintain our world leadership in developing the services, products, and an open and competitive market that lead to deployment of the Information Superhighway. Research and development will be an essential component of its sustained evolution.

In charting a course to meet these goals for the Information Superhighway, the Advisory Council identified what it believes are four critical issues that must be addressed and must be addressed early:

- What are the key areas of American life and work that will be impacted?
- What is the role of universal access in the digital age?
- What are the rules of the road regarding intellectual property, privacy, and security?
- Who are the key stakeholders, and what are their roles?

Recommendations

A. Impact on Key Areas of American Life and Work

1. Electronic Commerce

   The Federal Government should identify and resolve legal/regulatory and policy issues related to the development of electronic commerce.

2. Education and Lifelong Learning

   The Federal Government should create an initiative to accelerate access to the Information Highway and facilitate the integration of information technology and lifelong learning while stimulating market development of educational products and services.

3. Emergency Management and Public Safety

   Ensure that these Emergency Management and Public Safety needs are met.

4. Health

   The Federal government should resolve legal/regulatory barriers to delivering healthcare information across state borders. It should also establish a funded project to evaluate issues of cost, access, and quality in Telemedicine applications.

5. Government Information and Services

   All governments agencies should use the new communications technology in delivering services and performing daily functions.

B. Ensuring Access for All

1. Information Superhighway Deployment

   Commercial and competitive forces should be the drivers for establishing the Information Highway. Critical to remove regulatory disincentives to developing the Information Highway.

2. Universal Access and Services

   The definition of Universal Service must evolve with technology advances.

3. All Americans Must Have Access to the Information Highway Including Disabled Americans

4. Government's Role

   The Federal Government should step in when the competitive marketplace fails to achieve universal service and lead by example in using the Information Highway.

C. Rules of the Road

1. Intellectual Property

   All levels of government should promote public education re: intellectual property issues. The Federal government should ensure that other counties implement consistent intellectual property policies.

2. Privacy

   Federal government should pursue privacy issues reviewing existing laws and practices.

3. Security

   Federal government should not inhibit development or deployment of encryption in the private sector and should raise the issues of security and "foster mechanisms to promote private accountability for proper use of security measures."

4. Free Speech

   Government should stay out of regulating content of the Information Highway.

D. Key Roles

1. The Private Sector Must Be the Builder (of the Information Highway)
2. Communities Are Key to Access and Learning
3. Government Has a Critical Role as Catalyst
4. Individuals Must Take Charge (U.S. Advisory Council 1996)

In contrast

Let's now review a knowledgeable governor's observations and concerns of the impact of the Telecommunications Act on rural America in response to an FCC inquiry:

Impact of Telecommunications Act of 1996
on Rural America—South Dakota:

William J. Janklow, Governor, South Dakota

South Dakota's Telecommunications Needs. South Dakota is the most rural state in America. There are 58 American cities that have more people than our entire state. Of our 308 incorporated municipalities, 62 percent have fewer than 500 people. Of our 230 public and nonpublic high schools, 53 percent have fewer than 100 students. Nationally, 27 percent of public schools are classified as "rural." In South Dakota, that number is 72 percent.

South Dakota, as with the rest of rural America, has struggled in recent years to overcome the lack of a robust, public, switched information handling (voice, data, video) telecommunications infrastructure. We are proud of our successes but will not be satisfied until future upgrades make us fully competitive in the new global telecommunications arena. Vast physical distances in rural America lead to a remoteness from jobs, markets, educational opportunities and cultural resources. Telecommunications has been identified as a singular important tool in reducing and removing this isolation, offering fiscal, cultural, and educational opportunities and benefits to rural Americans. There is a need across rural America in our schools, libraries, hospitals and businesses (large and small) as well as in our state and local governments, for a communication infrastructure capable of supporting applications far beyond the current capacity of the public voice network as it exists today. Applications such as video conferencing, distance learning, telemedicine, and data networking require a vastly new and enhanced narrowband, wideband, and broadband network infrastructure with many specific technical and functional requirements. Such a network must be:

- Ubiquitous—universally accessible
- Feature rich[1]—beyond mere transport, the network must provide certain features and functionality
- Robust—able to sustain the rigors of growth and extensive public use
- Secure—the network must physically preclude unwanted access to information prior to the addition of encryption
- Survivable—the information network cannot go down!—must meet and exceed standards set by today's telephone network
- Addressable—the ability to connect and communicate with a specific person easily and securely on a dial-up basis
- Switched—circuit, packet, channel type switching—each suited to specific application requirements
- Symmetric—two way, equal bandwidth both ways
- Cost effective

This new network would operate at the *narrowband* rate of 64 Kbps to 144 Kbps, *wideband* rates of 144 Kbps to 45 Mbps, and eventually at *broadband* rates of multiples of 50 Mbps.

---

[1]Data features included broadcast, multicast, delayed delivery, packet interleaving, byte interleaving, code conversion, polling, inquiry facility, three-attempt limit, low error rate, data collection service, high/low grades of service, standard interface, bit sequence independence, short setup, auto callback, redirection of calls, speed/format transforms, abbreviated address calls, closed user groups, short clear-down, manual/auto calling and answering, data service classes, barred access, remote terminal identification, and multiaddress calls.

As information becomes the product of the Information Age, the telecommunications infrastructure becomes the railway and highway, offering access to data content and providing a means of delivering information goods to market. Thus the very real danger of creating a society of information haves and have nots very quickly transforms itself into economic haves and have nots. An advanced telecommunications infrastructure could enable rural America to grow and flourish. This same technology could also destroy rural America if remote communities are denied access to it—much the way towns away from the railroads and later the interstate highway system quickly perished once those infrastructures were established!

If rural America is to realize the salutary benefits of the Information Age, a robust Telecommunications Infrastructure must be ubiquitously established, offering all Americans equitable access to these new narrowband, wideband and broadband services. This infrastructure is the essential key, the foundation supporting the free, open and competitive information marketplace.

Impact of the Telecommunications Act on South Dakota. In reviewing the impact of the Telecommunications Act of 1996 on South Dakota, we have identified several areas of concern that must be addressed to establish an environment which protects the existing voice network and enables an enhanced public information network to be established.

The Telecommunications legislation breaks up the local telephone monopoly—opening the existing voice telephone network infrastructure in the hopes of encouraging competitive free market forces to establish a proliferation of new enhanced network service offerings as well as of lowering the cost of these and existing telecommunications services. However, the current voice telephone network, in its present form, is ill-suited to support the host of advanced narrowband, wideband and broadband voice, data, and video services that today's information users require. The accelerated "deployment of advanced telecommunications and information technologies and services to all Americans" envisioned by the Telecommunications Act cannot be realized until an enhanced network infrastructure capable of supporting enhanced services and applications is first established. Indeed, the free market will likely perform well at overlaying a host of new innovative services at competitive prices once a robust underlying public infrastructure has been established. But to achieve such an infrastructure—history shows us that as with the canals, railroads, telephones, and interstate highway system, government must provide the appropriate direction and incentives within the competitive environment to formulate the *right* infrastructure *everywhere*—even in rural America. This must also be done with the realization that competition does not always work the same way in rural areas as it does in the urban environments. To promote merely a series of disjointed, limited, autonomous networks would be a disservice to America and would not achieve the Information Highway required to meet our application needs, nor those of our children.

Some people fear that the legislation lacks proper incentives to encourage such an upgrade of the existing infrastructure and the deployment of new advanced infrastructures. They contend open access for resellers will likely act as a *disincentive* to local service providers who would incur the large cost of upgrading their network—only to see resellers *cherry pick* the key early entrant customers that have traditionally been used to offset the initial cost of the upgrade. Safeguards in the Act should be implemented to prevent their activity. The ability of resellers who, under the Act, seek to buy existing services at wholesale rates and *cream skim* the local market is a great concern. This may in fact have a detrimental effect on the price of local phone service for most consumers without providing any appreciable new enhanced services. Again, safeguards should be implemented.

Indeed, in the new competitive arena, the incentive to own existing POTs networks is diminished—as evidenced by the flight of RBOCs/LECs who are in the process of selling off their remote exchanges throughout rural America. They have chosen to focus on the most densely populated (and thus the most profitable) regions in state, out of state, out of region, out of country.

This flight of traditional LECs, coupled with stringent collocation, resale, unbundling, and interconnection requirements, may quickly result in a world in which America has lost its *Keepers of the Network*. Those who assume the role (and significant expense) of becoming this protector of the network must ensure that the basic (lifeline) infrastructure does not go down. This includes arduous testing, integrating, maintaining, and operating tasks as they interconnect with other networks and is especially true where collocation with other network elements/systems is involved. Who will bear this cost now—especially in the rural arena, which is left with the least cost effective area to serve?

Interconnection, collocation, and unbundling can represent a very real danger to the existing network if appropriate rational limits and boundaries are not established to prevent abuse of the voice telephone network. While the technology arena does offer nearly limitless potential, individual technologies do have very real limits. (As with the carpenter's tools, each has a specific purpose and it would be an abuse of that tool—say a screwdriver—to use it for a task for which it was not designed—e.g., chiseling with a screwdriver. The result is a poor job chiseling and damage to the screwdriver, inhibiting its subsequent ability to perform the task for which it was designed!) So too with the voice telephone network, which is increasingly being asked to function as a data network with its short holding times, and multiple attempts or extraordinary long holding times for example Internet access. The Internet (a data network) too is facing abuse from voice and real time video conferencing applications. Such abuse often results in brownouts, loss of service, and dramatic increase in the cost of operating the network so inefficiently. These costs are ultimately passed along to the customer.

Collocation offers its own special dangers to a network. As complex pieces of equipment interconnect—how will blame and liability be established as one system causes the other to fail? In a wide-open game of interconnection and collocation the network will be vulnerable to unanticipated situations resulting in possible catastrophic failures. Again, appropriate safeguards must be established to ensure the survivability of our public infrastructure. As hospitals and businesses move their data handling applications fully "on line," the cost of such a catastrophic failure of the network in human and financial terms escalates dramatically.

Conclusion. The FCC's final recommendation must address and resolve these issues and concerns, as they consider their course of action dealing with:

1. Protecting the existing infrastructure for life line services.
2. Ensuring that an enhanced Information Infrastructure is established for rural America.
3. Managing the Universal Service Fund to ensure that all participants contribute appropriately and money is fairly distributed to rural America.

Games, Arenas, Technologies

In considering what games to play, with what technologies, in what arenas, several insights come to mind:

- Enhanced infrastructures are essential for establishing enhanced services.
- Each technology is like a tool. Use the right tool for the right job.
- There is no magic technology pill; one size does not fit all.
- Games cannot be played without appropriate playing fields (arenas) supported by sustaining technologies.
- Establishing a new network infrastructure remains a nontrivial task.

With these insights, providers have several options to play and several arenas to play in over the late years of the twentieth century and early years of the twenty-first century (see Figs. 2-11 and 2-12).

Players can continue to provide dial-up analog modems that use the existing voice network to interface to more and more sophisticated terminals, connecting them to a router-type, limited, low-speed data networks such as the Internet or simply allow their LANs to connect directly to an Internet service provider. Alternatively, they can successfully span services by adding additional higher-speed digital communications data-handling capabilities to the basic copper distribution plant and then augmenting existing services by upgrading to a fully switched fiber-based network. To these, they may then connect various forms of wireless services: personal phone, wireless data, satellite packet switching, and direct broadcast services. Next,

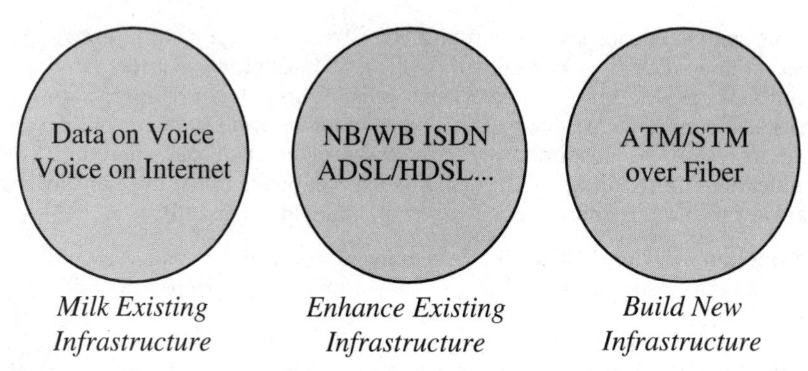

Figure 2-11. Three games to play.

networks service nodes can be attached, providing access to both local and remote unique or common databases.

Once the last-mile-blocking bottlenecks are removed, opening local transport offerings, both sides of the interconnection, customer and network, will expand and expand again as the customer-premise side engages in fully distributed communications throughout the campus, apartment complex, or business highrise, enabling not only voice but data and video interactive conversations. Similarly, the rising appetite for access to new services of information will encourage a proliferation of service nodes, thereby requiring an expanding and exploding, new, universally available, internetworking, interserving network infrastructure. This is achievable once the addressing, directory, network management, and operational issues are resolved to achieve dial-up, fully switched connections for data and video calls over expanding digital communications facilities, similar to the 100-year-old analog voice network achievement.

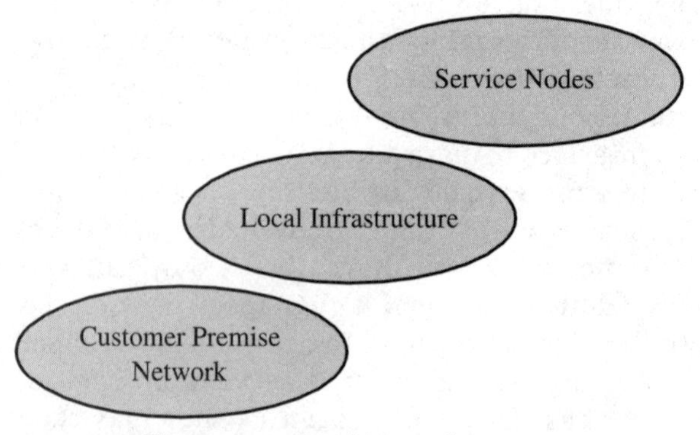

Figure 2-12. Three arenas in which to play.

Technology trends

As is readily apparent from Tables 2-2 and 2-3, the various providers and suppliers will be participating in a variety of technical domains, both terrestrial and nonterrestrial. The sky is the limit as both wireline and wireless services compete for providing solutions to this expanding arena. In fact, the internetworking of these offerings and interconnections will be essential as both rural and urban applications merge in the national and global information marketplace. Digital communications will play an ever-increasing role in these endeavors, so let us first pause and review the continuing progress and role of *digital communications.*

Digital communications

Throughout the 1930s, '40s, and '50s, communications blossomed in both the wireless and wireline worlds of radio and telephone and, later, television. The predominant technologies were based on various analog modulation/demodulation techniques as voice, data, and video signals were superimposed on high-frequency carriers and multiplexed together to be transported. As high-wattage radio antennas were established on high-rise towers to broadcast radio frequencies long distances, the telephone took the wireline route, as telephone poles were established down

Technology Trends & Alternatives

- Technology Evaluation Criteria
  - Addressable/ Switched vs Broadcast / Routed
  - Secure: physical access vs encryption
  - Survivable:
    - Vulnerability to: natural hazards, equipment failure, human disasters ...
    - OSSs
    - Physical Architecture (Ring vs Tree & Branch...) & Ability to do Physically diverse routing
    - Size of Failure Groups
    - Duplex systems
  - Asymmetric vs Symmetric
  - Contention / Blocking
  - Ability to Handle Success
  - Capacity: shared vs dedicated
  - Delivery Restrictions:
    - line of sight - physical requirements: size, environment conditioning... - cabling/wiring - conditioning lines
    - radius of turns for cable (fiber) - topology (tree & branch vs ring...)
  - Bandwidth: narrowband, wideband, broadband
  - Features: voice, data, video, multimedia, image, graphics, animation

Table 2-2. Technology evaluation criteria.

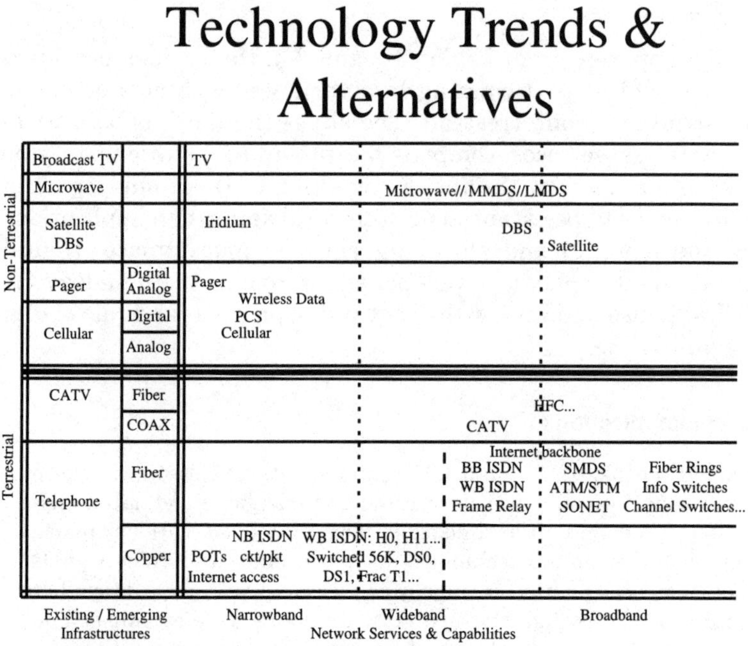

Technology Trends & Alternatives

	Existing / Emerging Infrastructures		Narrowband	Wideband	Broadband
Non-Terrestrial	Broadcast TV		TV		
	Microwave			Microwave// MMDS//LMDS	
	Satellite DBS		Iridium	DBS	Satellite
	Pager	Digital Analog	Pager Wireless Data		
	Cellular	Digital	PCS Cellular		
		Analog			
Terrestrial	CATV	Fiber			HFC...
		COAX		CATV	
	Telephone	Fiber		Internet backbone BB ISDN WB ISDN Frame Relay	SMDS ATM/STM SONET Fiber Rings Info Switches Channel Switches...
		Copper	NB ISDN POTs ckt/pkt Internet access	WB ISDN: H0, H11... Switched 56K, DS0, DS1, Frac T1...	

Network Services & Capabilities

Table 2-3. Technology alternatives.

country lanes, while urban cable vaults blossomed as they terminated hundreds of thousands of copper pairs, bringing analog voice conversations to large central switching offices. (In time, past the middle of the century, after Sputnik, satellites were universally deployed commercially to distribute wireless television signals to remote locations.)

Unfortunately with regard to wireline, there was considerable expense in dedicating each conversation to a single wire pair and considerable maintenance time required to hook up and hand-wire each new request for service within the cable distribution plant and main distribution frames of the central office. Stored program controlled (SPC) switching and service systems helped speed up the customer service changes enabling numbering changes, number translations, and new routing table updates, as well as the enhanced capability, to provide new features more easily in a timely manner. But there was still a serious need to reduce the cost of the distribution plant, improve the quality of service, and reduce the proliferation of new central switching offices requiring new blocks of the dwindling resources of seven-digit local codes.

To address these needs, especially for long distance carrier transport, AT&T introduced digital communications in the early 1970s, based upon pulse code modulation techniques developed in the 1930s by an earlier pioneer, ITT Laboratories. Here, AT&T's T1 carrier systems, operating at 1.544 million bits per second (Mb/s), were able to transport 24 voice conversations over a pair of copper wires, where 8 bits of each voice conversation were sampled 8000 times so 64,000 bits were sent together with specialized synchronization (bits) information to formulate the T1 1.544 Mb/s transport system. Later, capabilities were expanded to T2 (6.31 Mb/s) and eventually T3 (45 Mb/s) capabilities.

It was noted in the mid-1970s by GTE engineers that their central offices in remote rural towns could "home in" on a single large office located usually at the county seat. Hence, they extended the digital long distance transport to the local community, as remote switching units were deployed in remote locations to "home" on centrally based units, thereby establishing digital clusters of 15 or so small towns and villages with the county seat. Thus were born integrated digital networks (IDNs).

Though there were administrative economies of scale and number group savings in deploying digital to the voice world in this manner, much, much more could be achieved by integrating digital voice with the world of data. This was the purpose and charge of integrated services digital networks (ISDNs). Their initial task was to enable the customer to send and receive voice or data from their locations over the copper pair, and to provide the capability to take immediate advantage of shared digital transport and switching facilities. In this, the user was provided a 2B+D interface, with the B channel having the capability of sending/receiving 64,000 bits per second and the D channel enabling 16,000 bits per second of signaling and control information. Thus the user is able to send digital voice in one B channel, as well as simultaneous digital data in the other B channel, or have two separate voice conversations in the two B channels or digital data in both B channels, at total rate of 128,000 bits per second, while having 16,000 bits of out-of-band signaling and control information in the D channel. This D channel is also able to send data in a packet form (with information packaged in a variable or fixed number of 8-bit bytes, having a header and a tail to differentiate the beginning and end of the message; here, the D channel's transport rate for packet data was 9600 bits per second).

Hence, this 2B+D interface was called the basic rate interface (BRI), enabling voice or data to be mixed together or allowing two voice conversations or two data conversations to be transported. These narrowband capabilities were then augmented by a high-speed (wideband) ISDN primary rate interface (PRI), enabling 23 B channels of 64 Kb/s and one D channel having 64 Kb/s of signaling and control information, equating to the T1 transport rate of 1.544 Mb/s.

In time, broadband capabilities will be available, enabling the user network interface (UNI) to deliver 155 million bits per second (Mb/s) and 622 Mb/s to the customer over fiber-optic local loops. These optical carrier rates are multiples of OC-1 (51.8 Mb/s). It should be noted that American and European systems will be in step at the internationally agreed standard of OC-3 or 155 Mb/s, with Americans evolving from the T1 rate of 1.544 Mb/s and Europeans from the E1 rate of 2.048 Mb/s.

Here, information will be digitally transported, using the synchronous optical network (SONET) transport capabilities, enabling voice, data, text, image, and video to be dynamically modulated and multiplexed over gigabit facilities. This then will be switched using asynchronous transfer mode (ATM) fast packet-switching technologies, as well as synchronous transfer mode (STM) circuit-switching capabilities. These techniques will later be complemented and extended by photonic multiple-frequency (colors) optical-electrical switching transport systems handling terabits of information.

Using these capabilities, as fiber is delivered first to the office (FTTO), then to the residential curb (FTTC), and later to the home (FTTH), customers will be able to see four or so high-definition video channels and communicate to the world at the 155 Mb/s rate, paving the way for high-definition videophone, high-speed computer-to-computer data traffic, and the ability to dial high-quality musical and sporting events, etc. In this manner, narrowband, wideband, and broadband ISDN digital communications will establish the communications infrastructure for a new society—the Information Society in the twenty-first century (Heldman 1996).

The Open Competitive Information Marketplace

So what does this mean? What does all this add up to? What is the result for the user? Where are we going? Where do we want to go from here? How can we get there?

First and foremost, it is important to realize that there are two distinct models—one from the perspective of building everything on the broadcast entertainment market and transport facilities, the other of constructing the future from the interactive telephony-based communication market facilities, be it for voice, data, or video. Those with the former perspective view the latter as being a shrinking, diminishing infrastructure, as voice services shift to upstream cable facilities overlaid with some data inquiry/response-type mechanisms in the belief that interactive video will come much later as facilities are upgraded to 500 channel capacities, thereby embracing the Model E construct for their evolving future.

Alternately, using Model T, the traditional carriers and suppliers, which have made billions enabling people to talk to each other, recognize the need to continue this "interactive" process in not only the data world for business but the video world for residential and business communities of interest. Here, many believe that in the not-too-distant future, people will become quite accustomed to seeing each other to relay ideas and express concerns and opinions. As they find the feedback benefits in actually being able to see slight changes in expression (worry, mounting anger, happiness, fatigue, fear, and so on), this feedback becomes invaluable in achieving a heightened level of communications so that talking into a "dead phone" will soon become an unrealistic, unappreciated alternative. As this happens, better resolution, higher quality, and nonblocking interactions will be demanded. Continuing "shared limited transport" will not be one's choice, thereby requiring fully switched, fully addressable, high-quality, high-bandwidth, secure, personalized facilities. The providers' future will be based on having networks that offer these high-quality, fully-switched interactive communications for voice, data, and video calls. These are achieved by expanding the existing copper wire plant, overlaying it with fiber, and augmenting it with wireless capabilities.

So in reviewing the Telecommunications Act and its implications, as noted in Fig. 2-5, as well as considering the many different and diverse views of the various players (Figs. 2-6 through 2-10), it is interesting to note how the multiple approaches (in the different arenas, playing the different games, using either Model E or Model T as a baseline) sometimes concur but many times conflict. As seen in the following observations, conclusions, and recommendations, these issues become especially noticeable as we consider the realities and ramifications of the various infrastructures in terms of success, growth, support, and collapse.

What's happening?

Opening the local loop of the local monopoly for access and competition at the curb, the neighborhood access box, the multiplexer, the concentrator, the remote switch unit, the remote line stage, the intermediate distribution frame, the main distribution frame, the channel bank, the switch database, the translation routing table, the tandem trunk, the ANI-STP/SCP/STS, or at any tariffed interface results in the following complex, uncontrolled, changing marketplace, displaying the appearance of having:

- A continuous flow of programs and projects by RBOCs as they attempt to compete (after complying with the "list of openness") within their region with their interLATA long distance services, their manufactured products, their information content programs, their telco or cable network, and their service nodes.
- A continuous flow of interexchange carriers' (IXCs, VANs, ATPs, CAPs, AECs) offerings to local users, providing local transport and information-handling services requiring interconnection, number portability, customer menus for carrier selection, and service selection directories advertising the various ISPs, ESPs, DBPs, ASCs, and specialized client server service nodes.
- A continuous flow of congressional bills, laws, and legislation to protect information integrity and content decency.
- A continuous flow of FCC and PUC rules and regulations ensuring access to anyone, anywhere, anytime, at any price. . . .

In reality, there are multiple areas of crisis and conflict, resulting in questions asking:

- Who really ensures access, access, access?
- Who really ensures security, survivability, and privacy?
- Who really ensures performance objectives are established and met?
- Who really allocates and prioritizes capacity, availability, and call-handling capabilities?
- Who really monitors the resale and unbundling of the local plant—the FCC, the PUC, the courts? (See Appendix A, FCC Pricing Rules.)
- Who really provides the rules, regulations, judgments, and decisions? (See Appendix A, RBOC Appeals, Court Rulings, and Upholdings)
- Who really is/are the keeper(s) of the network?

Where do we end up?

Just think what it truly means to achieve an open competitive arena in which everyone is doing anything they want to do to anyone, anywhere, at any time. Do we really want this result?

Just think what it really means to access anyone's database, load anyone's network with bursty short holding time calls or numerous long continuous bit rate calls, to violate and bypass security mechanisms, to have free transport to send advertisements to 1 million, 2 million, 10 million, 1 billion, 10 billion mailboxes. Just think what it means to try to sort, intercept, and redirect 10 messages, 100 messages, 1 million, 10 million, 10 billion messages from 1, 10, 100, 100,000, 1 million advertisers.

Just think—no decency regulations, no blocking, no audits, no consequences. Just think of what it means to have transport or service features that require interrupting call processing in a switch for 1 call, 100 calls, 1000 calls, 1 million calls, every call.

Just think of having 100, 200, 300 such successful features, provided by as many providers interrupting the call processing, requiring rerouting to specialized service nodes in real time. Just think of customers selecting each aspect of a local carrier loop: local exchange carrier, local service provider, state long distance provider, national long distance carrier, international long distance carrier, specialized service provider, encryption provider, and time-of-day least-cost route selector, with or without menus and directories, on each call, each call type, each application, for each priority level, for each class of service.

Just think of past wishes and desires of having cheap electricity and look what happened to unchanneled, uncontrolled nuclear energy projects, resulting in nuclear mishaps (also note the overreaction resulting in no nuclear energy, causing expensive power, no electric cars, etc.), or the desire for cheap water for mining and irrigation, resulting in contaminated and limited drinking water or soil salt contamination.

Just think of the past evolution of the automobile transport systems—the lanes, the roads, the highways, the superhighways, the traffic problems, the congestion, the unmarked police cars and helicopters, the high-speed chases, radar, the diverse driving habits of too fast, too slow, tailgating, the tickets, the courts, the prisons, the types of cars, their rising costs, the price of gasoline, construction repairs, driving weather considerations, the rules of the road.

Now think of transporting, accessing, searching, browsing, storing, manipulating, packaging, and presenting information in text form, graphic form, image form, vision form, at low speeds, at high speeds, with low resolution, with high resolution, to 10 users, 100 users, 1 million users, 1 billion users.

Just think . . . just think . . . just think. . .

What should or could we be doing?

What should we be doing or could we be doing in terms of conceiving, identifying, defining, designing, testing, deploying, verifying, modifying, and enhancing, both transport and service mechanisms and operational capabilities to handle success, success, and more success? Is it not a new age of information—the Information Age? Is it not a new information marketplace in which hundreds, thousands, millions of providers and suppliers offer their services to millions and billions of users?

Providers do not need to build something that collapses under the weight of its own successes. Users need:

- A structure for layering services, both transport and content.
- A structure for offering higher and higher speeds of transport.
- A structure for protecting privacy, offering security, ensuring integrity.
- A structure that is robust and survivable.
- A structure for sharing transport, so streets are not plowed up every other day to accommodate a new carrier, where customer premise buildings do not collapse under the weight of layers of cable wire, or where the people are not baked by voluminous numbers and intensities of microwave frequencies.
- A structure addressing rural as well as urban needs.
- A structure enabling local access to global arenas and vice versa.
- A structure serving unique communities of interest over common shared technologies.
- A structure of controlled access—entrances, exits, gateways, portholes.

- A structure supported by firefighters, firewalls, and fire control protectors.
- A structure designed for success in the global information marketplace.

Players need to:

- Readdress attempts to play in every arena. It is impossible to be successful in all arenas.
- Reconsider financial strategies. There are not enough dollars to do it right in every arena.
- Readdress short-term versus long-term incentives and rewards.
- Change mentality to build versus rape and pillage.
- Limit and temper expectations to grow and expand over time.
- Construct infrastructures to withstand the test of time and usage.
- Establish long-term research and development programs for establishing next-generation product lines with next-generation services for next-generation users.
- Support extensive operations, administration, maintenance, and provisioning (OAM&P) for present and future offerings.
- Distribute offerings throughout the community, town, cities, regions, both urban and rural, local and national, around the globe, universally, ubiquitously.
- Protect and monitor users, punish offenders, and reward good behavior, not bad.
- Reassess short-term financial incentives versus long-term rewards for executives in the formation of mergers, acquisitions, and partnerships.
- Foster the expansion of product lines.
- Encourage the rebuilding of society with a new information economy, supporting new higher-paying jobs, jobs, jobs.

It's a new game with new rules of behavior for the new users, so as we look to the future of superhighways interconnecting supercomputers, serving the new users, the computer arena needs:

- Rules, regulations, and rights.
- A layered networks' layered services infrastructure.
- Freedoms and restraints for the open arena.
- Protectors for security and privacy.
- Robustness and survivability.

- Addresses, addresses, addresses.
- Access, access, access.
- Interconnecting, internetworking, interserving mechanisms.
- A new, open, competitive information infrastructure to support and sustain the new, open, competitive marketplace (see Fig. 2-13).

Building Blocks for the Future

In considering the preceding conclusions and recommendations, it is essential to note that to successfully construct an open, competitive information marketplace, the right building blocks must initially be put in place. If the blocks are improperly placed, it becomes extremely difficult to remove unsatisfactory blocks and still keep the structure from collapsing, especially if a large edifice has been quickly deployed on a totally inappropriate, nonsupporting foundation. This becomes more and more apparent if an abundance of unrestricted players are all doing their own things, adding more and more services, any way, on the wrong infrastructure.

With this in mind, it would be fruitful to carefully consider the full ramifications of the issues raised here and then proceed to the potential solutions provided in the remaining chapters of this book. First let us move to a different vantage point for a somewhat different

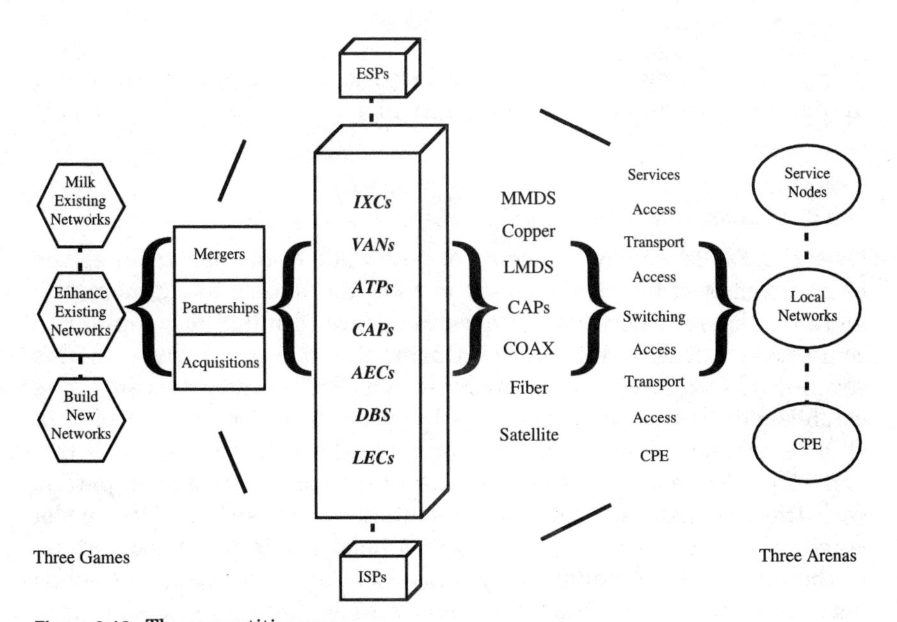

Figure 2-13. The competitive arena.

Figure 2-14. Building blocks for the future.

perspective on how to open and protect the local telecommunications arena. Let's begin by reviewing the history of the local monopoly (Fig. 2-15).

The local monopoly

Over the 1990s we saw a closing of both ends of the spectrum as the users attempted to become closer to their databases, using more and more transparent network services to ensure terminal-to-computer to computer-to-terminal interaction. During this period, wideband ISDN's seven-level Open Systems Interconnection (OSI) standards were being established. In the late 1980s, the Department of Defense decided to shift its strategy from supporting its past protocols, such as LAN 802 and TCP/IP, to ensuring that its future protocols are more compatible with the International Telecommunications Union's (ITU) OSI model standards. As we look at the layering of networks from the local arena to the global, this standardization is essential. Otherwise, the public network in the United States becomes an island unto itself, requiring extensive gateways to the world. It also will become more internally complex as numerous autonomous layers of private networks require front-end processors to bridge or gateway across it to each other.

As we change from predominantly voice to integrated voice, data, and video information interaction and interexchange, the public infrastructure compatibility issue will become worse. If it is not resolved, the public network will not facilitate the full utilization and interconnection of the many private networks. These needs and concerns have subsequently required the computer industry to establish specialized bridging and gateways through specialized processor databases, network controllers, network integrators, and access servers to achieve specialized interconnection, interprocessing, and interservices across these private networks, thereby bypassing the local monopoly. So what's new?

Even though Judge Greene (DOJ) had allowed a videotex offering, the issue of information content manipulation was still key to implementation of new services by both the RBOCs and the other players in the information marketplace until the new freedoms of the Telecommunications Act of 1996. However, as late as 1995, the judge had raised valid concerns by noting the past aggressive role of the Bell Operating Companies (BOCs). As indicated in Fig. 2-16 (page 59), he suggested that to resolve the bottleneck issue, there needed to be some major changes in the form of technology, risk, and ability. That is to say, we need major changes in technology that enable firms to better compete in a changing marketplace, which reduces risk to new carriers and players. These changes become essential in a fully open arena, especially as the RBOCs, as a result of the Telecommunications Act, are allowed to manufacture their own equipment. The changes in the opened monopoly must limit the ability of the RBOCs to:

- Discriminate in the purchase of new equipment.
- Inhibit the proper dissemination of information to other manufacturers and providers, such as delaying interface specifications to the network until RBOC product lines are in production.
- Charge higher prices for basic infrastructure offerings to subsidize other new services.

Having had further concerns about RBOC manufacturing, the judge considered the impact of their possible removal of major parts of the network product manufacturing market by controlling internal purchases, forming special alliances to exclude competition, and providing enhanced offerings in concert with new basic services, as well as increasing prices of basic offerings due to excessive increased costs in the manufacturing of products for their enhanced services. These concerns were reflected in the Telecommunications Act of 1996; it freed the RBOCs to manufacture their own equipment, but also attached contingency considerations to address these needs.

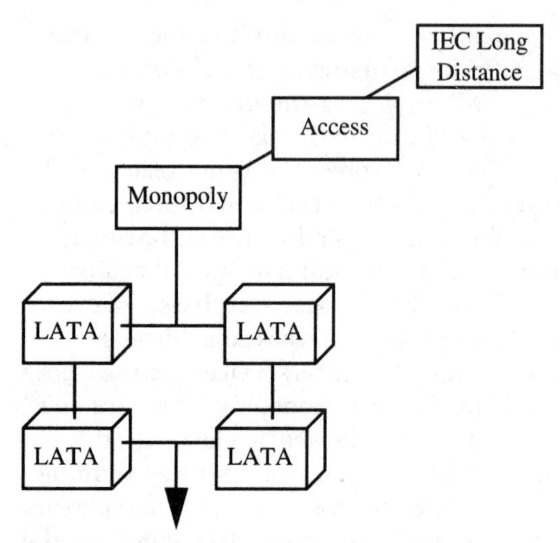

Figure 2-15. The local monopoly.

Finally, with regard to interexchange services, the judge noted the RBOCs' ability to block access to the local users, require market information from potential new providers, etc. "If they did have interexchange (long distance) capabilities this would give them a clear edge over others by having the ability to build 'total networks' across the nation to the exclusion of other carriers". This, he noted, destroys one of the two pillars that he put in place in 1984 to encourage a new competitive arena.

To overlook these concerns or pretend that they will go away is unrealistic, especially in a world in which there is limited trust in the RBOCs to establish a competitive marketplace on their own incentive. In the past, the BOC people have been known to be very competitive, both externally and internally, causing the judge to question if their mode of operation would now change in dealing with outside competition.

So what can be done to limit or remove these concerns to enable us, as one person said at a recent ISDN conference, "to get on with it"? Concern also existed in Congress for the lack of positioning of the United States in the information arena. One representative said, "We need to break out of these self-imposed restraints or we will find ten years from now that we no longer have a strategic asset in our telecommunication networks, as compared to other industrial countries." Similarly, Peter Huber once noted in a 1987 panel discussion at the Humphrey Institute that we are quoting the French as being so far ahead of us, when 20 years ago their phone network was less than desirable. Thus it is time to reassess and perhaps consider a revolutionary and politically controversial approach to resolving these concerns to get on with the game. After the Telecommunications Act of 1996, this becomes essential.

So in considering what's new, we cannot help but observe the intense change and new opportunity provided by gigabit transport and switching. Like it or not, this expansive technology will greatly transform our industry. We can choose to ignore it and pretend it is not happening, as extensive pricing continues to exist for handling small amounts of data. For example, data transport of 9.6K, 19K, and 56Kbps is still considered as having such extensive differences that considerably higher prices are required for 56K over 9.6K. Look at the pricing for frame relay at 56K, switched at 56K, and ISDN at 64Kbps.

Also, the close integration of computers with communications cannot be ignored as the next new wave of partnerships transform our daily mode of operation. Finally, U.S. firms must realize and accept that this new competitive arena is now to be the only field available to play the game, but, as noted, it must have a level playing field, and that was indeed the judge's concern, as well as that of the RBOCs' and the RBOCs' competitors.

> Many times it is extremely painful to share one's home with strangers, especially if it has taken many years of loving work and care to build it. But as we all know, new relationships can develop, as well as new incomes, when one allows an old house of many empty rooms to be filled with new life and activity, by transforming it into a boarding home. This added income may also enable the original owner to construct a new home of a different type and shape, that would be more economical and functional for a new age.

As we look at the world of internetworking, interprocessing, and interservices, we cannot help but note that many new types of services will become layered on top of each new offering. The numerous interconnected services will require an extensive billing system to keep track of each contribution. However, before any offerings can be provided, we should realistically determine what form of revenue-sharing scheme will keep the price of the total offerings within the budget of the users, at a level to still attract them to the public network, as well as remain on it?

One consumer advocate at a state regulatory review questioned why voice-only users should pay for new data users' networks. We cannot help but shudder at this lack of understanding in her attempt to freeze current services and offerings at this point in time to a somewhat half electronic (digital), half electromechanical (analog) voice world. This would limit the numerous technical advantages that could be obtained from the newer technologies, as well as sentence many of the current users who are served by older systems to, at best, a future of government-subsidized services, or, at worst, a removal from the future information marketplace.

Why? The future for "voice-only" users can be appropriately described by the following metaphor: Assume users are currently transported about town by large diesel oil-burning and belching buses and into more remote areas by different types of reconditioned vehicles, such as old school buses. One new technical economic alternative is to construct a large bullet train that can handle all types of passengers, providing a long chain of cheap seats. Without it, if we continue to use only old buses, many users will move to other fancier private transports that provide new services to meet more of their needs, especially as many of these services can be more economically obtained from the new transport capabilities as they become cheaper. The net result will be fewer passengers on the old buses, requiring higher and higher revenue per occupied seat. As time advances, more will leave the bus until the near-empty buses become totally uneconomical to run. Then, with local government support, the remaining users will be shuttled around by a fleet of reconditioned old buses, new minibuses, or, in some cases, taxi cabs. This low-volume, high-expense overhead operation further reduces any chance for shared mass high-volume movement discounts.

Alternatively, some customers who require only basic service are unable to pay expensive prices. It is quite feasible that the new high-speed train with its many seats would be able to accommodate their special needs. Here, if many riders are using the public facilities, these disadvantaged customers may obtain more economical transportation, perhaps even a free ride, where their small additional cost burden is covered by the large volume of paying customers who remain on the public facilities. Unfortunately, if the public arena clutches to old technology or does not take advantage of the new technologies' tremendous potential, it will lead to many more parallel trains (networks) provided in the private arena, which will, by the way, pick up more and more passengers from the public network as they pass through neighborhoods and business communities along their way. This then reduces the capability to cover the cost burden of transporting the disadvantaged riders.

CARS, CABS, and TRAINS

With this in mind, why not take advantage of the new technology and use it to catapult ourselves into the new information age? Why not construct bullet-type trains (networks) to cut across and throughout our local monopoly, enabling gigabit movement of information? If we allow these inroads and paths to ring and traverse across our communities, we will have established the basic new infrastructure needed for the Information Age. In addition, we need to construct it so that these ring transport vehicles can be accessed by both private and public users in a manner that enables easy entrance and exit. It needs to be priced to encourage dynamic usage and movement of information, not as traditionally provided in the past so as to make a lot from a few, but to make a little from many, which then becomes a lot from many (Fig. 2-16)!

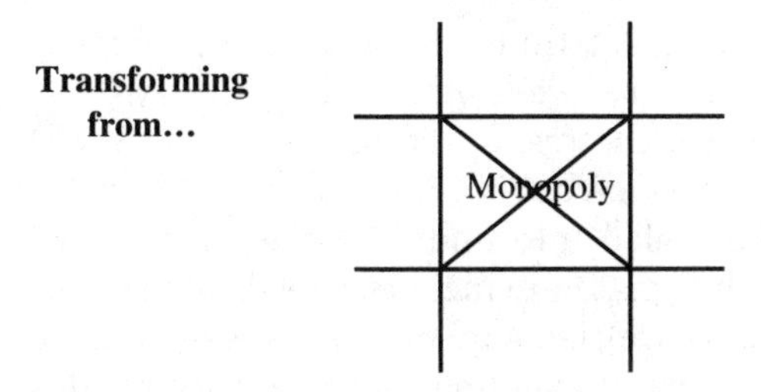

The Breakup of the Local Monopoly

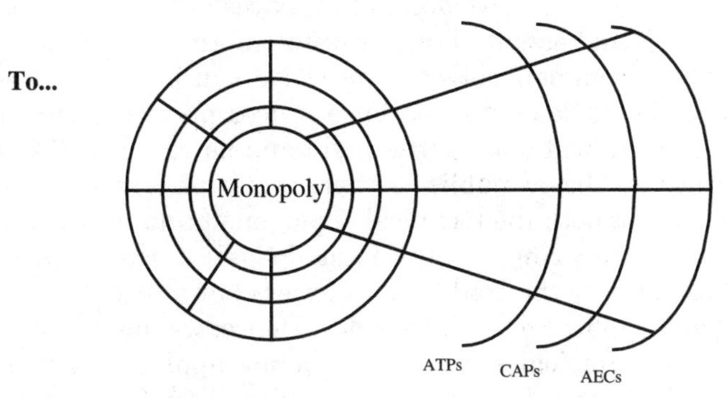

Figure 2-16. The breakup of the local monopoly.

These new common access bus systems (CABS) will enable new common access ring services (CARS) to be provided to both the public and private users. In this manner, a transfer ring access for independent nodes and systems (TRAINS) is readily available for all to use to formulate a transfer ring access to information networks and services (TRAINS). Additional transport manipulation and content manipulation capabilities can be provided at overlaid application service centers that provide specialized services for transport, information, and entertainment.

Can this concept become reality, or is it simply a figurative model that is not realistically feasible? Can it indeed be the vehicle to carry us forward into the Information Age? Is it an acceptable vehicle for the new post-Telecommunications Act marketplace, or should we look for something else? Does it successfully satisfy the judge's pre-Telecommunications Act concerns? To remove concerns of administrative blockage, we need appropriate open network architecture (see "Open network architecture" later in this chapter) and comparative effective interface (CEI) points of entrance and exit. Perhaps these access nodes and ring transports need to be managed by a separate division of the RBOC, which could have up to 50 percent ownership, or by a group of local providers. Sharing can occur if, for example, a block of ownership shares are allocated according to the size of the actual

LEC and ATP transport providers, not financial speculators—noting what happened when rural frequencies were obtained by lottery.

With these mechanisms in place, the transport ring network would then provide access to any providers' switching and database services, with revenue from the ring channeled into updating the entire basic infrastructure. Hence, it is also designated to meet the needs of the poor and remote rural communities, as well as support the access to basic public transport, where the common high-volume usage network resides for transporting voice, data, and video information throughout the local community. Thus all have access to services and features that are considered basic and essential to a growing information community.

Additional common network gateways can exist to enable further access to the various information services providers by splitting the network's technical and operational administration of the ring transport from the basic public services network. By using these new approaches for both the technical deployment and the business administration of a new ring network located closer to the users, we are able to access not only the traditional carriers but alternative carriers and specialized application service nodes. Hence, we have indeed been able to use new technologies to formulate a new approach to achieve a new, open, competitive information marketplace (Fig. 2-17).

Therefore, something new has indeed taken place to support the Telecommunications Act of 1996 that opens the arenas for all, while enabling LECs and RBOCs to provide "content" manipulation and presentation, as well as interLATA intrastate transport and manufacturing (Fig. 2-18).

As the Telecommunications Act of 1996 enables RBOCs and LECs to manufacture products, it is interesting to note that in reviewing past concerns of future telco manufacturing monopolies, many of the suppliers had noted that there will not be enough money for the RBOCs to both build these new networks requiring a full range of new services and spend the considerable sums necessary for personally designing and manufacturing the systems needed in the network. Table 2-4 (page 63) suggests that both endeavors are at the multibillion-dollar level. However, there is general agreement that the LECs need some R&D to provide the planning and requirements for these new systems, as well as operational support centers for controlling and administering their operations with some degree of new feature programming. Also, for some specialized products, the LECs may elect to design and manufacture a needed product. In a fully competitive arena, this choice should be left to the provider as a "make or buy" decision. Using the new competitive concept mentioned earlier, a new hierarchy will most likely develop in which the new access transport rings support the LECs' basic public network interfaces to their competitors' advanced service networks, which may support private services need-

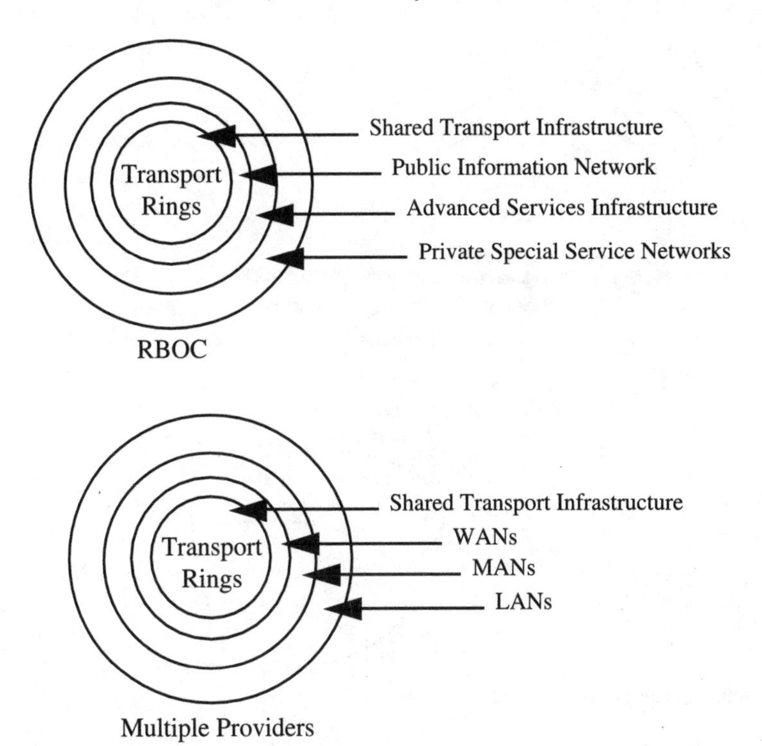

Figure 2-17. A multiprovider/RBOC approach to the new, open, competitive information marketplace.

ed for specific applications, such as hospitals, etc. The resulting array of systems and services using OSI standards can then be provided by many suppliers and used by many providers. Thus we will have established the needed infrastructure on which we can get on with the game and move forward into the information era, all participating in the growing, blossoming ISDN marketplace together!

Open network architecture (ONA)

ONA can be summarized as follows. The desire to open up a switching system to enable enhanced service providers (ESPs) to competitively provide their services to customers is indeed a noble objective. However, we must be very careful that what we do doesn't jeopardize a finely balanced structure. Advanced information networking (AIN) was mainly designed for a limited number and type of voice "enhanced database search" services. It has indeed tickled the interest of many ESPs to also provide similar services on an interrupt basis, but this challenges network integrity and feasibility as more and more services are provided by more and more ESPs in this manner. Some rapid service

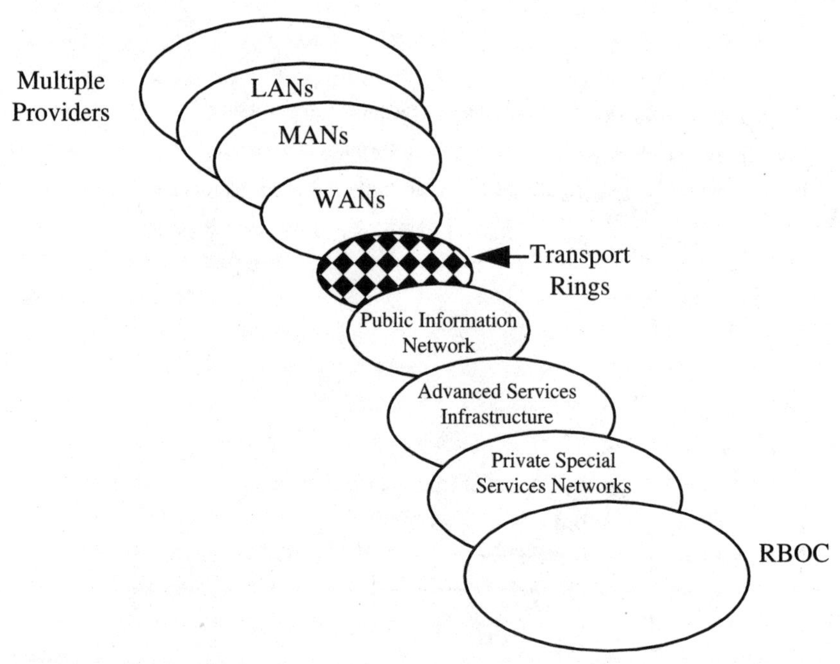

Figure 2-18. Sharing the local monopoly.

delivery approaches, using adjunct processors to provide inline software, have unfortunately opened the system to another degree of openness, which further causes concern for ensuring the integrity of the system. Also, these endeavors, depending on pricing of segmented offerings, will affect both public and private networking pricing for similar offerings. As RBOCs unbundle the machines, they must place themselves in the "if this, then what" situation to determine if they can indeed live and survive in the new scenarios that take place.

Some service providers have indicated that Bellcore's AIN architecture and the various RBOCs' versions of it don't provide for ESP involvement and software control of network switches' call processing. Therefore, some service providers have asked the following questions:

- In what direction is AIN leading the industry?
- What is (should be) the future network architecture that enables a competitive marketplace?
- Why can the RBOCs have access that ESPs cannot?

There's a view that new services can now be easily, quickly, and inexpensively added inline to call control either on an interrupt basis or by adding new software to the system in an adjunct processor. So it's quite natural that some providers would ask for the same interfaces as basic serving agreements (BSAs) with an internal set of basic serving elements (BSEs), like call transfer, to interrupt the call processing of the switch to bring the customer to their service platforms, where they can provide enhanced services (ESP) and information services (ISP). Pressure exists for immediate relief using AIN interrupts and unbundled pricing for numerous BSA and BSE offerings, but there's considerable danger that, as large expenditures occur for ubiquitously unbundling every aspect of the system, the net result could be a very expensive underlying transport mechanism that's extremely difficult to maintain and administer both economically and safely.

Pricing, collocation, number addressing, interface, and profitability issues are indeed the challenges of a new network architecture, as we shift from separate public or private networks to public and private internetworking. We must deal with the entire picture and not just a piece, on both an evolutionary and revolutionary basis, to resolve the considerations and observations noted.

Table 2-4. Multibillion-dollar Program Strategies.

Product suppliers	Networks provider	Service providers
Multiple prime supplier systems ■ Access switch R&D, $2 billion ■ Next-generation switch, $2 billion ■ Support centers, $2 billion ■ Alternative exchange centers, $2 billion ■ Info switches, $1 billion ■ Service nodes, $1 billion Total $10 billion R&D + $20 billion support = $30 billion program	■ Ring alliances for local transport access and control, $10 billion ■ Private-to-public internetworking number-translating and routing system, $3 billion ■ In-service gateways, $1 billion ■ International gateways throughout the global Europe, Asia, Africa, South America marketplace, $1 billion ■ Local switching and information handling of interactive voice, data, video messages, $10 billion Total $25 billion × 4 major carriers = $100 billion program	■ Service nodes for voice, data, video services Total $10 billion + $10 billion to support = $20 billion program

So the bottom line is $100 billion (networks) + $30 billion (products) + $20 billion (services) = $150 billion to get the game going, plus an additional $150 billion to finish rewiring America with fiber = $300 billion dollars.

Until now, the industry has not provided the world with a vision of where we're going and how we want to get there in terms of a feasible architecture for future networks and services that permit a competitive market to prosper and flourish. It's time to lead the way out of this confusion with these concepts for future narrowband, wideband, and broadband networks and services, in terms of the layered networks' layered services model. The industry is now at a crossroads.

Therefore, subsequent chapters attempt to look at the consequences of the actions from the different perspectives of what happens to the RBOCs' private offerings and public offerings from the network—service, pricing, network integrity, address availability, number portability, collocation, and overall feasibility and integrity—point of view, especially as fully switched networks not only deliver to the world voice services, but enter into the complex world of delivering data.

In the future, the integration of the layered networks' layered services (LNLS) model (see Chapter 6) with an appropriate layered information networking architecture (LINA) should hopefully provide more viable and feasible ONA access points. Here, access nodes located closer to the user provide direct connectivity to ESPs, as LINA offers traveling call-handling service requests to distributed processing platforms. Subsequent ESP direct access to customer transport call-control databases, as well as the use of ISDN D channels to communicate to Customer Premise Equipment (CPE) directly from Enhanced Service Providers (ESP) service nodes, will enable services to be more transparently provided throughout the central office (CO) network.

It might be feasible to open one's house for a few guests to walk around, but when 10,000 are attempting to do so, it's a different story.

A Future Look

As we look to the future and beyond, we cannot help but anticipate expanding growth of information usage, especially towards the end of the 1990s, as more and more inexpensive computer MIPS (million instructions per second) and communication bits become available to the user at reasonable prices to encourage this growth. On the other hand, if governmental restraints inhibit the LECs from providing an economical transport offering due to either too many fingers in the pie, too many cooks in the kitchen, or no incentive to cook the meal, then only the more specialized and focused offerings will be provided to the private world of closed user groups with little available to the general public. In this case, the ISDN marketplace will have shifted from the public to the private arena, where several integrated services digital networks will be provided in parallel to each other and accessed only via private tandem gateways and bridges for interexchange of information.

In any event, the ISDN marketplace is developing its form and expanding its shape. It will not be stagnant. It is much more than a narrowband, wideband, and broadband interface; rather it is an array of information networks' services offerings. Its growth will occur whether it is in a public or private arena, whether in the United States or Japan. As businesses become more interdependent, as segments of society become more interwoven, as the world becomes closer, there is an ever-increasing need for more versatile and complex "integrated networks' integrated services." This is indeed the vision and the scope of ISDN in the new competitive marketplace—the ISDN marketplace.

In Conclusion

The Telecommunications Act of 1996 promotes an open, competitive information marketplace interconnecting multiple, diverse providers' networks and services, or, presented another way, it becomes a "connect anyone, anywhere, to any service, any way, anytime" directive.

Unfortunately, many such straightforward, simple goals are not simply achieved. As implementation complications and complexities arise, this directive may become "just do it and worry about the ramifications later." Alternatively, it has been said by somewhat desperate manufacturers in a similar fashion: "Ship it and fix it in the field!" However, history has shown the actual ramifications of this mentality, as indicated in high cost-overrun military programs, such as in one new tank project. When the armor was decreased in thickness to reduce weight in a tradeoff for speed, the net result was getting the tank to the battle quickly so it could easily be destroyed by the first shell. A similar situation arises when telecommunications short-term versus long-term requirements are conflicting, such as the objective for an open, competitive information network interconnected to multiple providers and users, versus the more stringent requirements for a secure, survivable, robust, economical, personal, growable, integrated, etc., narrowband, wideband, and broadband interactive network.

So the new Telecommunications Act's goals for the new telecommunications age—to open up the playing field so it becomes one in which anyone can play any game, any way—can create conflicts as various implementation approaches arise. Here, reality must be addressed, especially as LECs carefully consider the full range of costs and ramifications of meeting the prime directive: "openness." Here "open, any way" means to many people "any place, every place" in the network, such as at the B box in the customers' neighborhood, at the remote line unit, the remote switch unit, the digital cross-connect, the intermediate distribution frame, and the main distribution frame, or at the initiating/terminating equipment of concentrators, contenders, multiplexers, and modems, as well as at the line stage, the switch translation tables, the customer databases, the lowest software subroutine, and at the STPs, SCPs, etc.

As "open" now applies to any and every existing tariffed service, it is necessary to reconsider how various techniques for achieving various services may have worked in the controlled environment of the telco having only a few internal overlays but will not work (survive) in the new, complex, open arena. This reassessment is similar to the need to readdress loading and boarding procedures to handle the new threat of hijacking and bombing of commercial passenger airlines. Checkpoints for entrances and exits needed to be established at key locations to ensure passenger protection by meeting new security requirements. So it is with the open competitive information marketplace.

LECs may desire to draw "lines in the sand" and then leave the playing field determination to the lawyers and appeals courts as costs and delays mount up for all parties involved. Alternatively, they may simply open everything up and wait for the catastrophic disaster, so they can assume the "see I told you it wouldn't work" position and then try to fix it, if it can indeed be fixed.

As the current baseline network is being "attacked" externally from all sides to be opened here, there, and everywhere, many internal programs are also in place or being proposed for applying service overlays and parallel offerings using emerging wireless technologies with various forms of additional use of (burden to) existing wireline capabilities. Together, the impact of both external and internal forces on the traditional network establishes an increased degree of vulnerability and support complexity that over time will grow and grow.

A Rare Opportunity

With all these changes and complexities, one could continue to paint a more and more dismal picture; instead it is time to step back and review the current situation with a more positive perspective by applying a more logical, controllable approach that addresses the many conflicting desires for "openness." It is time to offer a suitable way through the complicated, convoluted maze of realities, thereby providing a viable realistic solution that can be embraced by most parties (realizing, of course, that it is impossible to please all the people all the time).

In fact, if the future network is properly addressed with the correct answer to the mandates of the Telecommunications Act of 1996, not only millions of unneeded capital costs and human resources can be saved, but billions can be obtained from new revenues, as traditional providers develop the right sustaining infrastructure. Indeed, there is now a rare opportunity to provide "the missing link" to interconnect private customer-premise systems via a public infrastructure, as well as provide access to customer-selected alternative networks and database services. This link is a full-blown *operational access network*, located closer to the customer, providing *access switching nodes* that facilitate all forms (any way) of customer access, be it via analog modems, digital concentrators, or various types and speeds of circuit, packet, and channel transport vehicles.

This access node enables direct routing across the public network infrastructure, as well as paths to points of presence (POPs) of IXCs, ATPs, CAPs, VANs, and AECs and to ISPs' and ESPs' service centers. This platform is not a dirty, cheap, nonsupportable, interface box with limited cross-connect switching or packet-handling capabilities and controls. It has true network-quality switching nodal point capabilities for handling extensive growth in traffic of varying types and call mixes: voice, voice and data, data and text, text and image, image and graphics, graphics and one-way video, and fully integrated, two-way, symmetrical, interactive, switched voice, data, text, image, graphic, video conversations.

A new level

This node is actually a new level of the traditional five-level hierarchy that has evolved over the years since the initial 1934 Telecommunications Act, where the lowest entry level of the hierarchy is the traditional Class 5 central office (end office). However, this structure was based on analog voltage-net-loss (VNL) transmission principles employing the transport of groups and supergroups of coalesced voice calls. As conversations are digitized and quantized into 64,000-bit channels and transported over time-shared, nondedicated paths throughout the network, many new digital technical capabilities now overcome the previous analog transport restraints, thereby enabling distributed switching capabilities to be located closer and closer to the users. The initial applications of these technologies brought forth the remote switch nodes and remote line units for remotely selecting and distributing calls.

Once information is collected in the digital manner and identified as to call type, destination, and service needs, then there is no further requirement for traditional call-handling procedures and specialized physical mainframe connections. This call identification information accompanies the call in a traveling "Class Mark" package. As call types change from voice to data to video, it notes the required different and diverse forms of communications, using the effortless digital means of information transfer. However, by "homing in" on the singular local Class 5 from these initial remote switching systems and concentrators, they became simple extensions of the traditional central office, but by enabling selection of alternate paths to different Class 5s, POPs, or service nodes, these remote access nodes become highly functional Class 6 entities and can be dispersed any place in the network closer to the customer. They have limited front-end translation routing capabilities but extensive diverse customer-premise interfacing capabilities, using a full range of standard interfaces for not only narrowband, but wideband and broadband traffic, covering the full spectrum of information-handling characteristics, attributes, and service loads for the expanding applications of the blossoming information society.

Here, the new fiber-based access transport rings, with fully switchable Class 6 access switch nodes (see Figs. 2-19 and 2-20), enable private-to-public networking, customer choice of multiple providers, number portability translations, global addressing, POP access, customer provider service selection, entrance to private/public service centers, robustness, security, survivability, privacy, password verification, blockage, and audit trails, as well as preparation for future interfaces to the numerous forms and types of future CPE services.

Information switching systems

As Fig 2-19 indicates, private network nodes can bypass the traditional carriers or access network services via these new Class 6 nodes. Information switches can provide shared or specialized services above the network, with or without customer-premise systems. Extensive internal CPE networking may be established via new Class 7 entities that provide access to these new Class 6s for transport to traditional Class 5s, POPs, or service nodes and info switches.

Class 6 switch: What and why?

In considering a new Class-level switching node located closer to the customer, we have to ask the questions why, when, where, and, of course, what and how. When asked what is accomplished with the Class 6 switch, we might consider the following:

- Obtain survivable transport.
  Home on multiple base units.
  Survivable rings.
- Use the fiber to its maximum.
  Share facilities as close as possible to the customer.
- Reduce mileage cost to the customer.
- Enable direct access to other carriers' points of presence (POPs) without going through the entire public network.
- Achieve virtual private networking (VPN) under customer control.
- Enable dynamic bandwith allocation.
  Multirate – $n \times 64$ Kbps
- Provide narrowband ISDN/non-ISDN networking interface.

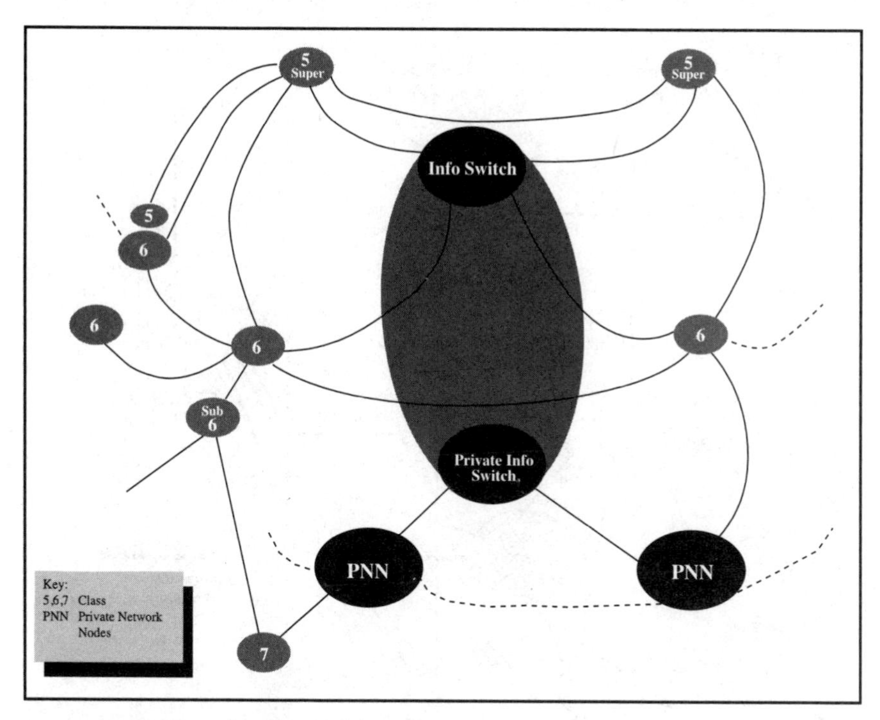

Figure 2-19. Public and private (P&P) network nodes.

- Provide wideband/broadband interfaces.
  Frame relay interface
  Switched Multimegabit Digital Service (SMDS) interface
  Fiber Distributed Data Interfaces (FDDI) interface
  ATM interface
  STM interface
- Channel-switching interface.
- Network management interface.
- Provide P&P (public and private) internetworking access points.
- Enable shared transport between nodes for private networking over publicly shared facilities.
- Provide quick access to info switch service nodes for advanced services.
- Control Sub 6 distribution plan fan-out for fiber to the home or pedestal (FTTH or FTTP).
  New fiber distribution plan
- Provide interface node for broadcast capabilities of high-definition TV (HDTV) or private communications services (PCS) network via Class 6 or Sub 6.

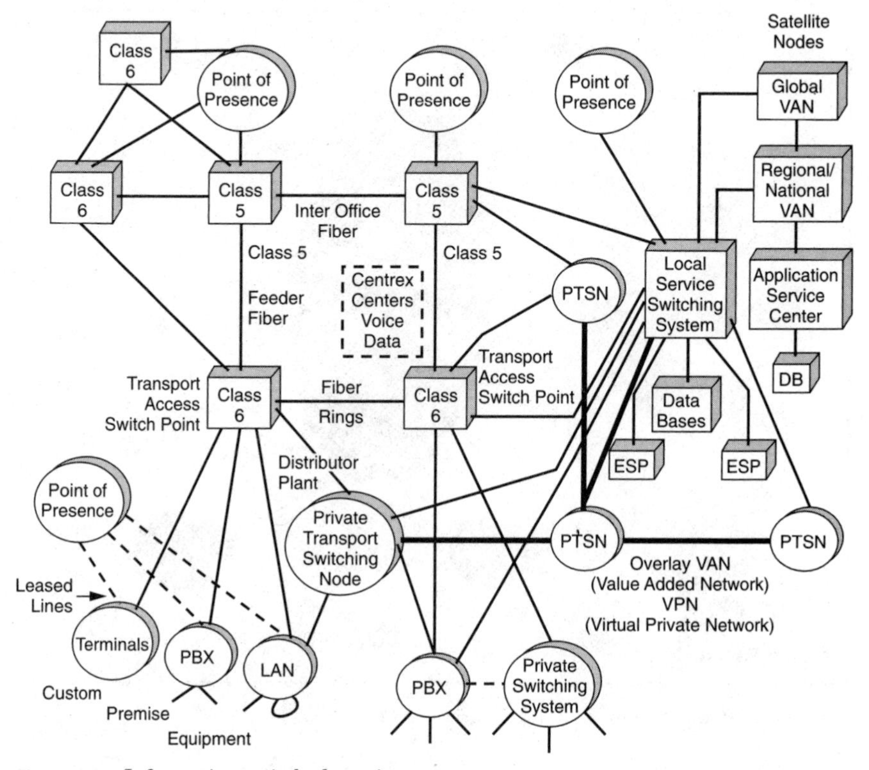

Figure 2-20. Information switched services company.

- Provide front-end translations and interfaces for LNLS addressing.
- Provide direct access to multiple Class 5s or via multiple Class 6s to multiple Class 5s. (Note: This is a key reason for Class-level access.)
- Provide front-end functions to interface over INA/ONA to a new Class 5 broadband superswitch, other Class 6s (to prevent "ring around the rosie"), and the new customer-premise Class 7 switch; also to provide INA/ONA network interface functions to interface with info switch-type service nodes.

Class 7 switch: What and why?

Over the years, many have questioned the difference between a PBX (private branch exchange) and a PAX (private automated exchange). The answer was quickly forthcoming and easy to understand: With the PBX, a customer can originate and receive calls from the network, while a PAX is strictly an internal system on customer premises. While PAXs usually don't have an internal operator, PBXs usually have an operator to intercept calls from the outside world and route them to an internal address. More sophisticated PBX fifth-generation systems have voice-answering capabilities that automatically ask the outside party to dial the internal address once the network is finished completing the call to the PBX.

Centrex-type services were initially provided on a network basis to internal user groups, where numerous customer campuses could obtain services from platforms shared with other user groups. Here, internal numbers were made available to the outside. Their terminals could be reached using a hunting mechanism that searched through available access lines. However, rather than use a single common number to the system, the individual internal number could be provided without requiring the intercept operator or announcement for inward dialing.

With the advent of data came LANs, cluster controllers, distributed network gateways, etc., to intranet data within the customer premises or internet between customer premises on a low-speed basis, on a wide-area basis (WANs), or around a city as metropolitan area networks (MANs). With the discovery of economical T1 (1.5 Mbps) and T3 (45 Mbps) transport capabilities came the private-to-private (inter) networking nodes (PNNs), located on selected customer premises, using leased facilities between nodes. Here, protocol conversion, bridges, routers, and gateways blossomed to achieve LAN-to-LAN internetworking as frame relay and FDDI protocols emerged and blossomed.

However, the need to internetwork more and more terminals to more and more locations focused attention on the asynchronous transfer mode (ATM) technology of the public network, as well as synchronous transfer mode (STM) circuit-switched technology. This technology combined with the power of the fiber to introduce narrowband/wideband/broadband switches in the customer environment. These switches can now provide *n* number of transport channels over direct fiber to any terminal (voice, data, text, image, video), and these switches will become the new PBXs or business-serving modules. But they'll be distributed internally and fully integrated with the outside networks. In this manner, they'll become the new Class 7s, by enabling direct network addressing and dialing to any customer terminal via direct open network architecture (ONA) interfaces. Hence, the platform enables direct access via Class 6 rings to the traditional network, to internal networks, to other private networks, as well as to other points of presence (POPs) or to various service-provider info-switch-type service centers.

Summary

Thus there is indeed a rare opportunity as a result of the Telecommunications Act of 1996 to deploy the new hierarchy, specifically designed for a diverse marketplace, using digital capabilities to achieve a fully distributed intelligent information transport and processing network. The network's access node serves as a *"firewall,"* protecting the traditional network from destruction but opening it to transport the customers' information to numerous alternative network and new service providers. There is no need to burden the traditional network's higher layers with traffic that can cause extensive blocking, overloading, and backhauling. In fact, with the new Class 6s at the entrance level with corresponding Super 5s at appropriate higher-level locations, both in ring-configuration structures, the network now takes advantage of gigabit fiber-transport capabilities. Traffic loading can be reconfigured appropriately, providing required local access to POPs and service nodes, as well as eliminating many of the traditional roles of local tandems. Network management mechanisms as well as operational controls can be extended to this new access level as they are fully deployed across a widening range of layered networks' layered services. In time, customer premises will have new Class 7 switching nodes as distributed switching capabilities fully populate customer campuses. Here again, these network administrative controls can then be applied deep within customer premises.

Hence, these fiber-based access rings with Class 6 level switches can be deployed under shared ownership to eliminate plowing up city streets for each new participant's fiber/coax network. They should be designed to not only meet existing voice services needs, but be growable to support an evolving body of new, fully integrated voice, data, and video services for the forthcoming information era (Heldman 1993, 1994, 1995). Now that the second shoe has dropped with a loud thud, with far- reaching ramifications, we have an opportunity to not only address this latter concern, but achieve the former; a rare opportunity, indeed!

References

1. Cauley, Leslie. 1996. *Wall Street Journal*. Cellular-phone spin-off marked start of slide leading to PacTel's deal. April 2.
2. Heldman, Robert. 1993. *Future Telecommunications: Information Applications, Services, and Infrastructure*. New York: McGraw-Hill.
   ———. 1994. *Information Telecommunications: Networks, Products, and Services*. New York: McGraw-Hill.
   ———. 1995. *The Telecommunications Information Millennium: A Vision and Plan for the Global Information Society*. New York: McGraw-Hill.
   ———. 1996. *Encyclopedia of the Future*. Vol. 1: Digital Communications. New York: Simon & Schuster/Macmillan.
3. United States Advisory Council on the National Information Infrastructure. 1996. A nation of opportunity: Realizing the promise of the Information Superhighway. January.

Part

2

The Competitive Information Marketplace

As we integrate
communications and computers,
as we layer
information networks' services,
we are indeed forming
a global village.

Chapter

3

The Customer

Customer, customer,
be ye:
aggressive or timid,
assertive or reserved,
frugal or fickle,
intelligent or ignorant,
rich or poor,
right or wrong,
ye be—the customer.

It has been said many times that the customer is always right, even when 100 percent wrong; but the more informed customers are, the more they are right. Similarly, providers and suppliers also need to be better informed about their customers. In this world of continual change, we are rapidly becoming a more "market-based" industry, as we shift from simply viewing our users as subscribers to seeing them as customers. Here, the key to success is knowing what the customer wants. This has caused a flurry of activity to identify customer needs that can be fulfilled by such and such a product or service. Once this is achieved, further analyses and surveys will then quantify this information in order to justify the expenditures for delivering the new services. The "numbers game" can be played to convince the more financially based management that there are indeed sufficient potential customers who actually have this or that need.

But, who are the customers? What are their needs? How do we differentiate real need from wishful greed, as we attempt to determine differences between musts and wishes? (See Fig. 3-1.) These potentially would-be/could-be buyers are being scrutinized from every aspect. They are being separated and grouped by market segment, by industry sector,

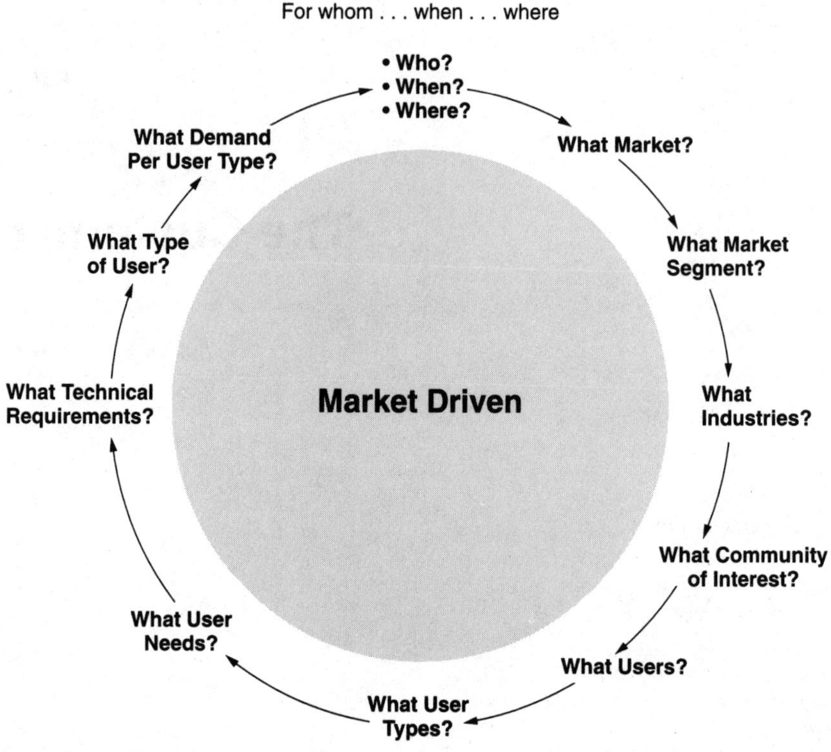

Figure 3-1. Market driven.

by geographies, and by demographics such as race, religion, age, and wealth. In the past, the traditional 20/80 percent rule noted that 20 percent of the people will purchase 80 percent of the items. Alternatively, in the real-world, Marketing-101 classroom, the top "street-wise" salesperson has often broken the market down into a less exact classification, using terms such as trash, mass, and class markets. We have often heard the words "mass market." Unfortunately, U.S. manufacturers over the 1960s to the 1980s have transformed mass market products into "classless" high-volume items. At the same time, "made in Japan" shifted to mean high-quality, low-cost items, while "made in America" changed to imply items of low quality and high cost.

Once import restrictions were imposed upon Japanese manufacturers, they took another tack. They recognized that the masses do appreciate being appreciated. As toilet paper became soft, buttons on car radios became smooth; seats, steering wheels, and shift mechanisms became "comfortable." This new emphasis on the customers' feelings and comforts has moved backroom, human-factors studies, which were previously performed after the design, to the forefront as the new science of ergonomics.

This has led to the "feature on the feature" realization. Here, for example, not only does the automobile radio become a stereo in the Toyota Camry, but the lower section of the unit pulls out with two cup holders located at just the right level to successfully hold just the right size cups to enable the driver (and passenger) to enjoy their morning "brew" while they listen to their digital disks on the way to work, or as they travel here and there.

So, we enter the exciting and complex world of user needs. We have needed to enter it for some time. We need only to look at the touch tone buttons on the traditional hand-held "princess" phone. Here you can note that there is no mechanical touch feedback to tell you that the tones have been sent. There is no snap to the buttons to indicate that there is no longer any need to continue to push. Since it is impossible to know, unless we hold the phone to our ear and push at the same time, we try to hold it as close as possible, and guess. There is also no visual display to show what numbers were actually sent to help us remember if we pause in our dialing. The results of various test studies indicate that there is a high potential (approximately 30 percent) for misdialed calls.

Similarly, call waiting of the custom-calling feature package denotes a lack of understanding and emphasis on customer needs. To demonstrate this, simply imagine waiting for two very important calls. One assumption could be that important calls cannot be missed. However, this should also take into consideration that many important calls cannot be interrupted (at least immediately). The original service was implemented in a manner that notified the called party that another call was waiting, but it did not notify the calling (waiting) party, that the called phone was in use. Hence, if the first conversation was not immediately interrupted, the second caller could surmise that no one was at home (or in the office), and prematurely hang up. A simple change would have been to provide a separate busy tone indicating that the called party was busy, but that the called party did know that another call was waiting, so the caller could simply hold on. This "twist on a twist" or "feature on a feature" would have made the world of difference in our daily lives. More people might have purchased the feature if the phone company had removed this major objection. Many times this is the difference between success and failure.

How many times have we seen such-and-such a product fail only to see it come back, usually by a different vendor with a slight twist or emphasis that makes it much more desirable and successful. This has been a successful, key strategy of the Japanese, and they still continue to browse through our technical patents and dropped market offerings. We have seen VCRs, as an underdeveloped American technology, capitalized by the Japanese, as they enhanced and perfected their operation.

In a market-driven marketplace, the customer becomes the consumer, the buyer, the user—the king or queen for the day. So, what does the customer want? What are the needs of the users (human and machine, basic and operational), as we proceed from product to feature to service to application, as we move from market to sector to community of interest? (See Fig. 3-2.)

The Customer

As we pursue our search for new services based upon applications, it is important to keep in mind both the human element and the machine element of the final solution. The providers of communication networks initially only concentrated on system requirements that made their network operation more effective. Their suppliers provided equipment for their customers—the providers—but not equipment for the providers' customers—the end users. Similarly, computer manufacturers initially only provided systems interfaces for machine-to-machine information exchange and processing, totally ignoring the need for "user-friendly" terminals. This caused a slow progression from machine-language assemblies to high-level language compilers in order to enable the user to program the systems with enhanced and versatile features, and reduce test-time errors. Similarly, today new data management systems are continuing their quest for greater ver-

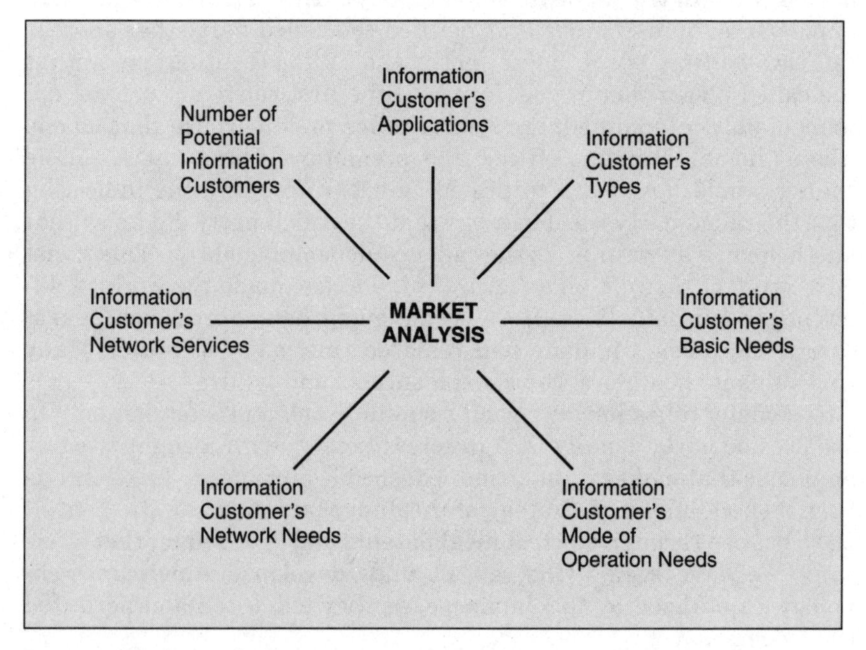

Figure 3-2. Market analysis.

satility, as new support languages attempt to enable immediate access to physically distributed information in multiple-vendor systems by using new SDL languages, relational databases, expert systems, fuzzy logic, neural logic, parallel processing, and data management techniques. Hence, we need to understand both the human and machine aspects of user needs as we consider information-service solutions for specific applications. (See Fig. 3-3.)

People

People have, in many respects, similar needs and desires; but in other areas, people are indeed quite different. The older we become, the more we realize how different we really are. In this world of complexity and choice, there is considerable effort to classify us in terms of political views, religious views, and personalities. Many times, for lack of a better yardstick, we use terms such as right, left, or middle of the road; but people are more complex than these general classifications. Figure 3-4 lists several attributes and personalities. One interesting exercise is to attempt to categorize your associates and family members. For example, consider who would be likely to go on a vacation to a distant or remote land, such as Papua New Guinea, or Hawaii? Who would go there in a cruise ship, or airplane? Who would go in a "tramp freighter?" Who would sail themselves? Who would stay a few days? Who would spend several weeks seeing its remote villages? Or who would spend a year or two learning the culture? So it is with new services; a few are leaders; some are followers; most are watchers, and a few are stone throwers.

Basic Human Needs
Human Operational Needs
Customer User Needs
People Types
Computer Types
Market Sectors
Current Services
Industry Tasks
User Types
Market Segments
Geographical Areas
Future Services
Information Market Applications

Figure 3-3. Customer needs/customer applications.

THE INFORMATION SERVICES MARKETPLACE

PEOPLE

ATTRIBUTES	CHARACTERISTICS
• Fearful	• Leader
• Resistant	• Follower
• Adventurous	• Watcher
• Daring	• Stone Thrower
• Innovative	
• Conservative	
• Affluent	
• Poor	
• Middle Income	

Figure 3-4. The information services marketplace, people.

Finding "base motives" is also a challenge. These must be understood and considered. Many times what appears to be is not so. Why people do something may be tied to nonapparent economic or politically motivated reasons. One potential user of a narrowband ISDN trial period was championing the event because he would be given CPE terminals for the period, which would be leased by the network provider to demonstrate the effectiveness of ISDN. In actuality, some believed that the user had quietly arranged to pick up the terminals at a "salvage value" of 10 percent of cost from the terminal manufacturer after the trial. This caused a demand for more and more unnecessary terminal equipment to verify this or that feature. Once the tie was removed when the provider shifted to another vendor, the user was no longer interested in the trial. Hence, participants in new ventures need to be carefully chosen to ensure that personalities and personal goals are consistent with utilizing new technologies and services. Many will use equipment on a trial basis, but when required to pay, they refuse to buy. Are they really players or watchers, or in it for something else such as price or base motives?

Basic Human Needs

We have many human needs. (See Fig. 3-5.) Maslow's hierarchy of needs puts the human survival needs of food and clothing at the lowest level and then moves up through security, work satisfaction, power and control, recognition, and self fulfillment and contentment. Human needs are constantly changing, and we should not expect a product or service to become a major success unless it satisfies a basic human need!

THE INFORMATION SERVICES MARKETPLACE

BASIC HUMAN NEEDS

• Wisdom	• Power	• Shelter
• Knowledge	• Control	• Food
• Self – fulfillment	• Enjoyment	• Health
• Self – achievement	• Love	• Faith
• Sense of Quality	• Security	• Hope
• Success	• Communication	• Satisfaction

Figure 3-5. The information services marketplace, basic human needs.

Human Operational Needs

Today's society has progressed from the somewhat safe but confining parochial/provincial village boundaries. No longer are towns and cities self-contained entities. People are on the move. Business is becoming more and more global as operational boundaries shift from state to region to nation to country. Raw material is shipped for refinement in one country, from there sent to another country, where it is used in the assembly of products that are purchased back in the first country.

In this global society, information will need to be current, extensive, and reliable. Access to it must be timely, efficient, and effective. Its form of availability and utilization must support our information-services-based activities, just as the bulldozers provided real value to the ditch digger, or as the printing press supported the book maker. Information will be the tool of the twenty-first century to enhance our lives. (See Fig. 3-6.) Various possible uses and benefits have been projected in televised "information society" ads that noted how information can help us manage time and lower stress in our daily (operational) lives.[1]

Customer/User Information Needs

As information plays a larger role in our lives, we need to better understand where, why, and for whom information should be available. ISDN was initially established in geographically separate islands. Local area networks only provided interconnectivity within a physically limited, specific area. Unfortunately, times today are indeed multilocation, if not multinational. There is a need to interconnect locations or islands together. Closed user groups (CUGS) were designed by the computer manufacturer to connect like termi-

[1]US West Communications, AT&T, MCI, and Andersen Consulting, 1990.

THE INFORMATION SERVICES MARKETPLACE

HUMAN OPERATIONAL NEEDS

- Data Input
- Time Control
- Effectiveness
- Efficiency
- Conflict Reduction
- Low Frustration
- Stress Minimization
- Senses Utilization
- Information Exchange
  - Analysis
  - Agreements
  - Contracts
  - Decisions

- Operational Delay Minimization
- Business Travel Comforts
- Operational Supports (while on the move)
- Current Information
- Reliable Information
- Less Data More Analysis/Implications/Conclusions

- Computer Aided Task Support Systems
- Direct Personal Operational Improvement
- Human to Computer Interfaces – Friendly
- Computer Operational Needs
  - Computer to Computer
  - Terminal to Computer
  - Terminal to Terminal
  - Speed
  - Accuracy
  - Availability

Figure 3-6. The information services marketplasce, human operational needs.

nals or common families of terminals and processors to their mainframes. By using specialized internal communication interface protocols, they enabled only their terminals to communicate with the computers. Even as front ends were remoted, they could only interact with systems employing like protocols. This was so until the early 1990s, when internetworking between dissimilar systems became the new focus.

Each of the needs noted in Fig. 3-7 needs to be carefully analyzed and considered before launching a new offering. These needs indicate

THE INFORMATION SERVICES MARKETPLACE

CUSTOMER USER INFORMATION NEEDS

- Universal
- Uniqueness
- Price
- Performance
- Convenience
- Cost Control
- Privacy
- Security
- Ubiquitous

- Remote Control
- Available
- Time Control
- Time Programmable Event
- Immediate
- Flexible
- Choice
- Access

- Transparant
- Less Complex
- Indexed Information
- Friendly Interface
- Ability to Generate, Store, Process Manipulate, Transport and Present Information to the Eye, Ear, Hand, Nose and Mouth

Figure 3-7. The information services marketplace, customer/user information needs.

and sometimes dictate the slight twist, feature on the feature, or emphasis that will make the difference. For example, "ubiquitous." A key reason for ISDN is to enable remote users to access services formerly only available to closed user groups. For example, the doctor at home wishes to access x-rays in the hospital, or insurance firms need to access police automobile accident records. Hence, the network needs to be "ubiquitous" within a city or state before a large group of users will begin using it. This alters the small build-up strategy, but it promotes an all-or-nothing approach, which many believe has been the Achilles heel of the ISDN deployment strategies of the 1980s.

Computer Needs

The old strategy of putting oneself in the shoes of another to better understand the other's problems and decisions still holds. So, let's not limit ourselves to being human beings. If I am a computer working in some application, then what do I need? This is a fun exercise to help us better appreciate a computer's needs. (See Fig. 3-8.) We must recognize that there are as many various types of computer/terminal modes of operation as there are various types of computers and terminals. (See Fig. 3-9.) Each is more or less equipped to do different things, especially as the computer leaves the strictly computational world of arithmetic-type functions and becomes more involved with information manipulation, processing, and presentation. So, we need to reassess what type of computer/terminal device is needed and what feature requirements are essential in order to achieve success for some particular mode of operation in the specific application in which it is being utilized.

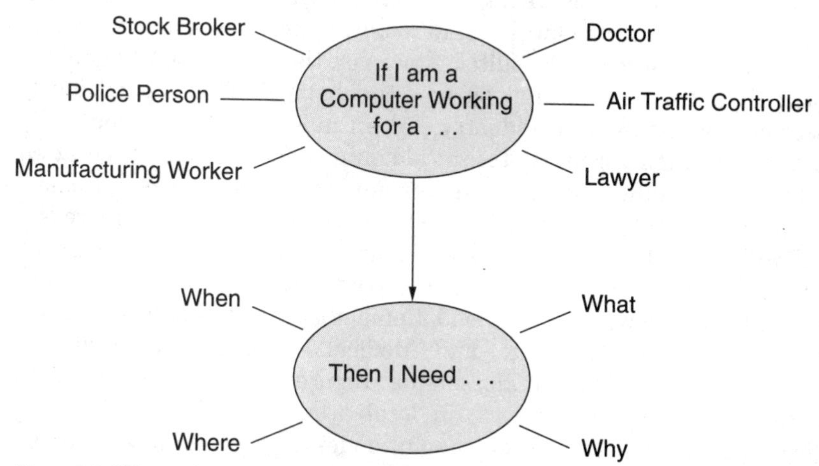

Figure 3-8. What computer needs?

Personal PC's	Cluster Controllers	Terminals
System PC's	Super Computers	Satellite Computers
Microcomputers	Communication Controllers	Information Switches
Minicomputers	Sensors	Protocol Converters
Mainframes	Work Stations	Display Systems
Specialized PC's	Parallel Processors	Front-end Processors
— Factory Automation	Distributed Processors	Intelligent Systems
— Tool and Die	Large Mainframes	

Figure 3-9. Computer types.

User/Computer Needs

User-applications analyses generally consider:

- Work at home.
- Remote shopping.
- Remote banking.
- Remote education.
- Access to entertainment.
- File access and management.
- Data manipulation and processing.
- Archiving.
- Presenting multimedia information.

(See Fig. 3-10.) This enables the medical, legal, financial, and other industries to exchange information in real time. Hence, time again becomes a major factor as we improve productivity by removing the "hurry up and wait" aspect of our lives. In today's society, more and more specialists are forever being consulted. However, the "real time" dilemma is getting them this or that analysis on a timely basis. This is particularly a problem for the medical industry. For an industry that is supposedly modern, with highly complex technical equipment, its mode of operation of moving the results of tests to the consultant is in many cases archaic. The more remote the doctors are from the hospital complex, the more isolated and alone they become. Similarly, as more and more people await trials or are involved in lawsuits, the traditionally slow legal system becomes bogged down with overload. Remember the old sayings, "Justice delayed is often justice denied," and "Medical delays are often fatal."

Yes, private networks are blossoming. Figures for the 1990s indicate continuous, tremendous growth in local area networks (LANs), but, unfortunately, that was only a nice solution for a few closed office groups. The interconnectivity problem has continued to grow.

• voice	• interconnectability	• survivable	• multimedia
• data	• internetworking	• securable	• multinetwork
• text	• interprocessing	• reliable	• multiservice
• image	• interservices	• portable	• multiapplication
• graphic	• interfaces	• maintainable	• multispeed
• video	• integration	• available	• multinational

Figure 3-10. User/computer needs.

Communities of Interest

This then brings us to a new approach. Why continue to build islands of users that also need to interrelate with numerous users outside their island, but can't? Why not construct networks to interconnect the total community of interest (COI)? In a sense, this has been attempted in the past for large business users, who can pass on the high costs of private facilities to their customers. In the 1980s, they began to construct expensive backbone networks to interconnect their firms together, using leased lines, microwave, radio or satellites. Autonomous internal networks were "bridged" or "gatewayed" together, using bridges, routes, or broutes to interconnect Ethernet, Token Ring, or Token Bus local area networks over metropolitan area networks (MANs), or wide area networks (WANs), as will be discussed later.[2]

However, these costly endeavors still solved only 50 percent of the problem. Many separate, small firms or customers needed to connect to the larger businesses' networks. This then leads us to further questions, such as: What about extending internal networks to the full spectrum of users? Also, what about extending both the internal and external users within a COI to other COIs? But, first, what do we mean by a COI?

Market Sector

We can break the potential "information marketplace" into areas such as business, residential, government, and education. Then, we can refine our breakdown by sectionalizing business into large and small; government into federal or state, etc. We can then segment these sections by function, further decomposing state government into judicial, police/fire, road, park, etc., to form a subset of functional areas that could be considered candidates for closed user networks. For example, a radiologist medical group would like to have the ability to interexchange x-rays in order for the various doctors to interconsult.

---

[2]See Chapters 4, 6, and Appendix B for further information.

INDUSTRIES

• Banks	• Law Agencies
• Federal Agencies	• Manufacturing
• State Government	• News/Magazines
• Education	• Wholesale/Retail Stores
• Investment Firms	• Information Services
• Entertainment	• Health Care
• Insurance	• Home Communications
• Transportation	• Operations/Utilities
• Small Businesses	• Large Business Complexes

Figure 3-11. The information services marketplace, industries.

All in all, major industries, such as those noted in Fig. 3-11, can be identified, sectionalized, and segmented. Here their subgroups interexchange information internally and send and receive information from an ever-increasing span of external business databases and customers.

Conceptual Planning Exercise

If we were to pause now for a moment and conduct a workshop to consider the various interconnectivity needs of these industries, we might determine that a much wider spectrum of connectivity is needed than what would be initially considered. So, if we played the "if" game—if I am a doctor within the medical industry, with whom do I need to communicate to obtain and exchange information? (See Figs. 3-12 through 3-15.)

MARKET SEGMENTS
(Community of Interest)

MEDICAL INDUSTRY

• Radiology	• Medical Journals
• Speciality	• Medical Library
• General Practitioners	• Medical Computer Search Firms
• Hospitals	• HMOs
• Pharmacies	• State Poison Centers
• Insurance Firms	• 911
• Drug Industry	• Federal Disease Control Centers
• Teaching Universities	

Figure 3-12. Medical industry (community of interest).

MARKET SEGMENTS
(Community of Interest)

LEGAL INDUSTRY

- Lawyers
- Local Courts
- State Courts
- Federal Courts
- Prisons
- Half-way House
- Incarceration Homes
- Superior Court Ruling System
- Law Libraries
- Legislative Laws
- Tax Rulings
- Real Estate Ownership Files
- Land Ownership Files

Figure 3-13. Legal industry (community of interest).

MARKET SEGMENTS
(Community of Interest)

FINANCIAL INDUSTRY

- Stock Exchange
- Commodities Market
- Exchange Rates
- International Markets
- Federal Reserve Network
- Funds Transfer Network
- Transaction Bank – Bank Network
- Credit Card Network
- Branch Banking Network
- Remote Teller Network
- Smart Card Network
- Bond Market
- Investment News Network
- Pension Funds Management
- Research Data Bases
- Tax Advisory

Figure 3-14. Financial industry (community of interest).

MARKET SEGMENTS
(Community of Interest)

AUTOMOBILE INDUSTRY

- Factory Assembly
- Parts Orders
- Market Orders
- Spare Parts
- Accidents/Collisions
- Theft
- Piece Parts Industry
- Tire Industry
- Repair
- Title
- Insurance
- Advertising
- Servicing
- Leasing
- Cabs
- Rental Cars
- Rental Trucks

Figure 3-15. Automobile industry (community of interest).

COI Networks

In addition, there are unique, specific COI networks for specific private industries or shared endeavors, as well as those with geographic community of interest (COI) emphasis in local, national, or international arenas. (See Figs. 3-16, 3-17A, and 3-17B.)

Private Networks

- IBM
- IGM
- GE
- RCA
- AT&T
- US WEST
- Notre Dame University
- FBI
- Department of Defense
- SAC
- St. Luke Presbyterian Hospital
- Chicago School System
- Mayo Clinic Network

Shared Networks

- Shared Tenant Services
- Credit Check Verification
- Retail Point of Sale
- Lost Child Network
- Fugitive Network
- Coin Dealers Network
- Library Book Exchange
- Mayo/Johns Hopkins/Georgetown George Washington Clinics
- Denver's Medical Network

Figure 3-16. Specific community-of-interest networks.

MARKET SEGMENTS
GEOGRAPHIC EMPHASIS

NATIONAL NETWORKS

- Washington, D.C.
- New York
- Boston
- Chicago
- Atlanta
- Minneapolis/St. Paul
- Omaha
- Kansas City
- Dallas
- Houston
- Phoenix
- Albuquerque
- Las Vegas
- Spokane
- Seattle
- Portland
- San Francisco
- Sacramento
- San Jose
- Los Angeles
- San Diego

Figure 3-17A. Geographical COI networks.

MARKET SEGMENTS
GEOGRAPHIC EMPHASIS

GLOBAL

- New York
- Stockholm
- Madrid
- Paris
- London
- Geneva
- Zurich
- Moscow
- Hong Kong
- Rome
- Brussels
- Bonn
- Frankfurt
- Auckland
- Tokyo
- Sidney
- Singapore
- Vienna
- Honolulu

SPECIFIC GEOGRAPHIC NETWORK

- Washington, D.C. Legal Network
- Wall Street Financial District
- Boston – Washington Corridor

Figure 3-17B. Geographical COI networks.

Private and Public Networking

Hence, there is a need to internetwork both internal and external users within a community of interest across local, regional, national, and global geographical areas. This, therefore, necessitates the involvement of more than one carrier or provider, and it shows the need for internetworking private and public communications, from both a transport and a service basis.

Additionally, as noted, there is a need to access other industries, some continuously, others on an occasional basis. Hence, the need exists for cross-COI information networking. Here, an accident patient's medical billing from a medical COI crosses to the insurance COI, or to the banking COI, while the insurance COI needs to access the police COI to obtain more detailed information concerning the particular accident.

Hence, all major COIs often need to interconnect at some point in time, usually resulting in the situation shown in Fig. 3-18. Figure 3-19 also indicates related areas that are less frequently accessed, or separately accessed from more predominant COIs. Eventually, there will be the need for the more universal public/private interconnectivity so that all the private networks can be accessed via a public entity. Hence, in time, the COI will utilize more publicly common networks, but can privately provide value-added overlays that ensure security, privacy, extended service access, data manipulation, and priority override capabilities for those within the COI. (See Fig. 3-20.)

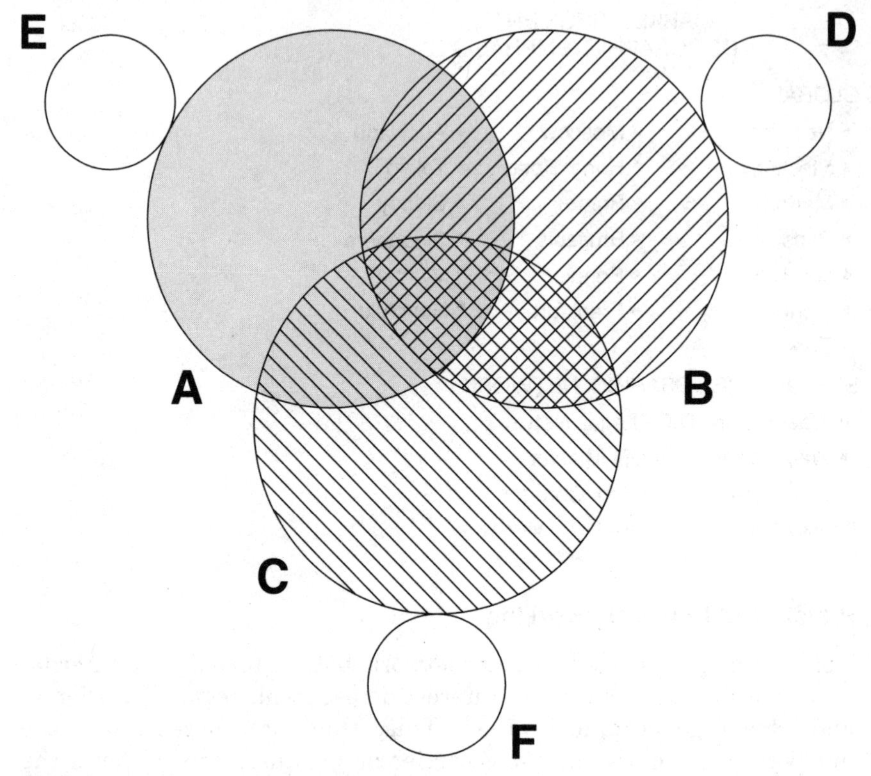

Figure 3-18. Private community-of-interest access and overlay.

User Types

As noted in the preceding analysis,[3] we can analyze each major industry to determine what type of networks or services best fit their present and future mode of operations. This is done by first identifying several voice, data, and video categories of information movement and management (IM&M), such as inquiry/response, data collection, data distribution, remote documentation, and display, etc. Then we can review each major operational task of the firm to determine which form of information movement category is (or could be) used to accomplish the task. Once this is identified, a range of technical values, such as data speed, connect time, holding time, can be applied to each endeavor to more closely approximate its real-world application, using current and future technology.

In this manner, we can determine various types of users per industry. Next, we can relate them across the industries to determine those

---

[3]See *Future Telecommunications,* McGraw-Hill, Chapter 3.

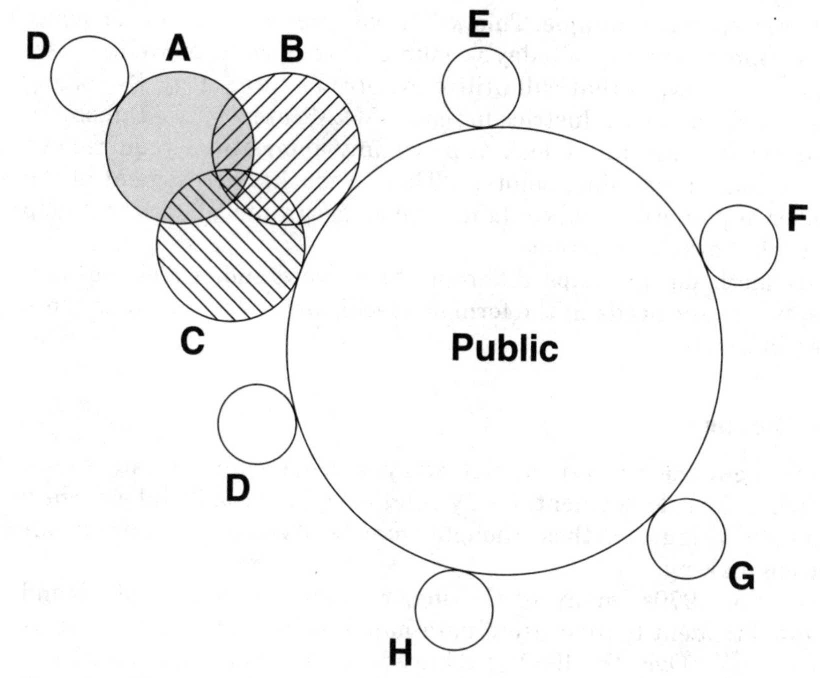

Figure 3-19. Transition—part public, part private.

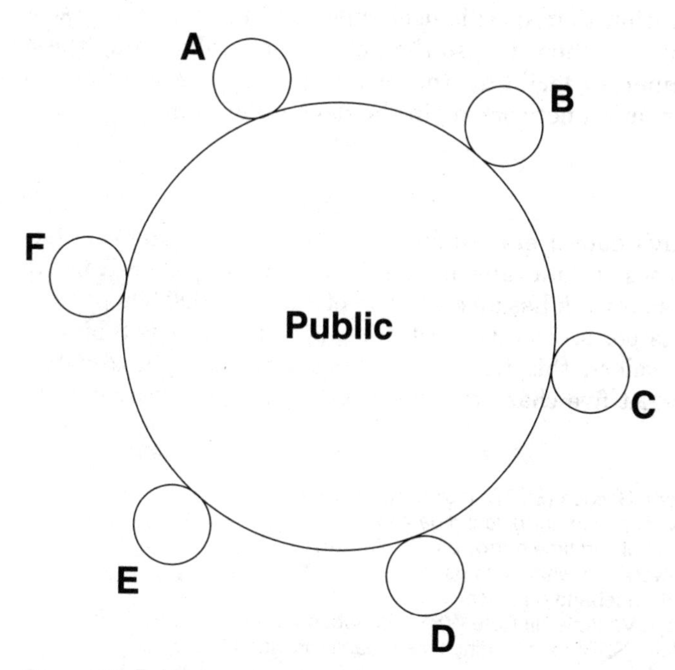

Figure 3-20. Public/private networking.

that are common or unique. This will leave a common set of user types, which represent many of today's communication users, as well as a set of future user types that will utilize tomorrow's technology. By sorting lists of tasks across industries for each IM&M category, we can obtain a somewhat quantitative look at how many applications require such an information-handling solution. Then, a market assessment of the number of potential applicants in a given geographical area will help establish market projections.

This methodology helps differentiate narrowband, wideband, and broadband user needs in the form of specific user types, such as those noted in Fig. 3-21.

User Groups

As we begin the analysis and strategy sessions over the late 1990s, involving fiber deployment, CATV relationships, broadband switches, data networking, etc., these thoughts may be of some assistance in our decision making.

Over the 1970s, many of the major industries were visited and reviewed to identify their users' data-handling needs for internal operational tasks. Over the 1980s, we have had numerous opportunities to rethink these analyses and update them due to changes in the marketplace, technology, and the regulated/nonregulated arena. This has led to the realization that specific user types do indeed apply to specific applications, but there is also the need to generally group these users in a manner to facilitate the evolution from narrowband to broadband information networking in the new millennium.

Data History

The bulk of today's data users use dial-up facilities to interconnect terminals to databases at data rates from 1.2 Kbps, 2.4 Kbps, 4.8 Kbps, to 9.6 Kbps. Today's network has an error rate of one in 10,000 bits per second, at 4,800 bits per second data rate. A single character is 8 bits; so with message headers, this translates to one error in approximately 1,000 characters. At five characters per word, this is 200 words; at 10

- Plain Old Telephone Service (POTS) narrowband voice user
- Interactive inquiry/response narrowband data user
- Fast connect/short holding time narrowband data user
- Private virtual network wideband data user
- Wide area network wideband data user
- Burst interexchange Variable Bit Rate (VBR) broadband graphic display user
- Continuous Bit Rate (CBR) long holding time broadband video user
- Etc.

Figure 3-21. User types.

words per line, this becomes one or two errors per page. However, errors usually come in bursts in actual usage, such as a white line or jumbled characters in every other FAX message. To increase data rates, conditioned point-to-point special service circuits enable better computer-to-computer throughput at 9.6 Kbps, 14.1 Kbps, 19.2 Kbps, 56 Kbps, etc. However, these specifically conditioned lines are expensive and do not meet the general needs for switched services, such that one terminal can access many computers or so computer power can be distributed across many systems.

To enable data users to better communicate within an office, LANs were established with bridges, routers, and gateways to interconnect them. Today, computer firms such as DEC have decided that internetworking their systems to other firms' systems is the way to increase usage and sell more computers. Such firms would like to communicate between systems over a public network. Alternatively, other firms such as IBM have in the past been quite happy to construct their own separate, private networks.

Another major issue currently facing the RBOCs is how to encourage many of the dial-up users to move off the flat rate voice network to a new data network. Many customers are unhappy with the current mode of operation but have few alternatives or incentives. Without removing flat rate or experiencing increasing delay in the voice systems' transport time (due to a high number of data-user accesses), many small business/residence users will continue to send data over the voice dial-up network.

Today, as the computer enters every facet of the workplace, there is now a trend for large to medium-sized business users to cluster their terminals with LANs and channel their traffic over separate private networks, using T1 and T3 leased lines between CPE nodal switches. However, continuing personal computer usage growth opportunity represents a yet untapped market for many residences and small businesses. They are currently unconnected to either the large businesses or each other, except perhaps through the dial-up voice net from their off-line systems. Alternately, many still use manual transport and operational methods. As the information exchange now expands to every aspect of the market for data, image, and video, how we will transport and process it will be quite challenging. Currently, many of the RBOCs are simply encouraging dial-up data use of the voice network or selling transport capacity.

Solutions

Let's divide future voice, data, image, and video users into ten major groups:

Group 1. Will continue to remain on the current voice network in the dial-up mode, using cheap access to Internet, even when an alternate public data/voice ISDN network is available.

Group 2. Will remain on LAN-type private networks, even after public switched data/video services are available.

Group 3. The bulk of today's narrowband data users (16 out of 26 potential data-user types) can be nicely handled by an overlay data network that provides fast-connect, medium-speed connections at reasonable circuit and packet-handling rates over new Basic Rate Access (BRA) ISDN standard interfaces.

- This is an emerging new market with totally new revenue as users change from manual methods to online computer usage.
- This is the reason for deploying ISDN to obtain the B channel digital data interface of 64 Kbps or 2B at 128 Kbps.

Group 4. This group of wideband data users dynamically demand access to a variable number of 64-Kbps channels. Fractional ISDN (f-ISDN) or fractional T1 (4, 6, 8, 12, etc., 64-Kbps channels) are now available to the user on point-to-point drop-offs. There is a growing need for a variable number (N) of channels for not only the DS1 rate of 1.5 Mbps but up to the DS3 rates of 45 megabits. This capacity will initially be accessible to the user from new digital cross-connects, as main distribution frames are automated, restructured, and moved closer to the user. These services will later be provided by the new transport access switching nodes (Class 6) for both wideband (group 4) and broadband (groups 5 & 6) users.

Group 5. Broadband-services users will be tied to CCITT B-ISDN interfaces and Bellcore SONET (STS/OC) rates, as multiples of approximately 50+ megabits. The industry appears to have picked CCITT H4 (135 Mbps) and Bellcore's OC-3 (155 Mbps) as the megabit rates that will meet the needs of compressed advanced quality TV and radiology imagery. Therefore, OC-3 or 155 Mbps will be one grouping for variable-bit-rate (VBR) broadband users that can be served by new integrated ATM voice /data /image systems. Extended wideband/broadband services up to the 45-megabit (T3) or 50.+-Mbps (OC-1) rate will be initially handled by overlay broadband/wideband switching fabrics added to current switches for selected Group 4 & 5 users. Hence, wideband may, by the turn of the century, extend from 1.5 Mbps (T1) on up to 51.84 Mbps (OC-1) in multiples of T1 (DS1) rates or 64 Kbps (DSO) channels.

Group 6. Super-broadband services users at rates above OC-3 (155 Mbps) from OC-12 (622 Mbps) to OC 192 (9.9 Gbps) can be handled by future circuit and channel STM-type switching modes of operation. Somewhat compressed, continuous bit rate (CBR) high-definition TV is still around 600 Mbps (compressed 155 Mbps, very compressed 20 Mbps), which may determine the upper transport bound. These higher speeds and more ubiquitous broadcast offerings for CBR videophone and high-definition TV will require the new broadband urban/metro's (Class 5 and Class 6) switches at the turn of the century, as well as new internal CPE broadband switches (Class 7). In this manner, these new systems will eventually meet the needs of the full spectrum of voice/data/video user groups.

Group 7. MAN-type nonswitched broadband services, enabling high-speed data interexchange, up to 100 or 200 Mbps, using products such as SMDS and FDDI.

Group 8. Broadcast broadband services are usually nonswitched but sent directly to the user via passive optical fiber networks (PONs), satellite Vsats, radio, microwave, cable, or Direct Broadcast Systems (DBSs).

Group 9. Cellular network services for car, air, train, boat, and remote rural users.

Group 10. Personal Communications Networks/Services (PCN/PCS) for mobile urban users.

Applications

In reviewing the different needs of these separate groups of data/video users, note that each group can be satisfied with different technologies. Circuit switching versus packet switching versus channel switching versus nonswitching will again be reviewed as we go to higher and higher megabit rates and higher usage. Before we get to the where, when, and why of deployment strategies, we need to ask what services for what users at home and in business, and how we can best meet their needs. (See Fig. 3-22.)

The Global Customer

As one executive once said, "My firm does not stop at a Local Access & Transport Area (LATA) boundary." Yes, we need to provide complete network solutions, be they local, regional, national, or international.

As we enter the time of the "stateless corporation," the multinational Europe 92, Rebuilding the Eastern Block, Japan 2000, Singapore Financial Network, Caribbean Data Net, French Videotex, and the third world "missing link," we see a shift to emphasizing information movement & management (IM&M) as a strategic asset.

Group	*User*	*Rate*	*Application*[1]
Group 1	Dial-up data users	1.2 – 9.6 Kbps Voice grade	Inquiry/response
Group 2	LAN data users	1.2k – 10 Mbps Nonswitched	Data collection Data distribution
Group 3	Narrowband data users[2]	64 Kbps – 128 Kbps Switched data grade	Transaction Videophone I[3]
Group 4	Wideband data users[4]	Variable 64k 64k – 1.5m – 45 mbps	Data manipulation Videophone II
Group 5	Broadband users[5]	50 – 155 Mbps OC1 – OC3	Data presentation Videophone III ATV SMDS
Group 6	Super-broadband users	Super channel OC3–OC12–OC24 155–2.4–9.9 Gbps	Videophone IV HDTV
Group 7	MAN data users	100 – 200 Mbps	SMDS FDDI
Group 8	Broadcast broadband	50 – 600 Mb/s	ATV HDTV Work station Multimedia
Group 9	Cellular	Radio spectrum	Mobile phone Mobile data set
Group 10	Radio	Radio spectrum	Personal phone Mobile data set

[1]Note that each application will have each type of data collection, distribution, storage, manipulation, access, transaction and presentation offerings. These are only presented as representative examples.

[2]Group 3 users can be handled by overlay data networks using existing digital switching systems with ISDN basic rate interface (BRI).

[3]Picturephone/videophone/viewphone will have ever-increasing resolution as bandwidth (data rate) increases.

[4]Group 4 users of selected wideband offerings will be limited to specific user groups using overlay broadband fabric switching, as well as remote access nodel switches that provide digital cross connects, ATM, SMDS, FDDI and frame relay interfaces and Class 6 functions.

[5]Ubiquitous offering of videophone over the entire area, via distributed megabyte facilities to the home/business, requires the new broadband urban switch, using the full range of B-ISDN interfaces.

Figure 3-22. The ten user-group applications.

From data entry capabilities in the Caribbean to the financial clearing houses in Singapore, we need to globally enter, transfer and process information. No longer must all aspects of complex financial transactions be physically performed within New York City or Tokyo. Location has become incidental. Normal economic forces are taking place, where economic considerations dictate where and how information is handled. Just as the flow of goods from raw material to refining, assembling, and purchasing spans many countries, so will the handling of information. Many third-world countries see this as an opportunity to pull themselves into the post-industrial economy. They may not have the expensive physical plant to manufacture high-technology items, but they can readily provide data entry from lower-cost operators. The system for distributing calls to remote clusters of telephone operators was often used to move "information please" calls to more remote communities to take advantage of less expensive pay scales.

So what are the needs of these global stateless corporations? What are the needs of their global customers? Can long-distance transport continue to be expensive as we drop more fiber across our oceans and insert more satellites in the sky? We are now able to provide inexpensive high-speed data circuits to remote areas of the globe. This enables economical transport "dumps" from remote third-world front-end information processors back to mainframe computers. These remote processors perform all the time-consuming entry, compilation, and record building activities that are often too labor expensive to be performed in the more industrialized nations. These tasks are completed at considerable cost advantages and in a timely manner in these distant locations. (We should also note how important information has become. Cable News Network demonstrated this during the war in Iraq. As CNN newscasters told the world audience where the bombs landed in Israel, the Iraqi gunners used this information to realign their targets.)

The needs of the world's customers will be differently time phased. Just as semiconductor assembly moved from California to Japan to Taiwan to Mexico to Korea to Asia in search of cheap assembly, so information "assembly" will move from country to country. On the other hand, few countries have given up their technology once they have become accustomed to using it. They have subsequently created new offerings using their new ability to "assemble." So it will be with information "assembly." Assembly for others soon becomes assembly for itself as the country begins to utilize the technology it learned while providing services to others. Soon the country becomes its own best customer and also begins to compete in the global information marketplace.

In any event, it is important to recognize that the simple fiber can provide a fruitful marketplace at the end of its pipeline, just as water brought to some desert location encourages a growing oasis. This opportunity is fostered by extending the COI of a particular industry to any location in the world. As these communities embrace information as a new economic opportunity, the world becomes a little smaller, a global village housing the global customers.

Observations

We have defined several specific techniques for identifying and categorizing customer needs, from considering their basic human needs to understanding their computer system's needs, to subdividing customers in terms of user types that require specific communication capabilities to fulfill job-related tasks. Next a different view was suggested that considered users in terms of their community of interest (COI), both internal and external. This indicated the need for both private and public network solutions, both integrated and layered together. Finally, specific user groups were established to help classify users in terms of their information transport service needs. These separate and varied approaches help formulate a better understanding of the total picture as we begin to obtain a reasonable model depicting both the opportunities and the solutions that identify and satisfy our current and future information customers' needs and expectations.

In this we have taken the first
step down the path to meeting
the information networking
challenge of the 2000s–2020s
for the global customer
of the forthcoming
information society.

Perspective: The Customer

On reviewing the customer needs for each market segment, sector, or functional area, we may make the following observations concerning customer needs and market opportunities.

The information marketplace opportunity

- Over the 2000s, information communications customers will need totally new services to meet their personal, business, and family needs.
- Communications firms will meet this challenge by using the existing and new fiber-based plant to provide a growing array of new

evolutionary and revolutionary narrowband, wideband, and broadband information services.

The information customer's basic needs

- Increase personal knowledge and skills.
- Enable our customers to be successful in:
  a. business
  b. family life
  c. personal life
- Reduce stress and increase performance.
- Manage time.
- Access information on a timely basis.
- Tie dispersed communities of interest together.
- Enable small firms to improve performance and productivity.
- Enable large firms to improve performance and productivity.
- Reduce travel time.
- Access and exchange information.
- Improve the quality and timeliness of decision making.
- Improve the quality of life.

The information customer's operational needs

- To see information:
  a. in text form
  b. in image form
  c. in graphic form
  d. in video form
  e. in person
- To control information movement and services.
- To rapidly exchange information.
- To access remote databases.
- To translate data into graphics.
- To store and retrieve images.
- To dynamically move varying amounts of information, requiring dynamically changeable transport capabilities.

- To enable the internetworking of private and public networks to encourage usage and growth of shared public facilities.
- To enable interoperability of computers of different capabilities.
- To provide secure and survivable information transport throughout the network from customer to customer.
- To encourage the growth of new services as features are added to features to meet the new and changing needs of our customers.

The information customer's network needs

- To transport voice, data, text, graphic, image, and visual information:
  - a. terminal to terminal
  - b. terminal to computer
  - c. computer to computer
- To ensure the ease of growth of information movement with
  - a. more and more usage
  - b. higher and higher speed
  - c. no real-time transport bottlenecks
- To ensure the ease of growth of information movement
  - a. locally
  - b. regionally
  - c. nationally
  - d. globally

The information customer's market needs

The information customer needs a structure:

- From which to offer new services, both the regulated and nonregulated.
- From which to offer enhanced voice services.
- From which to offer data, image, and video services.
- That ensures secure and survivable (S&S) transport of information.
- For meeting private internetworking needs of large business users.
- That enables access to the full community of interest across small and large businesses.
- That fosters the movement and management of information (IM&M) in all aspects of the marketplace. (See Fig. 3-23.)

Telecommunications Management Planning Cubic

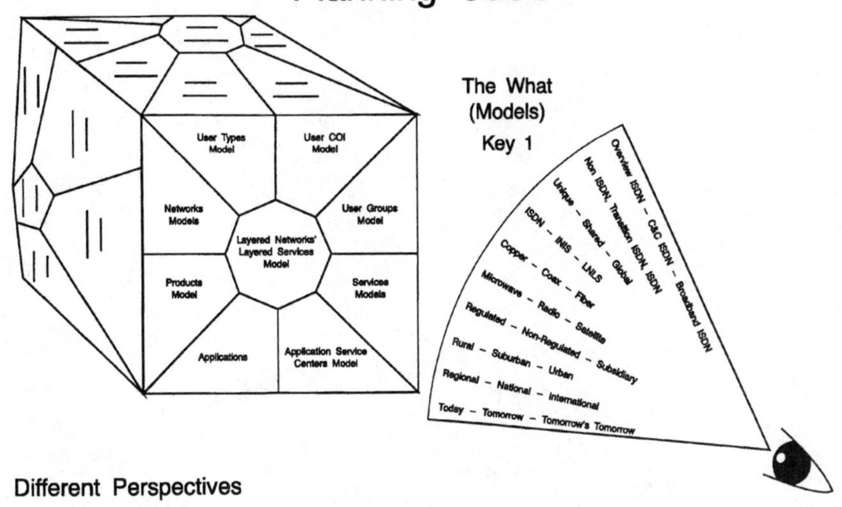

Figure 3-23. Models.

Chapter

4

The Services

Caveat Emptor: *Let the buyer beware.*

Marketing and technical planners and designers approach the subject of services from different directions (application or technology) as they attempt to identify and define what services providers' networks should provide, or what features suppliers' products should offer. Traditionally both the computer and the communication industries were basically technology driven. (Even though the computer industry had its highly sophisticated user groups, they were usually only changing or modifying variations of a compiler's syntax or semantics.) However, with the advent of the Apple computer with its Scottish Macintosh "mouse," the computer industry shifted into a more customer-focused, market-based, user-application approach. Here, graphics and imagery replaced number displays as input-output mechanisms become more "user friendly." There was new-found recognition that the customer had many different, dissimilar systems that needed to be "networked" together, thereby promoting a renewed attempt to obtain interconnection, interprocessing and interservices.

As more and more potential users embraced the faster, easier-to-learn windowing techniques, computers left the accounting field to spread across the industries to every conceivable application, with many quite inconceivable applications yet to come. Technologists could say, and rightly so, that if the personal-computer chip technology had not progressed to the point that economic computer power and sophisticated input-display mechanisms were available, then market wishes would not have been so readily fulfilled; nor would the market application planner even have known the potential of this new technology to satisfy so many new market needs.

As we look at the history of mankind's progress and advancement, technology begets technology as usage breeds more usage. We have

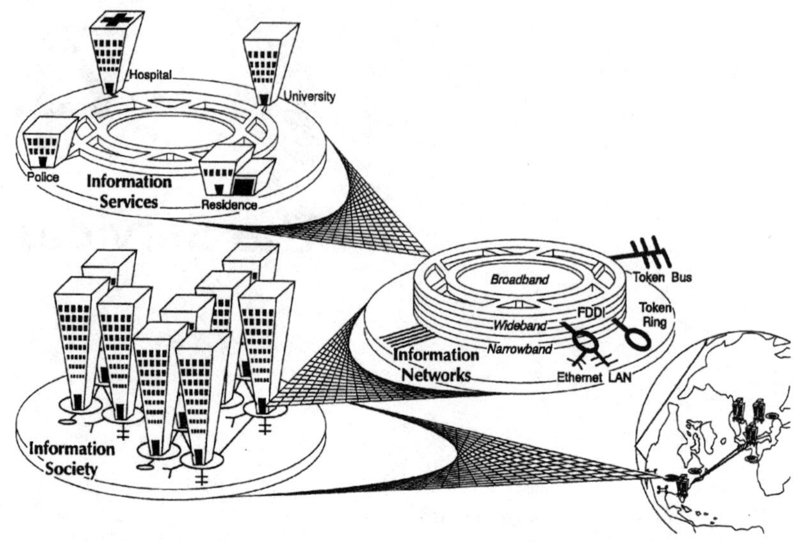

Figure 4-1. Global markets.

seen home radios shift to car radios and "walkmans," record players to stereos, one-prop mailplanes to super passenger jets. The more we use technology, the easier it is to see more uses for it. (See Fig. 4-1.)

So it also was with communications. However, in the past, the ability to broadcast video images directly to a remote receiver over the radio spectrum caused an artificial separation of wire-based interactive voice services (telephone) from wireless broadcast video services (television). As these massive industries developed quite autonomously from each other, a deep chasm developed between them; within this chasm was data. Deep down at the base of these towering mountains, following its own winding path, was the fledgling world of data, prospecting for its particular form of gold to "strike it rich."

For many years data users were quite content to explore their own way as they proceeded from cavern to cavern, creating their own local domains, quite oblivious to what was happening above them. In this manner, data used what it could of whatever fell down around it from both the wire-line and wireless worlds of voice and video. During the 1960s through the 1990s, quite an industry emerged of noninterconnected domains; but little of it saw the light of day, as the growth of the data kingdom was shadowed by the towering regions of voice and video.

Changing Markets

As noted in Fig. 4-2, computers found their initial applications in the financial industry, as they migrated from their birth as calculators to

Figure 4-2. Changing markets.

various aspects of the accounting and stock market industries. Here, they communicated between each other and to the user at ticker-tape rates of 60 and 100 words per minute to spell out financial successes or failures. While the stock market collapse in 1929 was largely brought about by highly leveraged margin calls, more recent reports note that the "almost" collapse some seventy years later in 1988 was based upon sophisticated computer trading, where computers took advantage of differences between the "big board" in New York and the futures market in Chicago. What the aftermath study noted of the free-fall drop in prices in 1988 was the actual "collapse" of the communications networks. They could not transport quickly enough the massive amount of information between these two exchanges and the brokerage houses. The delays further aggravated the differences, causing further computer trading based upon the widening disparities resulting from these delays.

This earthquake-like communication upheaval was followed by the Hinsdale fire, the New York City software bug, and the Hinsdale cable cut. This left those who had become quite dependent on autonomous private networks' leased lines or dial-up WANs to suddenly be very concerned with security and survivability (S&S), especially as more and more of the business operations became more and more dependent on the integration of C&C (computers and communications).

The 1990s also introduced a new factor into the equation, as the decade unleashed the buying and selling of huge firms at huge prices, causing administrations to search for more economical ways to cut costs and increase profits. Firms became more and more global as they searched for the most economic labor force to provide different pieces of their "widgets." They also introduced "low or no inventory" goals, as pieces were ordered electronically to be delivered "just in time" for assembly. So the world changes as we shift to more and more data dependency.

It's especially important to determine what products are possible from the complex, different technologies and to identify what features the products can deliver. It's equally essential to determine how to package these technically possible features into appropriate services that can be applied in specific applications to meet the varied and changing customers' needs.

Alternatively, customer needs in specific applications can point to services that can be obtained from packaging various features from current and yet-to-be designed products. In reality, both approaches take place as technical possibilities are translated to market opportunities and vice versa.

Hence, the real change for the communications industry is not the approach of being technology-driven or market-driven, but the financially backed shift to market-based management, where renewed emphasis is placed on the forgotten customer, especially the lonely data "subscriber" who had to purchase expensive transport to move low-speed "bits and bytes" to Internet databases.

With the discovery of the customer (Chapter 3) and a better understanding of technology (Appendix B), let's see what features and services can best meet new needs.

In reviewing numerous industries in terms of the tasks that are performed in each and then denoting the type of communication network that helps accomplish the task, it became quite apparent that communications customers' needs fall into three distinct classifications: narrowband, wideband, and broadband. Narrowband communications are tied to the "baseband" twisted-pair bandwidth capabilities defined by the 3- to 4-kHz transport boundaries that limited analog communications to the traditional 9.6-Kbps data transport rates. With the advent of IDN (Integrated Digital Network) with ISDN basic rate interfaces of several 64 Kbps, we have improved the transport range for narrowband to efficiently transport 160 Kbps in information. Similarly, wideband capabilities now extend up to 1.54 Mbps (T1) and 6.3 Mbps (T2) on copper. Some believe wideband range should include T3 or 45 Mbps delivered by fiber or coaxial as the first entrance of broadband, where broadband services will be provided from SONET's 51, 155, and 600 Mbps transport-package envelopes.

In any event, an interesting exercise may be to reassess the twenty-six user types (noted in an earlier book, *Future Telecommunications*), in terms of narrowband user types, wideband user types, and broadband user types. This will help determine how many potential information users from the eighteen or so industries would be best satisfied by narrowband, wideband, or broadband communications to help them perform their specific applications' detailed tasks.

As data users demand more S&S (secure and survivable) transport, as information movement and management (IM&M) becomes more and more distributed, as data networking of computers and communications becomes integrated, C&C becomes the primary emphasis of the 1990s. New and expansive private networks will see considerable growth in data usage and application.

This driving force has become quite evident as PC networks, mainframe clustering, and paralleled database management track fully distributed parallel processing. Similarly, file transfer access and management (FTAM), non-ISDN and ISDN addressing, routing, message sequencing, delayed delivery, storage, and data manipulation and processing are requiring more interconnection in the form of internetworking, interprocessing, and interservices. Here, we interconnect throughout the domain of C&C over the range of P&P, as both private and public services are layered together to achieve the desired customer solutions. (See Fig. 4-3.)

The Information Marketplace

Let's take a look at the forthcoming information marketplace in terms of present and future voice, data, video, text, and image services that can be delivered from the features offered by global narrowband, wideband, and broadband telecommunications networks' products.

Once understood, we will be able to determine what the various common and special carriers (IXCs and LECs), service providers (SPs), information providers (IPs), agents, integrators, database sources (DBS), other equipment manufacturers (OEMs), and third-party software providers (OSPs) can deliver to the consumer, government, residence, business, education, and military market sectors noted in Chapter 3. (See Fig. 4-4.)

Narrowband services

Voice and low-speed data can be ubiquitously provided over today's narrowband facilities. ISDN provides the ability to offer the customer integrated voice and/or data transport facilities. Using the basic rate access interface, two 64-Kbps channels and one 16-Kbps signaling or 9.6-Kbps packet-handling channel are available to the user. This has been widely noted as the 2B + D BRI ISDN interface. Over the 1990s,

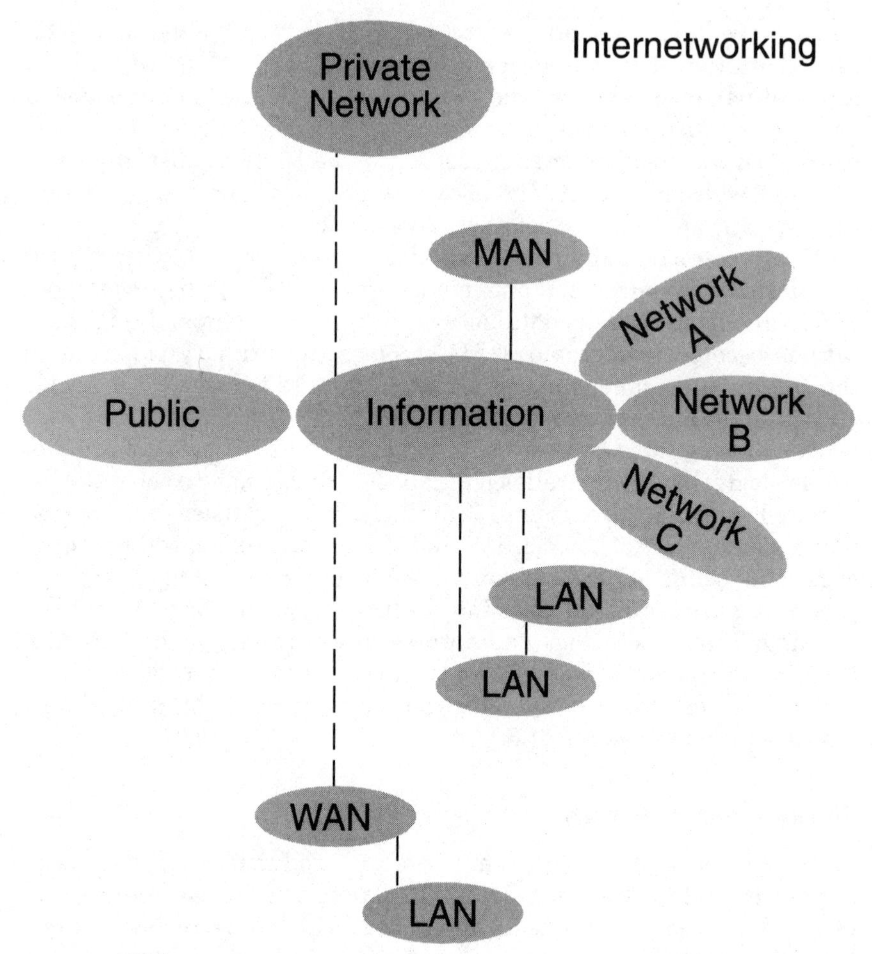

Figure 4-3. P&P

this was mainly sold as an ISDN interface to obtain two voice channels and a slow-speed 9.6-Kbps data packet channel. (Packet implies segmenting the message in a series of fixed-length or variable-length payloads with a destination address header and tail attached to each payload.) Ease of station movement was noted as a key feature of ISDN, plus the ability to know the calling-party ID. This is due to the parallel technology called Common Channel Signaling System #7 (CCS7) that was also being deployed with the new Integrated Digital Network (IDN) switching system. As a result, the CLASS family of features and services were added to the traditional telco's offerings noted in Fig. 4-5A and 4-5B.

The ability to know who is calling whom enables a whole host of new and exciting features under the umbrella of CLASS services. Here, also, repeat calls can be periodically redialed at the distant end with a

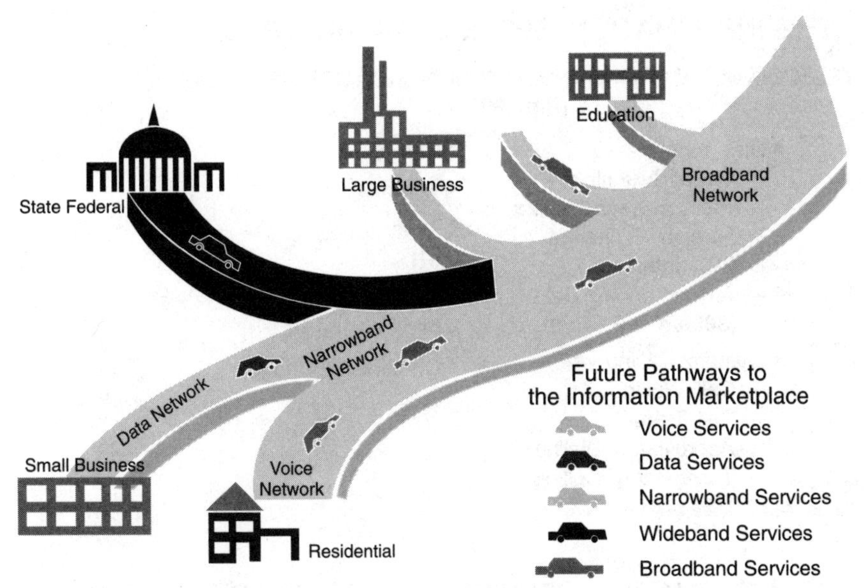

Figure 4-4. The information highway.

special ring to note when the party has been reached. Return call enables customers to easily return the last call or a call from the received-call list. Priority call enables a busy person to choose which call they wish to receive. In the future, using D channel signaling, the calling party may be able to request preferential treatment. Selected call forwarding enables desired calls to be routed to selected locations such as the home, car, voice mail system, or to voice/text systems, where the contents are changed to a data message that might be routed around the world. Other features such as call block and call trace enable specialized call handling on specific calls. Finally, Identicall-type features allow the calling party number to be displayed, or even the ability for the phone to announce by name who is calling.

Other voice-based services can be added to existing systems, provided above the network, or outside the network, depending on regulatory/nonregulatory decisions. Here, we see voice mail or voice messaging as viable offerings, as well as automated voice attendant services. Similarly, enhanced centrex, 911, enhanced 911, 800, 900, customer line identification database, operator call assistance, conference calling, customized services, telemarketing, database access, area wide centrex, shared tenant services, PBX automated attendant services, ACD, personal payphone services, high fidelity audio, intelligent CPE phone, service creation, and network management services expand the traditional world of voice-based services. (See Fig. 4-6.)

In these offerings, voice messages can be stored in one's mailbox or routed for specialized handling. Alternatively, a single voice message

THE INFORMATION SERVICES MARKETPLACE

TELECOMMUNICATIONS INFORMATION FEATURES/SERVICES
(Pre 1990)

- Dial Tone
  - Voice Switching
  - Interexchange Access
  - Foreign Exchange
  - PBX Interface
  - Asynchronous Data (300b/s – 9.6kb/s)
- Custom Calling
  - Call Waiting
  - Call Transfer
  - Abbreviated Dialing
  - Three–way Calling
- Centrex
- Weather
- Time
- 911
- Pay Phone
- PBX Interface
- Conference Bridge

Figure 4-5A. Voice services.

THE INFORMATION SERVICES MARKETPLACE

- NNX Code Special Handling
- 800 Services
- 900 Services
- 976 Blockage
- 976 XXXX Blockage
- Billing
  - Flat Rate
  - WATS
  - INWATS
  - CALL PACK
- White/Yellow Page Listing/Advertisement
- Feature Group A–D
- Data Transmission
  - DDS Rates
  - T1
  - T3
  - Conditioning
- Network Management
  - Code Blocking
  - Alternate Routing
  - Access Interexchange
  - Carriers

Figure 4-5B. Voice services.

can be sent to numerous destinations within an internal, closed user group. Or it could be broadcast to many locations throughout the world. Operator services can be expanded from traditional calls such as person-to-person, collect, third-party, calling-card, multiple-language, and directory-assistance calls. Locator services, virtual private network directory assistance, and telemarketing, where databases are accessed while conversing with the customer, can also be provided. Similarly, intelligent networking enables gab lines, advertising, and 800/900 service bureaus for entertainment, polling, promotions and personal messages, as well as wake-up calls, security and screening, and access to databases to help establish virtual private networks. Network man-

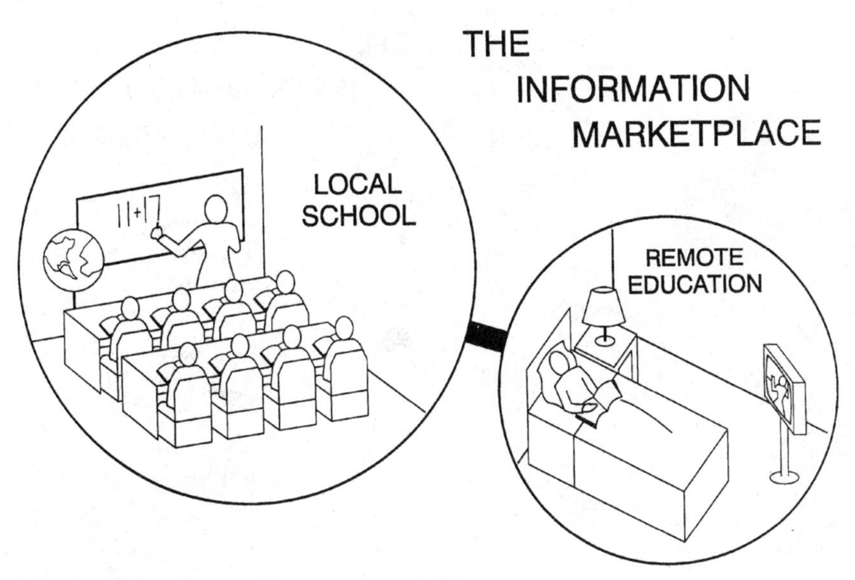

Figure 4-6. Education services.

agement services for routing, rerouting, accessing, blocking, disaster recovery, time variant bandwidth, or service options can also be provided to the customer along with high fidelity (7 kHz) stereo quality communication. (2 "B" voice channels is equivalent to 7 kHz.)

In this manner, numerous versatile services are available to the customer, features such as call screening, priority call override and reroute, hands-free calling, desk-to-desk messaging, and special calling directory lists using personal identity numbers (PINs). (See Fig. 4-7.)

Data

Narrowband data services are provided by traditional circuit-switching isochronous data streams at dial-up network rates of 300 (V.21), 1,200 (V.22), 2,400 (V.22BS), 4,800 (V.27), and 9,600 (V.32) bits per second, or ISDN rates of 64 Kbps or 128 Kbps, which can contain 14.1-Kbps, 19.2-Kbps or 56-Kbps (CSDC) data traffic. These switched data streams enable distant modems and Group 3 (2.4 Kbps, 9.6 Kbps) or Group 4 (56 Kbps, 64 Kbps) facsimile machines to communicate with each other asynchronously, where they use their own end-to-end protocols to send and receive data messages, using their own methodologies to determine error rates and retransport and sequence the orderly arrival of message segments. In this technology, leased lines for 14.1- and 19.2-Kbps services no longer require special expensive Cl, C2 conditioning, but these options have remained available in the 1990s if direct point-to-point analog service is desired. However, as noted, permanent or semipermanent

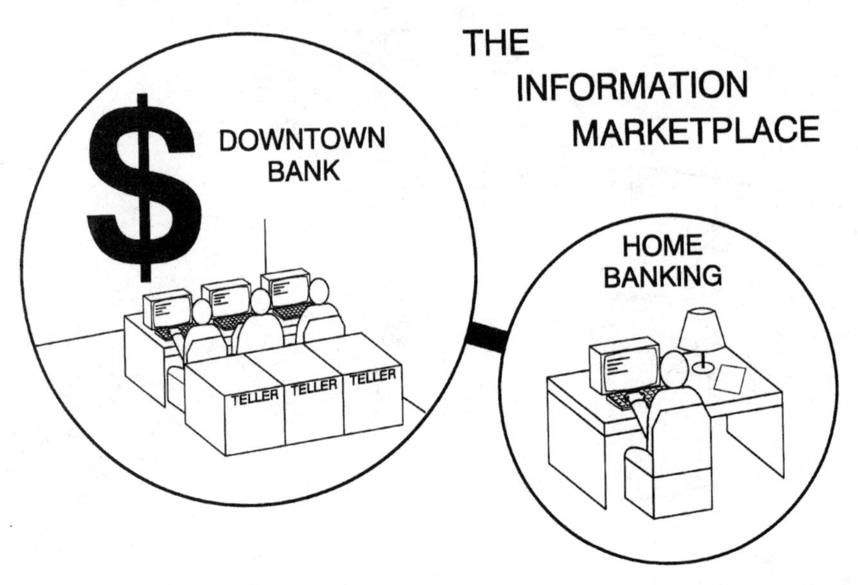

Figure 4-7. Banking services.

connections are now available on digital systems at the higher transport rates, so these earlier offerings will become academic.

Data may also be moved in the 64K, 128-Kbps packet mode for connection-oriented (the network remembers for all subsequent packets the path to the initial destination address) and connectionless (each datagram establishes its own path through the network) packets. Similarly, as noted earlier, the "D" channel of ISDN provides for 9.6-Kbps data packet movement.

In these circuit-switched or packet-switched networks, traditional data-handling services can be offered, such as broadcast, polling, delayed delivery, priority messages, protocol conversion (for example, asynchronous to X.25), code conversion (for example, EBCDIC to ASCII), file transfer, file access, packet sequencing, error detection and correction, security checks, audit trails, terminal verification, ID verification, closed user groups, open user groups, alternate routing, three-attempt limit, multiple addressing, rerouting, multiple carrier access, CO-LAN (19.2 Kbps) interface, X.400 electronic file-transfer messaging, X.500 electronic directory, PC-to-PC networking, TCP/IP protocol interface and control, encryption, electronic meeting phones, and shopping-bag gateways, as well as gateway services to distributed databases such as the Internet. (See Fig. 4-8.)

Sophisticated E-mail systems provide message transfer confirmation, forwarding, annotation (editing, adding, inserting), and distribution lists. They provide special features for recording messages, saving messages, replaying messages, checking incoming messages, and selectively filing or ordering messages. (See Fig. 4-9.)

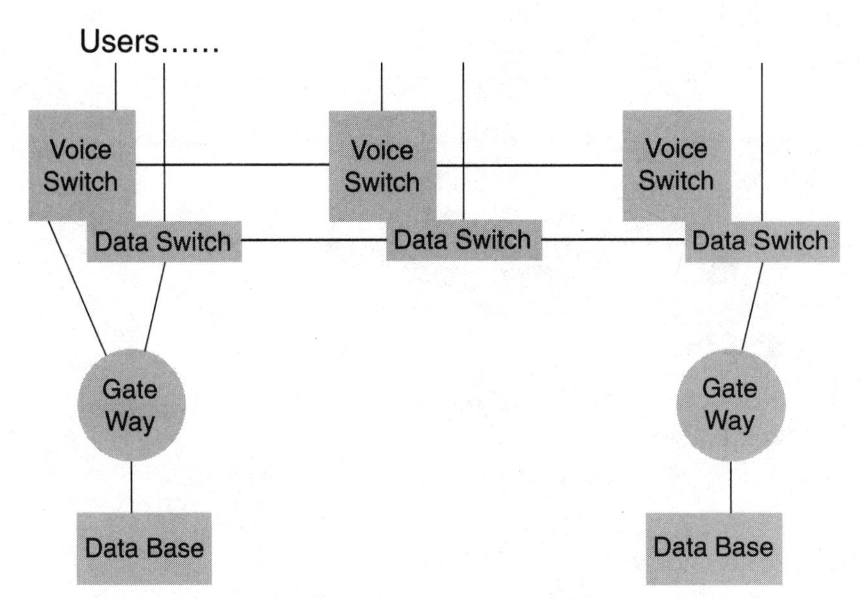

Figure 4-8. Voice/data switching.

In this manner, considerable narrowband services can be established by providing public voice and public data narrowband networks ubiquitously within a community of interest and throughout the community. (See Fig. 4-10A.)

In reviewing the data user needs (Fig. 4-10B) and the ISDN Feature Matrix (Fig. 4-11), we see that various levels of services, such as data manipulation and data processing, can be performed on customer premises. These figures also show how the Department of Justice MFJ restriction (for RBOCs to not provide "information services" where the content of the information is touched or changed) has caused services such as E-mail to be quite limited when provided in the regulated side of the business by the RBOC. Hence, in the early 1990s, after seven years of divestiture, the judge was requested to reassess his very restrictive rulings on information services, especially as this restriction was added late in the divestiture negotiations to satisfy the publishing industry. At that time, the full impact of the initial decision to limit the RBOCs from information services was not fully understood. Many years later, the RBOCs had not yet established public data networks, indicating their resistance to offering only limited data transport services. It should be noted that there is still a considerable and justifiable base for a public data network, as several RBOCs noted in their mid-1990s announcements that they planned to pursue public narrowband data network services. Now that the 1996 Telecommunications Act clearly removes any restrictions, opening up the arena to any and all, it will be interesting to watch the public data network bloom and blossom across America.

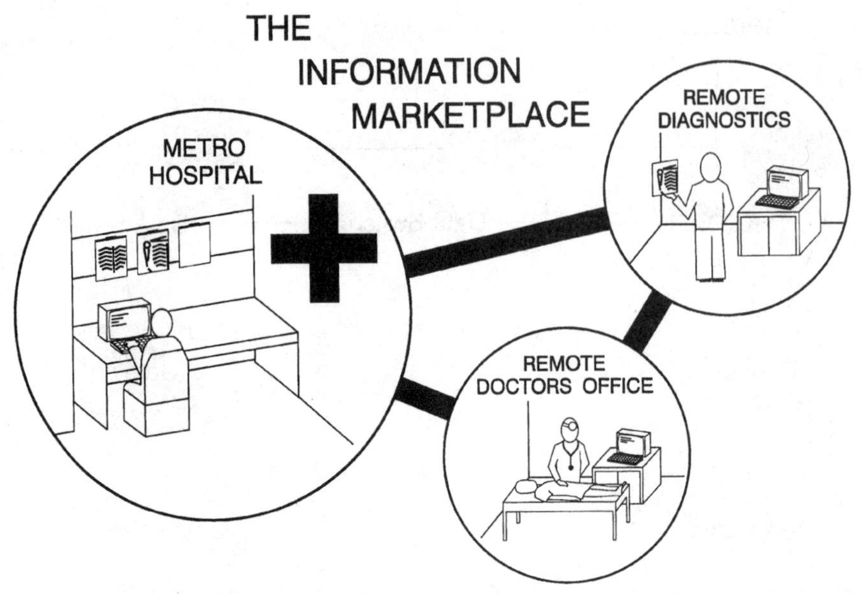

Figure 4-9. Medical services.

Wideband services

Narrowband services continue to advance, with speech recognition (speaker dependent and independent), talking yellow pages, junk mail blockage, and dial dictation. Thus, both large and small businesses have found the need to concentrate more and more of their internetwork traffic on higher-speed transport mechanisms, as data communication systems require more and more survivable routing. Hence, private networks took to establishing T1 (1.5 Mbps) and T3 (45 Mbps) hubs, as they formulated private network grids to accommodate their internal network needs. (See Fig. 4-12.)

Wide area networks (WANs) were being deployed to interconnect local area networks (LANs) using high-speed digital transport. Extended Digital Subscriber Loop (EDSL), known as Primary Rate Access Interface (PRI-ISDN) provided access to twenty-three 64-Kbps channels and its 64-Kbps "D" channel for signaling. In addition several PRIs could be concatenated together using a single "D" channel to achieve twenty-four 64-Kbps T1 rates of 1.544 Mbps. This and other variations led to Public Primary Rate ISDN and Digital Cross Connect (DCS) systems for T1 and T3 hubbing to enable DS0/DS1/DS3 (64 Kbps/1.544 Mbps/45 Mbps) channel switching at remote nodes closer to the customer. (See Fig. 4-13.)

Private networks saw similar needs and opportunities to lease fiber with or without electronics (dark fiber) and provide alternate carrier hubbing arrangements to internetwork large business enterprise networks.

- **Personal Needs**
  - **Less Time in Queues**
  - **Less Time in Work Commute**
  - **Flexible Work at Home**
- **Remote Access to Data**
  - **Banks**
  - **Stock Exchange**
  - **Financial Information**
- **Education at Work/Home**
  - **Specialty**
  - **Hobby**
- **Entertainment/Amusement**
  - **Interactive Games**
  - **Movies**
  - **International T.V.**
  - **Past Events**
  - **Travelogs**
- **Research Searches**
  - **Medical**
  - **Government**
  - **Legal**
- **Childcare Assistance**
  - **Work at Home**
  - **Remote Access to Doctors**
- **Remote Shopping**
  - **Goods**
  - **Foods**
- **Business Improvements**
  - **Increased Effectiveness**
  - **Faster Access to Info**
  - **Increased Personal Efficiency**
  - **Better Decision**
  - **Efficient Operations**
  - **Reduced Staffs**
  - **Cheaper Operations**
  - **Quality Products**
  - **Faster Order-Delivery Cycle**
  - **Better Planning**
  - **Deeper Market Penetration**
- **Computers**
  - **Inquiry**
  - **Search**
  - **Retrieval**
  - **Processing**
  - **Manipulation**
  - **Analyzing**
  - **Presentation**
  - **Instantaneous**
  - **Cheap**
- **Communication**
  - **High Resolution**
  - **High Quality**
  - **Reliable**
  - **Available**
  - **Accessable**
  - **Cheap**
- **Information Exchange**
  - **Voice** - **Video**
  - **Data/Text** - **Image**

Figure 4-10A. User needs.

Broadcasting
Delayed Delivery
Packet Interleaving
Byte Interleaving
Bit Interleaving
Code Conversion CCITT Codes
Polling
Inquiry Facility
Three Attempt Limit
Low Error Rate
Data Collection Service
High Grade Service
Standard Interface
Data Tariffs
Access to Lease Lines
Duplex Facility
Call Back (Automatic)
Redirection of Calls
Speed/Format Transforms
Multiple Lines
Incoming Calls Barred
Abbreviated Address Call
Packet Switching
Re-try by Network
Store and Forward
Short Clear-Down
Manual/Automatic Calling
Manual/Automatic Answering
Data Service Classes
Direct Call
Network to Subscriber Interface
Barred Access

Figure 4-10B. Data user needs.

	Voice	Data	Video
I/O	Terminal Access	Terminal Access	Terminal Access
Basic	POTS Billing DA	Data Transmission Circuit Switch Packet Switch Billing DA	Wideband Transmission Video Conferencing Megabit Switch
Level 1 Extended Data Transport	Custom Calling 911 Weather/Time Class	Security Error Rate Control Barred Access Async-X25-X75 Code Conversion	Megabit Transport Error Rate Control Quality Features
Level 2 Extended Inquiry/ Response	Delayed Delivery Voice/Text Customer Control	Broadcast Delayed Delivery Message Access Text/Voice Data Collection	Video File Access Picture Catalog Slow Screen Display
Level 3 Enhanced List Access Services	Inventory Control	Polling List Storage List Analysis Record Search	Slide Access/Display Video Record Search Video Entertainment
Level 4 Information Processing Services		Page Reformatting Program Packages Data Manipulation Data Processing Data Presentation	Video Games Graphics

Figure 4-11. ISDN Feature Matrix.

These private-to-public network nodal points provide logical interface points for frame relay (up to 2 Mbps) variable-length messages, SMDS (1.5 Mbps and 45 Mbps), and eventually higher-speed broadband platforms for FDDI-II (100 Mbps) and SMDS (140 Mbps) connectionless data-packet switching. (See Fig. 4-14.)

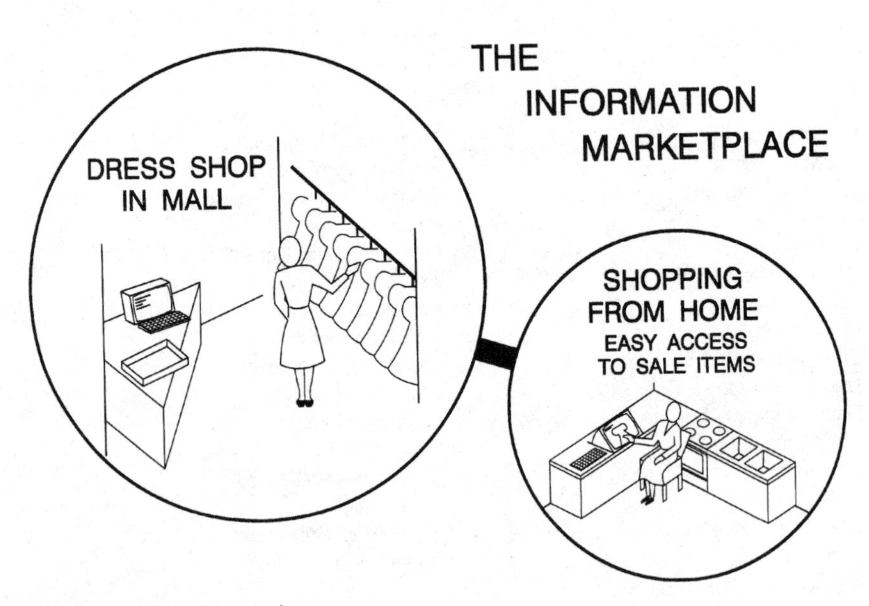

Figure 4-12. Shopping services.

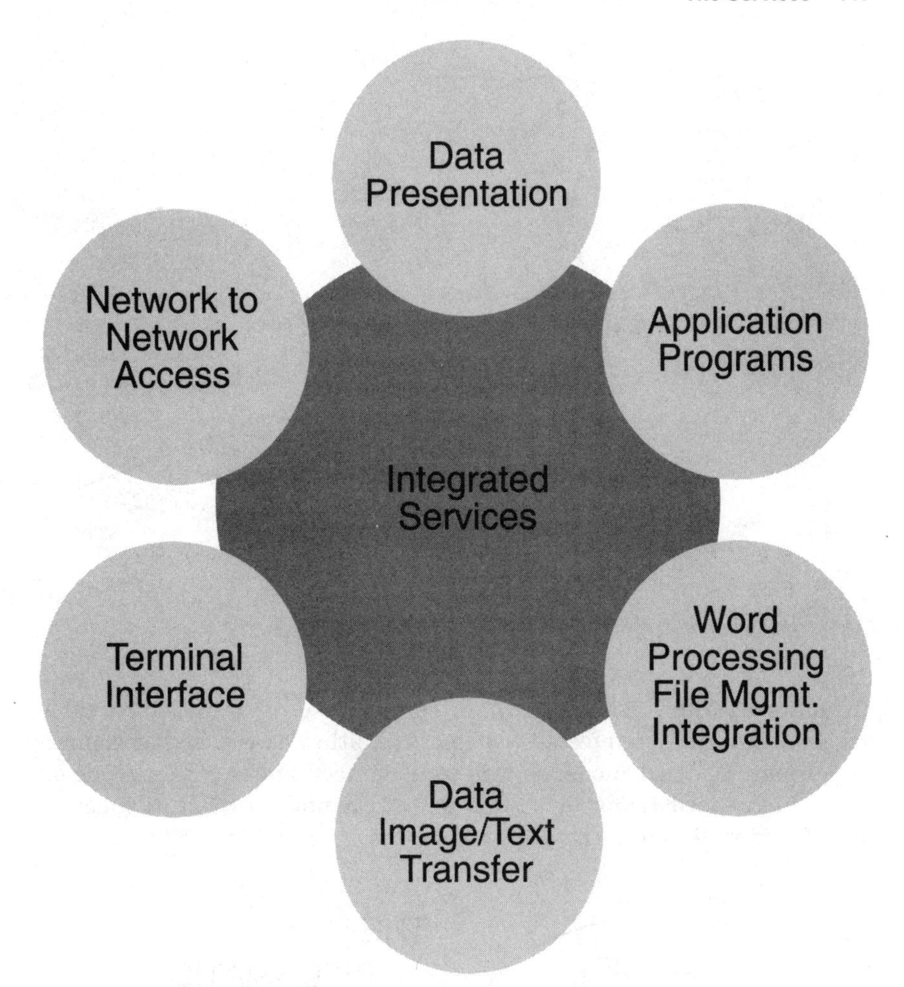

Figure 4-13. Integrated services.

As users demanded more and more control over their transport, they first requested variable bandwidth as needed. Here, fractional T1, fractional ISDN, or *N* number of DSO channels of primary rate ISDN could be obtained dynamically or predefined as required. One of the major problems of administering a "private network" is to keep operating costs down but still meet high-demand, busy-hour needs. So, if private network managers purchase too much unused, expensive transport, or if they do not provide enough capacity when their users need it, they do not meet one or the other conflicting objectives. Unfortunately, lead time to achieve new facilities was as high as 30 to 45 days, due to traditionally slow telco provisioning turnaround time. Hence, these large users first requested variable bandwidth, then dynamic bandwidth. Next they shifted to usage-based bandwidth on a permanent (call

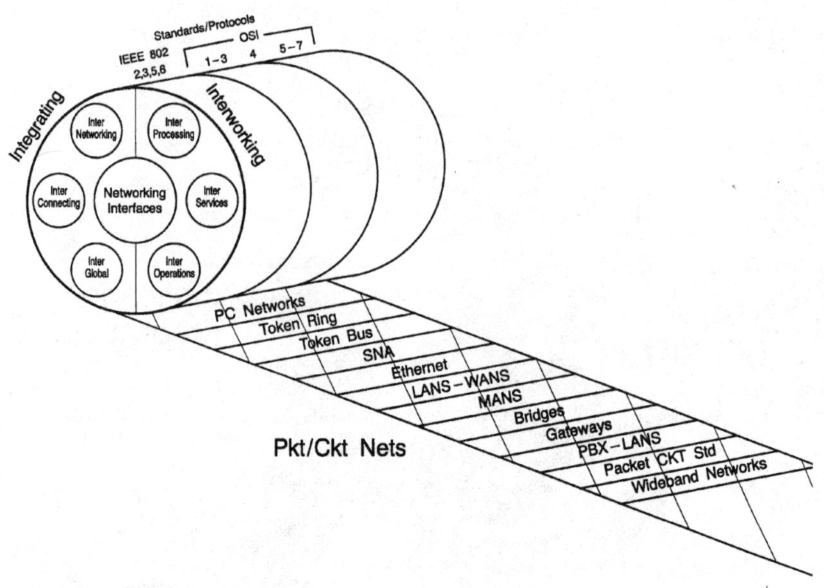

Figure 4-14. Networking C&C.

ahead for operator assist set up) or semiprivate basis (database table driven). Then they requested transport in either a switched or connection-oriented packet mode, as well as a connectionless packet mode on super high-speed transports such as SMDS and B-ISDN at megabit broadband rates. (See Fig. 4-15.)

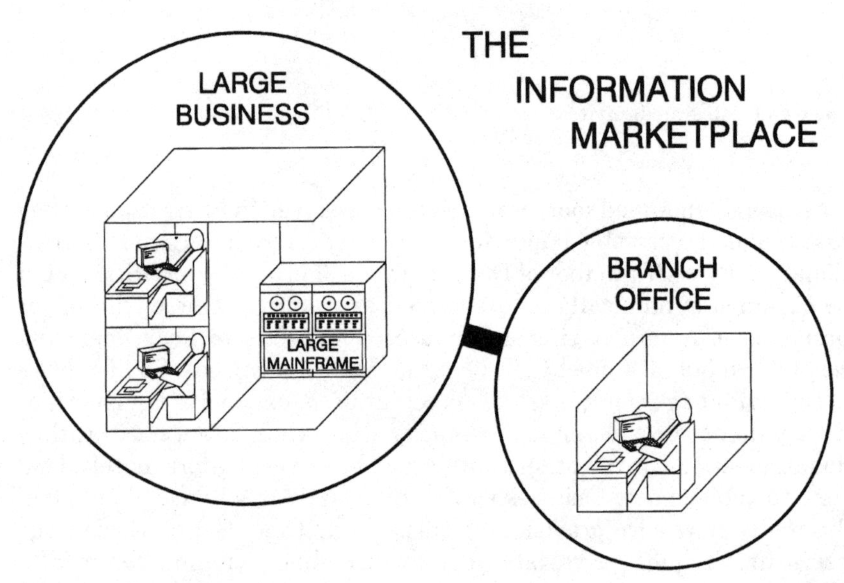

Figure 4-15. Business services.

Broadband services

This then leads us into the domain of broadband networks delivering a full range of broadband services. Broadband begins at STS-1 or OC-1 (51.84 Mbps), and continues in higher multiples to the agreed European/American standard OC-3 (155 Mbps), and then on to OC-12 (600 Mbps), OC-24 (2.4 Gbps), and OC-192 (9.9 Gbps) rates.

One may be wondering who needs these large amounts of bandwidths. It's important to keep in mind that users are concerned with resolution and response time, as multiple imagery-type medical systems send up to 100 images of patients or as high-resolution systems display information in 1,000 × 1,000 or 2,000 × 2,000 or 4,000 × 4,000 arrays, where 24+ bits of information for each pixel indicate not only the 16 levels of gray code, but also denote each color of the rainbow. Hence, several million bits of information are required to provide a high-resolution, accurate display of a still-frame picture. As noted, depending on the data transfer rate, the refresh time can be minutes rather than seconds. Information can be in the form of a continuous bit rate (CBR) or variable bit rate (VBR). Here, burst mode interactive picturephone information may compete for transport capacity with long-holding, truly continuous broadcast television channels. On the other hand, workstations may require periods of continuous high-bit-rate information exchange between the mainframe computer and workstation, followed by long periods of relatively little transport. (See Fig. 4-16.)

Compression techniques are continuing to advance, even though high-bandwidth capabilities are available. This need was originally

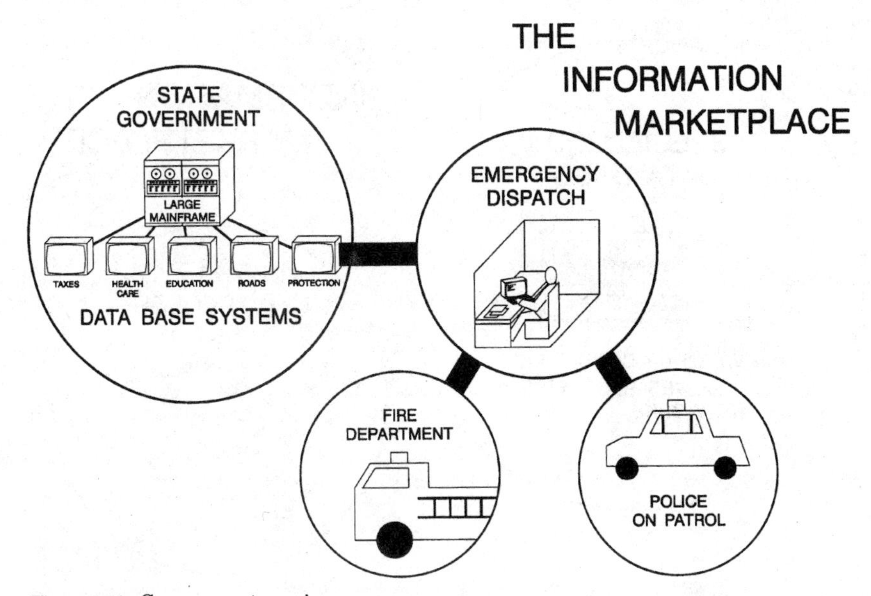

Figure 4-16. Government services.

due to the desire to reduce expensive transport. Unfortunately, transport will continue to remain somewhat expensive throughout the 1990s, as carriers (IXC and LEC) are reluctant to lose billions during the interim period between cheap "spring" transport that would enable an overflowing "fall harvest" of new services growth. Even with the entrance of VANs, only marginal transport reductions have occurred, as they also attempt to offset the costs of their new facilities by maintaining the high prices of transport. It will take more than one new carrier to drop the prices substantially. Eventually, this will happen as competition heats up similar to the PBX wars. The smart move would be for the new LECs to shift to service pricing for specific high-bandwidth ticket items, such as high-resolution picturephone (45 Mbps) or even 1.5-Mbps compressed picturephone.

Hence, compression techniques will still be needed as customers desire more and more high-quality images. High-definition TV originally was at the one-billion bit rate, compressed at 600 Mbps, further compressed at 155 Mbps, and is still targeted for the 6.3-Mbps NGTV level, perhaps reaching 20 Mbps in the early 2000s.

As we look at the other broadband video and imagery systems besides computer-to-computer data traffic, we realize that, for broadband, the applications are just beginning to be visualized. It is indeed the technology for the forthcoming information millennium. (See Fig. 4-17.)

The home of the future is envisioned to have a UNI interface of one 155-Mbps channel or four 155-Mbps channels to enable 4 to 6 HDTV programs, as well as the 2B + D and the 23B + D data streams. In this, passive optics can play a role for broadband-type systems, but

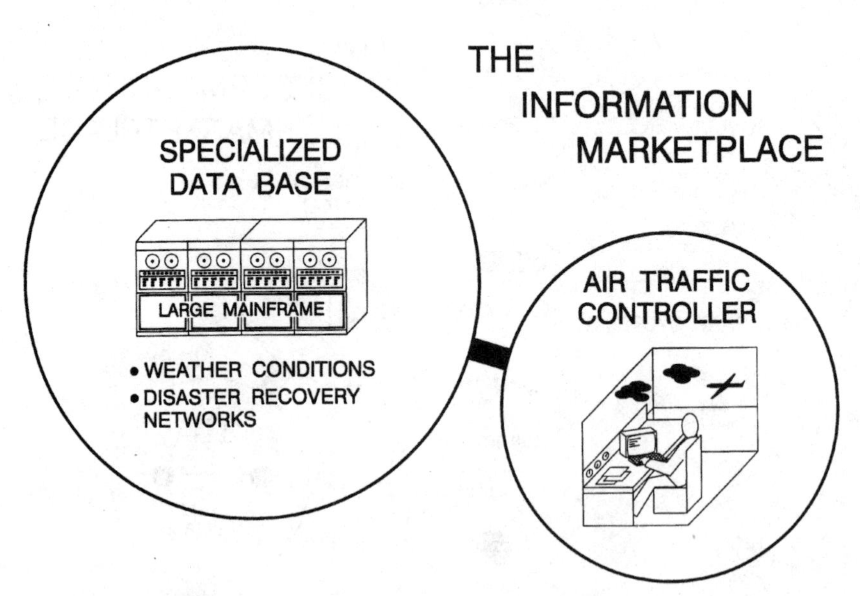

Figure 4-17. Air traffic services.

interactive movement can best be handled by active switched systems utilizing ATM or STM technologies. While the asynchronous transfer mode (ATM) system will best handle burst mode variable bit rate (VBR) traffic, synchronous transfer mode (STM) systems appear to nicely support continuous bit rate (CBR) traffic. Separate STM network fabrics can be designed for fixed throughput rates such as 64 Kbps, 1.544 Mbps, 51 Mbps and 600 Mbps. ATM appears to be nicely positioned at 155 Mbps, handling variable packetized voice and data traffic in fixed 53-byte cells. It simply keeps taking pieces (bytes) of the variable-length incoming data streams, so it really doesn't care how long they are, as long as it can keep up with the volume of input.

What does make the difference between these different systems and the need for high-capacity 9.9-Gbps transport facility services? As seen in Fig. 4-18, each form of data transfer will have a high number of services. All these successful services must be successfully transported. However, congestion analyses, querying theory, and traffic-distribution tables all show the need for more and more higher-speed bandwidth to transport all our lower-speed successes.

One bank, one hospital, one drugstore moving their own information may only require relatively low-speed and medium-speed transport buses. However, when 200 banks, 10 hospitals, 100 drugstores, and 1,000 small businesses exchange information, the need for capacity increases and increases. Hence, compression technologies will continue to have their role to play, during and after the transition period to broadband.

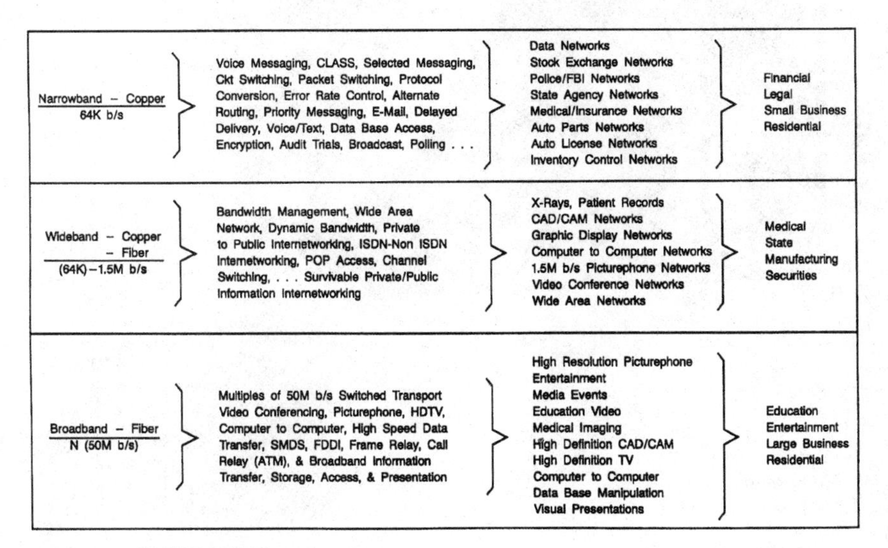

Figure 4-18. N-W-B ISDN services.

Applications

Besides computer-to-computer and CAD/CAM/CAE information exchange, broadband applications include: remote education classrooms; home catalog shopping; video mail; travel logs; dial-up entertainment (movies, sports, cultural events); video conferencing (1.5 Mbps, 45 Mbps, 140 Mbps); advanced TV (100 Mbps); high-definition TV (140 Mbps, 620 Mbps); x-ray; video storage; video services (shopping, finance, travel, entertainment, education, information); CATV dial-up, browse, and picture-in-the-picture; PACS (Picture Archives and Communication Systems); Remote CAT Scan (RCS); Magnetic Resonance Imagery (MRI); Computerized Axial Tomography (CAT); full-motion sports; electronic photographic storage; interactive TV, video security; high-resolution still graphics; white-board camera; and FDDI-II and SMDS high-speed transport. (See Fig. 4-19.)

Service nodes

There are several techniques for quickly providing services to the market; none of these techniques are cheap or easy. At best, all are com-

Figure 4-19. ISDN service game.

plex, but some are less complex and less dangerous than others. As we enter a world of interconnectability, interprocessing and interservices, we cannot underestimate the diversity and complexity of our new service environment. (See Fig. 4-20.)

To add a given feature and hopefully not bring the system down, we can collocate adjunct processors to existing switching systems and interrupt their call processing programs, containing upwards of 4 million instructions. Above the network, we can insert databases containing information such as virtual private network information, line identification, and 800 and 900 number and route translations, so that this information can be changed off line to more readily enable new network configurations or destinations. Alternatively, calls should be switched from the network to separate application centers, where advanced value-added services can be added to the call; or under switch control, these centers can dynamically access and change the switch's internal database parameters and values. Many of the new services will be required dynamically under customer direct or indirect interface, a nontrivial challenge.

ONA (open network architecture) requirements for RBOCs require network elements (NEs) to be accessible to OEMs third-party software suppliers, value added resellers (VARs), and other IXCs and VANs to

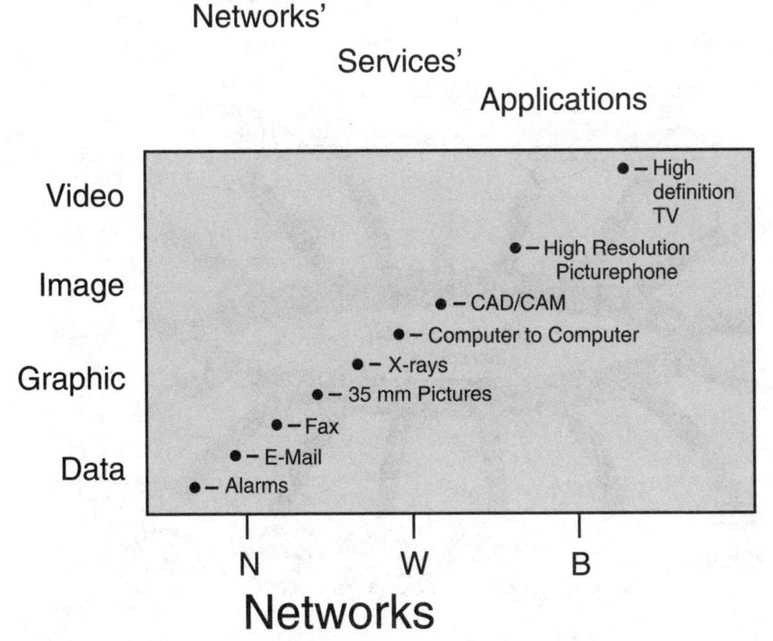

Figure 4-20. N-W-B.

ensure that a competitive playing field exists for all players. As we consider the cost of maintaining and protecting these interfaces, it's important to keep them simple. Hence, the objective of Bellcore's Information Networking Architecture should be to define the most appropriate architecture for not only ensuring a competitive arena but also ensuring that the network cannot go down as suppliers attempt to open up their switching systems for each of the types of service nodes noted here. Considerable analysis was devoted to their Intelligent Network 1+ and 2 efforts, as STP, SCP, SMS systems (noted in the technology chapter) are deployed for voice-based services. Now the complexity issues grow as we enter the combined worlds of data, image, and video services.

Similarly, the now "open" competitive arena considerations will help determine where best to offer services as the new networks become more layered to facilitate private-public internetworking, and as services become layered to enable application service centers, intelligent network nodes, and CPE application service centers to exist above or autonomous to the network. (See Fig. 4-21.)

Perspective: The Marketplace

Some say narrowband ISDN has missed its time, that the world will wait for broadband ISDN offerings. Users can do much more with a broadband channel than with a narrowband twisted pair; so why deploy narrowband ISDN? Why not wait for broadband?

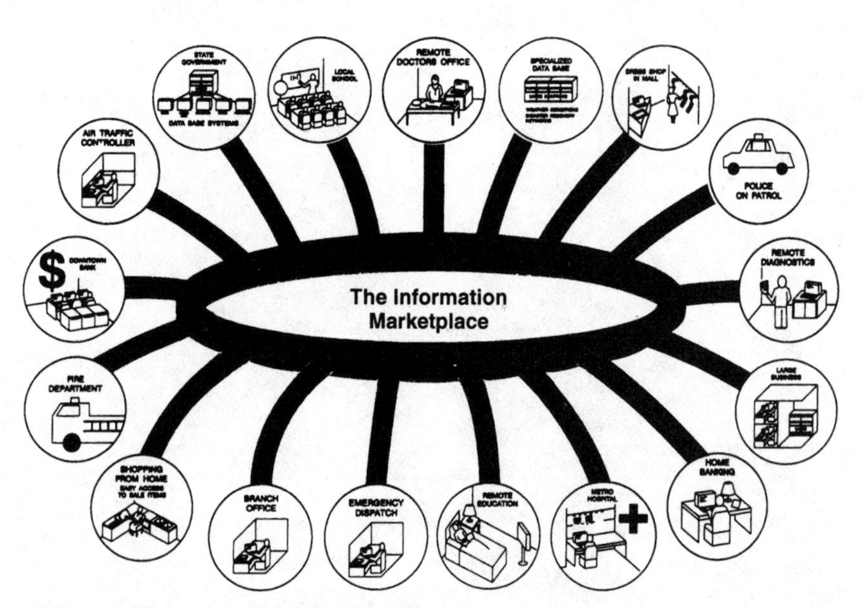

Figure 4-21. The information marketplace.

To answer these queries, we must step back and reassess who needs what, when, where and why. We have seen in an earlier analysis that sixteen or so out of twenty-six current data users from eighteen industries would be quite happy with a relatively error free (1×10^7 or better bit error rates (BER)) 64-Kbps terminal-to-computer interface, provided by a feature-rich data network that enabled numerous data-handling features. Yes, in the absence of the public data network, closed user groups have constructed their internal LANs and access the world through their WANs; some even ride across town on their MANs. But in looking at broadband, higher-speed networks, many are just carrying a considerable amount of lower-speed terminal traffic. Yes, workstations are indeed demanding higher-speed data movement, as T1 (1.5 Mbps) facilities are changed to T3 (45 Mbps), but this traffic is nicely handled by the wideband network and first-stage broadband. Video with high resolution and minimum compression will indeed require broadband. However, it will take time to deploy the fiber ubiquitously. It will require a major offering such as picturephone or HDTV to justify speeding up this implementation to obtain a full fiber plant in ten years or so. Similarly, much needs to be accomplished from a broadband-network-support capability to enable anyone to obtain any feature or service from any network anytime. (See Fig. 4-22.)

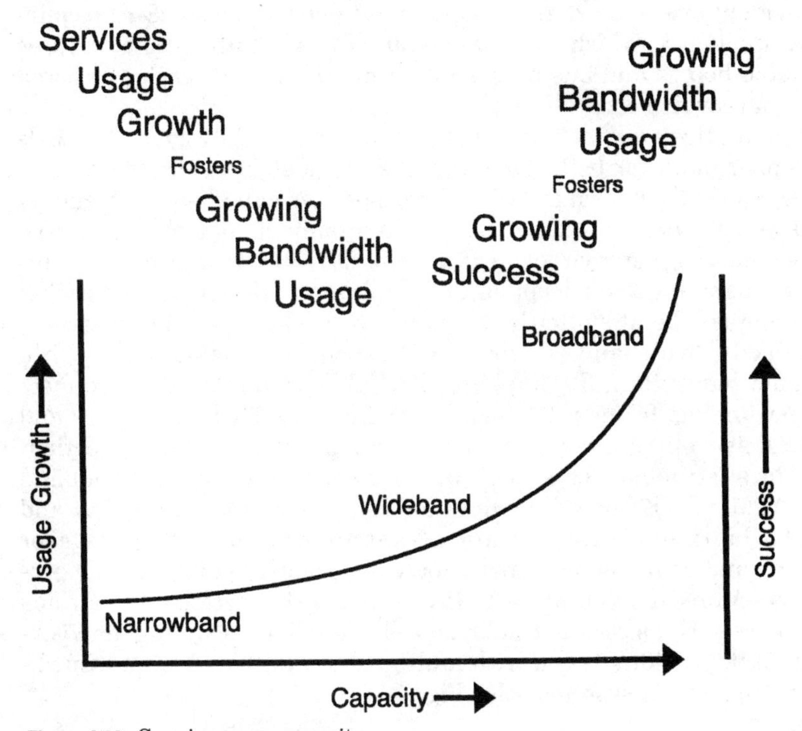

Figure 4-22. Growing usage capacity.

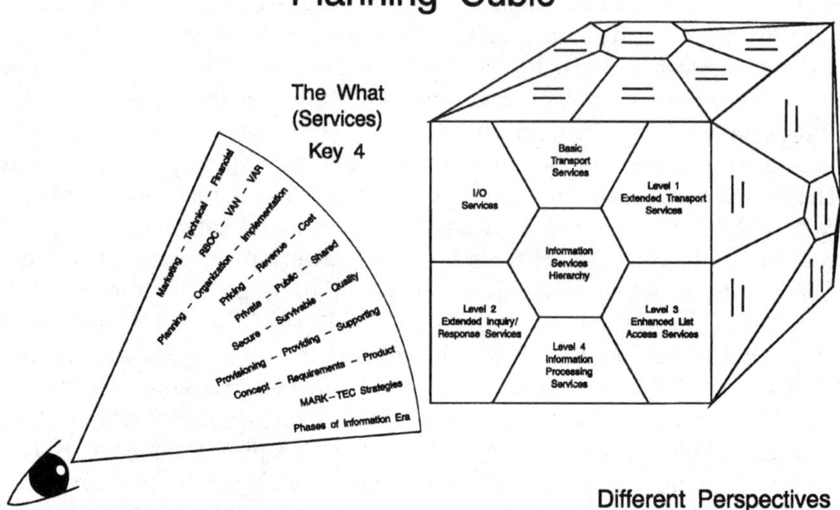

Figure 4-23. Services cubic.

On the other hand, narrowband ISDN does enable immediate deployment over the existing copper plant of a 64-Kbps/128-Kbps public data network, which can provide numerous data-handling features to enable homes and businesses to communicate with each other and access larger firms' databases.

All in all, there will be many, many new services, as we use the existing copper plant for both narrowband and wideband voice, data, text, image, and video offerings. "D" channel services will blossom as energy management, meter reading, remote environment control, alarm verification, message encryption keys, and audit trails are added to the out-of-band signaling and packet bus, and uninterrupted circuit and packet switching will be ubiquitously available to residence and businesses.

Similarly, wideband services will provide variable bandwidth, dynamic bandwidth, T1 hubbing, T3 hubbing, digital cross connect 3/1/0 switching, frame relay and SMDS interface for LAN bridges and routers, as well as interfaces for connection-oriented and connectionless data streams. This will be followed by overlay broadband STM/ATM SONET-based services for continuous bit rate (CBR) and variable bit rate (VBR) information. That, in turn, will set the stage for defining and utilizing new fiber deployment plans to ubiquitously provide 155-Mbps and 600-Mbps UNI user network interfaces for various user types such as residential, business, federal, educational, and governmental, providing them with multimedia, multivendor, multimedian, multiterminal systems. (See Fig. 4-23.)

Part

3

The Competitive Information Infrastructure

As roads connect,
bridges span,
rails track,
ships cross,
planes aviate
around the globe,
so "information networks"
must follow
here to there,
to everywhere.

Chapter

5

The Structure

There are many types of "building blocks," but, what do we want to build?

Over the 1990s we have seen many different types of structures, some stable, some unstable, some specialized. Some were dedicated to private entities; others were more public, more global. We have seen computer networks that provided their own unique protocols. Their designers spent considerable time and energy preventing "cloned" terminals from interfacing to their systems, in an effort to keep both the terminal and computer markets and to pursue vertical integration and ownership of all aspects of the application. Later, as cluster controllers, modems, multiplexors, concentrators, LANS, and application protocol interface programs were added to their offerings, computer manufacturers began pursuing every aspect of network transport, along with their input-output processing systems, to formulate total end-to-end solutions for their various customers' applications. (See Fig. 5-1.)

As a result, many specialized networks were constructed to provide a straightforward, immediate solution for networking a select group of computers and terminals. Hence, many closed vertical networks were established. Some were indeed very, very specific, so specific that nothing could be constructed on top of them. Any attempt produced a very shaky structure that usually came tumbling down at the first extensive growth of new users or new services. On the other hand, many computer structures became so closed that it was extremely difficult to add new, complex services on top of the transport mechanisms. Their lack of openness or inability to interface with multiple-vendor systems was such a challenge to the network manager that separated

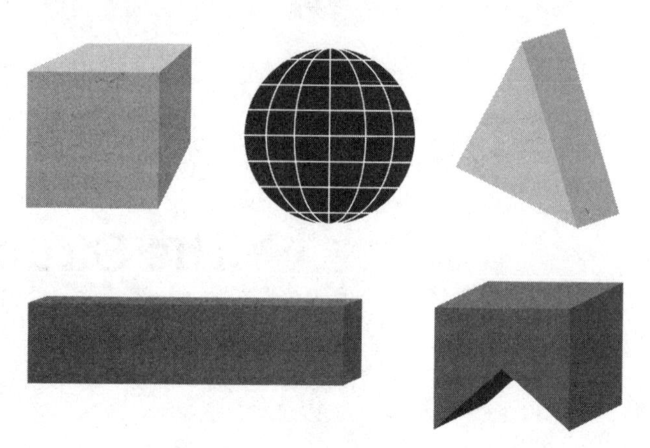

Figure 5-1. Building blocks.

networks were overlaid independently of each other. It became not uncommon to see two or three display terminals from different manufacturers sitting on a secretary's desk; by their presence, these redundant terminals displayed the need for a single terminal to be able to interface to noninterconnected databases. During this period of intense change, many firms spent considerable resources constructing information networks that, unfortunately, were many times obsolete before completion, as technology advances outstepped previous solutions. (See Fig. 5-2.)

Today, many ask, "What type of architecture will last far into the turn of the century?" We have seen the shift to integrating voice and data in private networks. Then there was the return of voice to the public network with the separated data on new high-speed T1 & T3 virtual private networks. There was a similar push for private networks to provide dynamically changeable, survivable bandwidth.

However, as time progressed, designers questioned how the excess capacity was to become available, if all the private contenders wanted it at the same time. Was it readily available? When idle, who was paying for its nonuse? As a result, many large private networks attempted to construct special offerings for other users in makeshift arrangements; they attempted to sell their idle capacity to smaller users who, for a cheaper price, would be willing to not use it when the larger firm needed it. Soon, they found themselves in the "information transport provider" business, with all its online network management headaches, as they attempted to keep their smaller users and themselves happy.

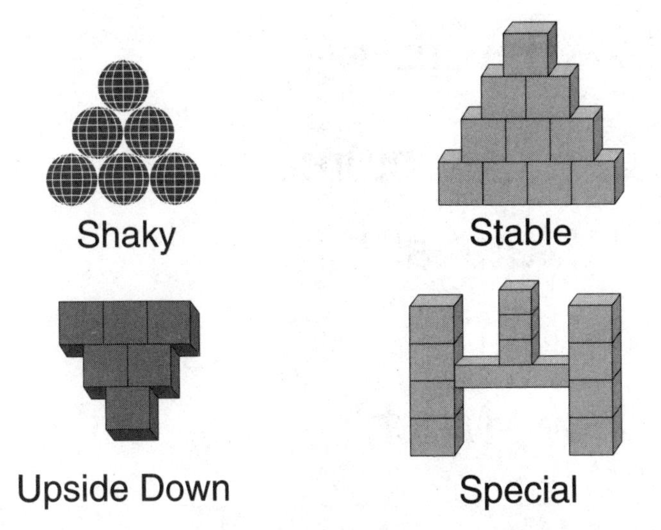

There can be Many Different Types of Structures:
Some Stable; Some Specialized, Some…

Figure 5-2. Architectural structures.

Building Structures

So lets take a moment to assess the structural issues. Indeed, what do we want to build? Let's step back and assess the opportunity. We could take the analogous view of constructing a physical building to meet our personal needs. Say, for example, we decide to build a garage addition to our home or even a new house. First we must ask ourselves, "What type of structure shall I build?" (See Fig. 5-3.)

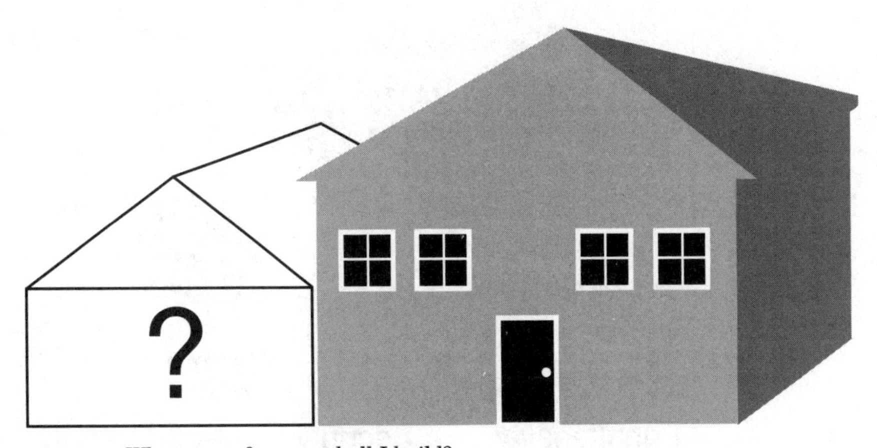

Figure 5-3. What type of garage shall I build?

GARAGE

- What Architecture?
- What Material?
- For Whom?
- For How Much?

Figure 5-4. Garage?

There are the usual questions to help determine not only the what and how, but also the why and for whom, as noted in Fig. 5-4. As we ask these questions, we begin to address the more basic issues of "use," as noted in Fig. 5-5. This then leads us to defining both the user features and the construction requirements. As we measure the height and length of possible cars and boats to ensure that they will fit within the structure, we must also determine the overall architecture of the roof design. (See Fig. 5-6.)

Next we must consider all aspects of personal usage requirements, including functionality. A structure can be many things to many people; it can be more than just a single purpose structure, unique and special; it can be more versatile, more functional. (See Figs. 5-7 and 5-8.)

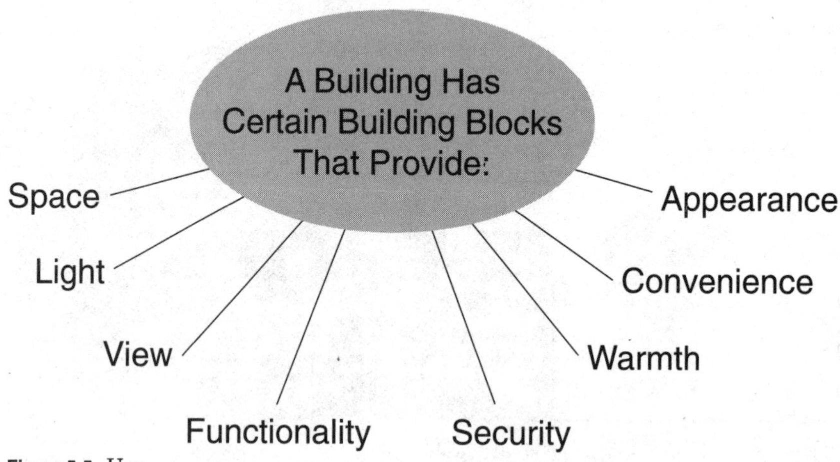

Figure 5-5. Use.

WHAT REQUIREMENTS?

- 2 or 3 Car?
- 22-24 Foot Width?
- 24-33 Foot Length?
- Single and Double Door
  or
  Three Single Doors?
- Asphalt or Shake Roof?
- Hip or Gable?

Figure 5-6. Requirements.

The Purpose For The Building Can Be To Provide
These Functions For These Users For This Use:

Functions	Users	Age	Use
Entrance	Children	Age 2	House
Living Room	Women	12	Office
Dining Room	Men	22	Store
Kitchen	|	32	Apt Complex
Bathroom	Rich	52	Orchestra Hall
Bedroom	Poor		Factory
Library	|		Jail
Offices	Educated		Outhouse
Workrooms	Non–educated		

Figure 5-7. Functionality.

WE ALSO SHOULD STEP BACK AND ASK. . .

What Else Can We Do with the Garage?

Figure 5-8. Observation.

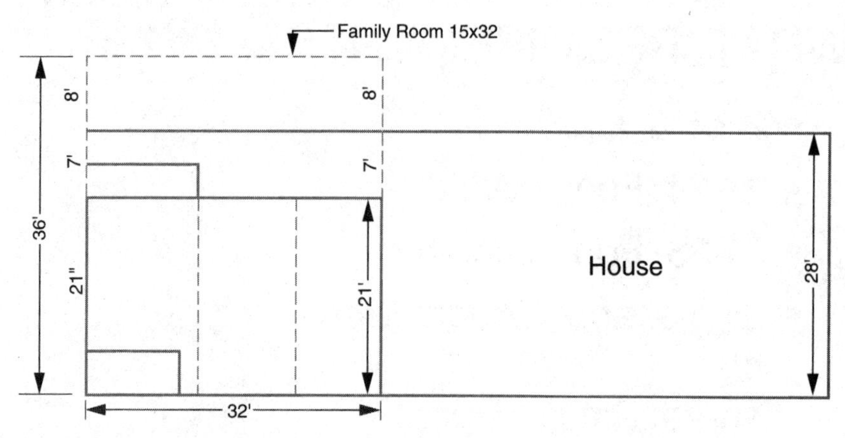

Figure 5-9. We could add a family room.

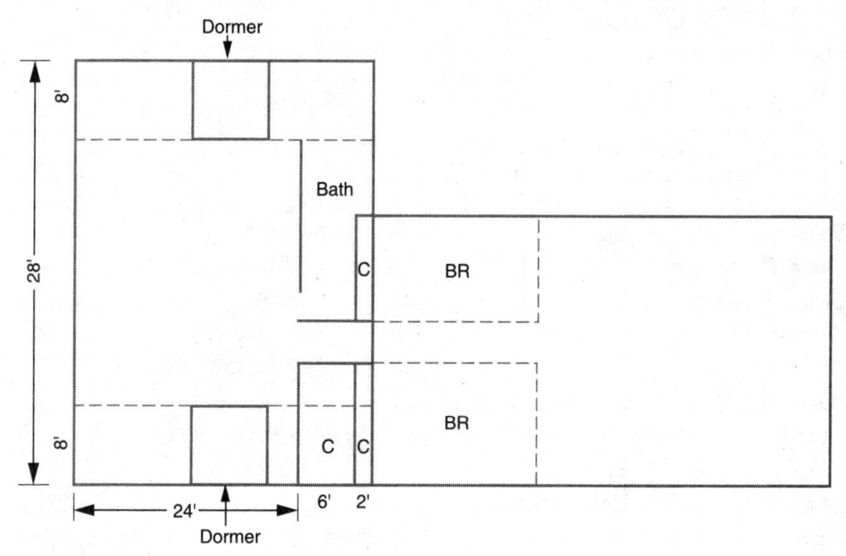

Figure 5-10. We could add rooms above the garage.

There are many other possibilities a structure could provide. We are already paying for a roof and outside walls and foundation, so why not add a family room behind the garage? (See Fig. 5-9.) But, why stop there? Why not use the space above the garage? (See Fig. 5-10.) Therefore, the multilevel garage could become a multilevel structure that takes economical advantage of every building opportunity. (See Fig. 5-11.)

Building costs vary, depending on materials. Real estate is usually sold by square foot or rooms. An interesting exercise is to see what the additional size actually costs in terms of labor and material, without selected enhancements such as type of carpeting or hardware, etc. The

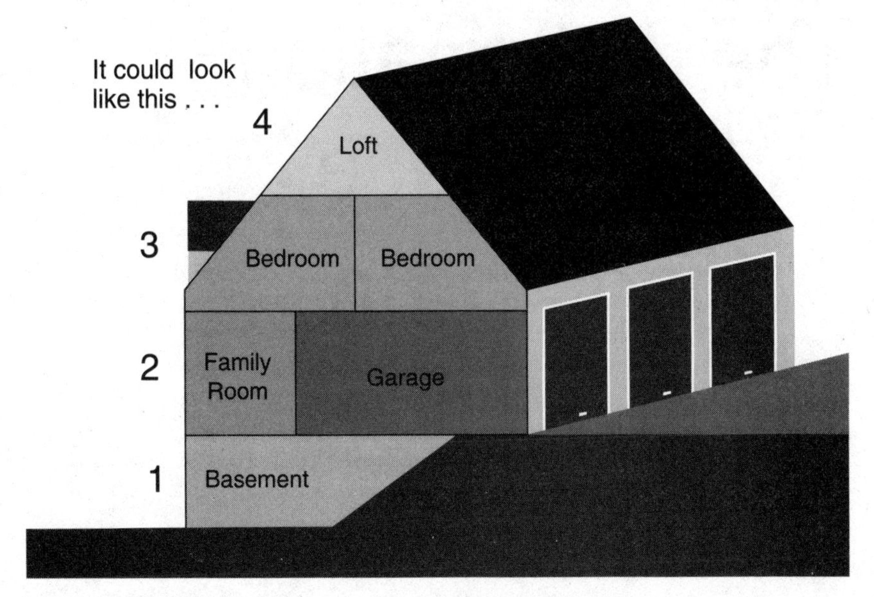

Figure 5-11. Multilevel.

basic unfinished building cost of additional space is really quite low, in the surprising range of $30 to $50 per square foot. Of course, finishing can be another $25 to $55 per square foot, depending on how it's to be finished. But as noted, the savings can be quite impressive if the owner decided to finish it off later, rather than simply pay for up-front market prices. Depending on how a building is constructed, what is done first will determine what can be added later. If done properly, the right structure can result in longer usage life, considerable savings, and many new and exciting features.

Telecommunications: Information Structures

Similarly, as we look at the future of telecommunications (see Figs. 5-12 and 5-13), we might ask similar questions. Pretending that the new structure is really an information structure, we can visualize similar considerations. (See Fig. 5-14.) Instead of a garage to house our transportation vehicles, let's consider communication facilities to support our use of information in homes, businesses, universities, and hospitals. (See Fig. 5-15.)

We are expanding from the traditional analog voice-based network with voice-only services, as noted in Fig. 5-16. We are beginning to use digital capabilities to better operate and expand the networks. This is similar to changing the garage door to a larger door with automated controls to enable two cars to reside more efficiently. In the communications world, these changes to digital technology represent more economical voice-call-handling capability. (See Fig. 5-17.)

The Information Marketplace

1990 – 2010

Figure 5-12. The information age.

- So What Are The Building Blocks Of The 1990 – 2010 Information Marketplace?
- What Do They Enable Us to Build?
- Who Are the New Users?
- What Are Their Needs?
    - People Needs?
    - Computer Needs?

Figure 5-13. Information marketplace questions.

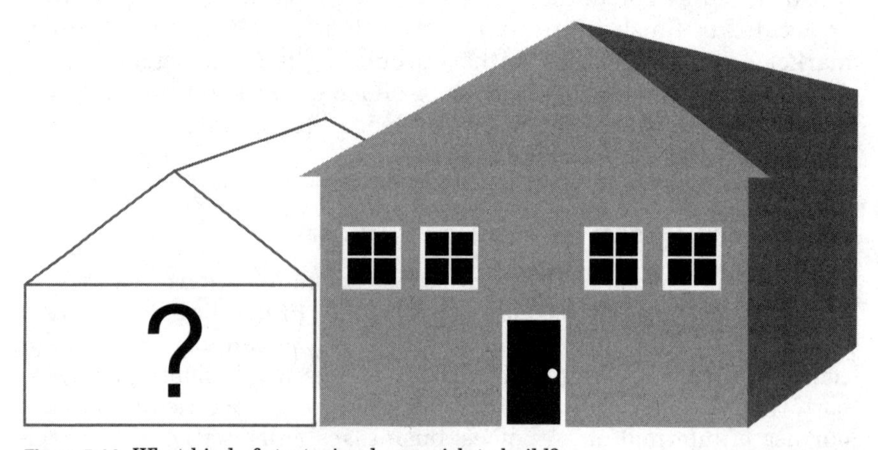

Figure 5-14. What kind of strategies do we wish to build?

Next, we can overlay data on our voice network using data-over-voice, dial-up data or separate leased-lines capability with a few data services. This has been demonstrated in the 1980s and 1990s by separate, expensive data packet switches, which attracted few users because of limited features and high prices. (See Fig. 5-18.)

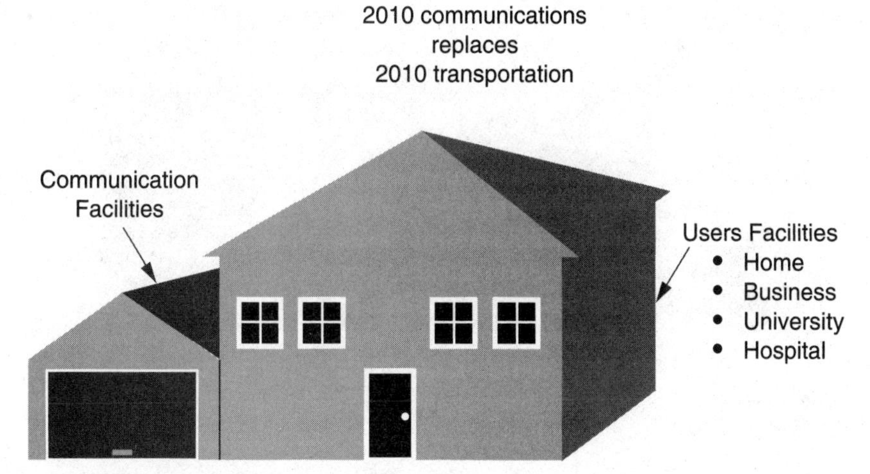

Figure 5-15. Communication facilities.

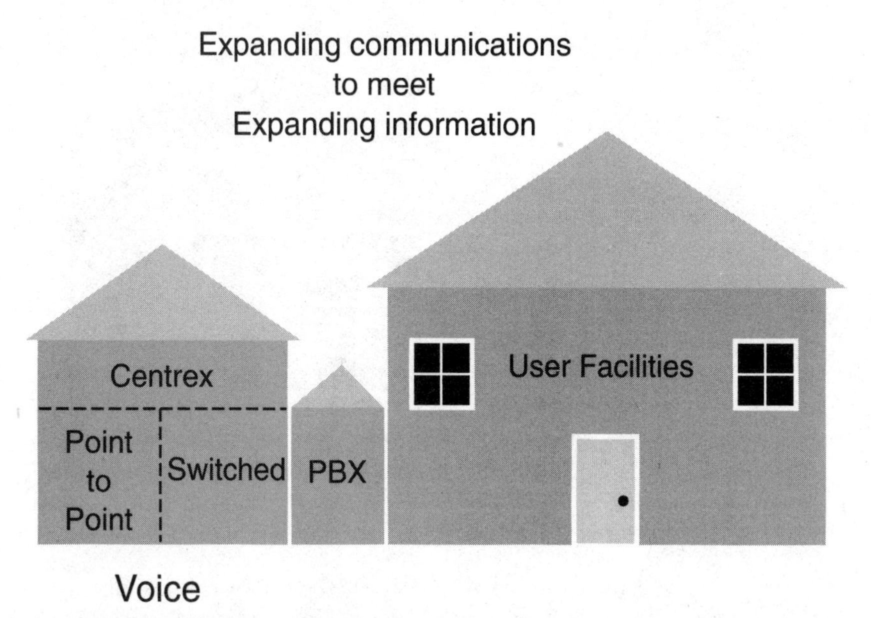

Figure 5-16. Voice communications.

If we again step back and consider a new multilevel structure, we could observe how we use different building materials for the different layers. As we used basement blocks, doors, bay windows, dormers, and skylights for different layers, we may wish to use fiber loops, access switches, data switches, gateway nodes, computer application programs, and ISDN interface services for new data, image, and video offerings. (See Figs. 5-19 and 5-20.) In this manner, the information

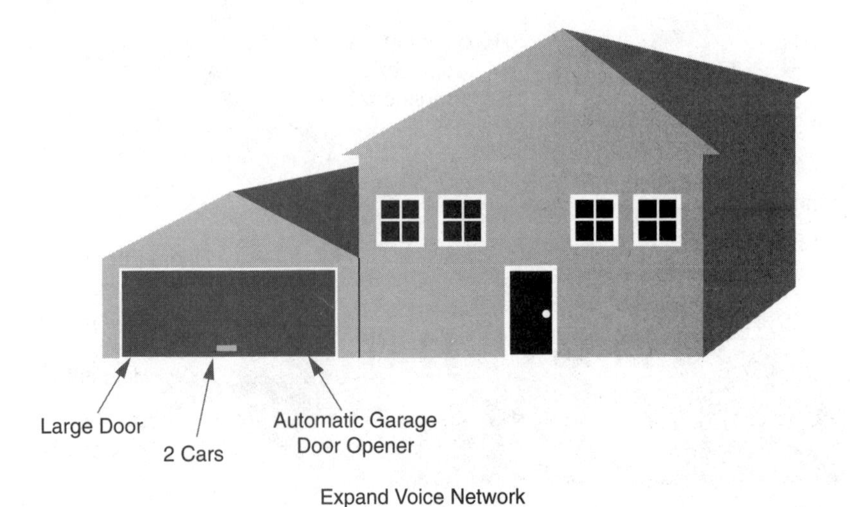

Figure 5-17. Digital overlay.

Figure 5-18. Data overlay.

industries' communication structure could begin to take shape from both a public-offering and private-offering point of view. (See Fig. 5-21.)

Layers on layers

As we begin to replace traditional public voice-only switches with new voice, data, and video metro-urban superswitches, the first step is to move new front-end information-access switches closer to the customer; this will enable private-public interconnection to survivable fiber rings that support shared information-transport services. From these ring switches, the customer can access unique offerings for a

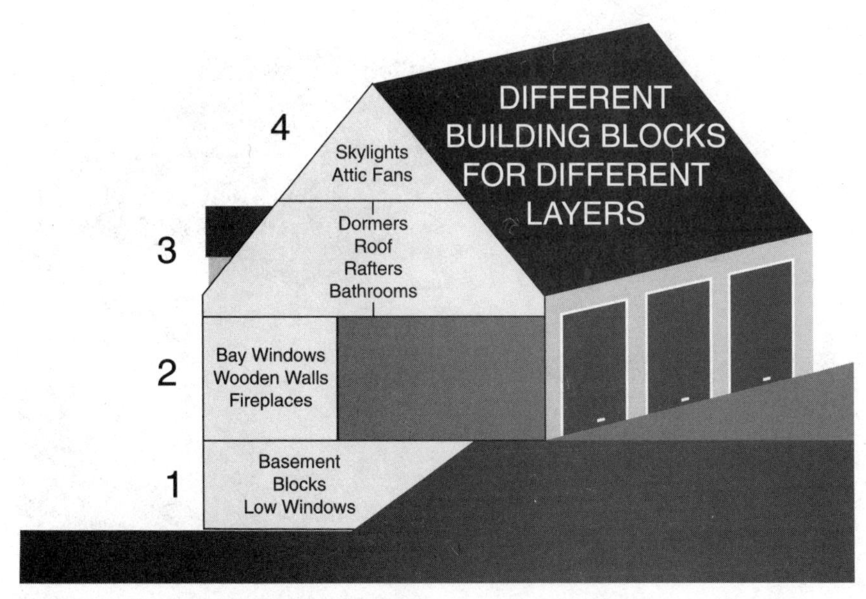

Figure 5-19. Construction building blocks.

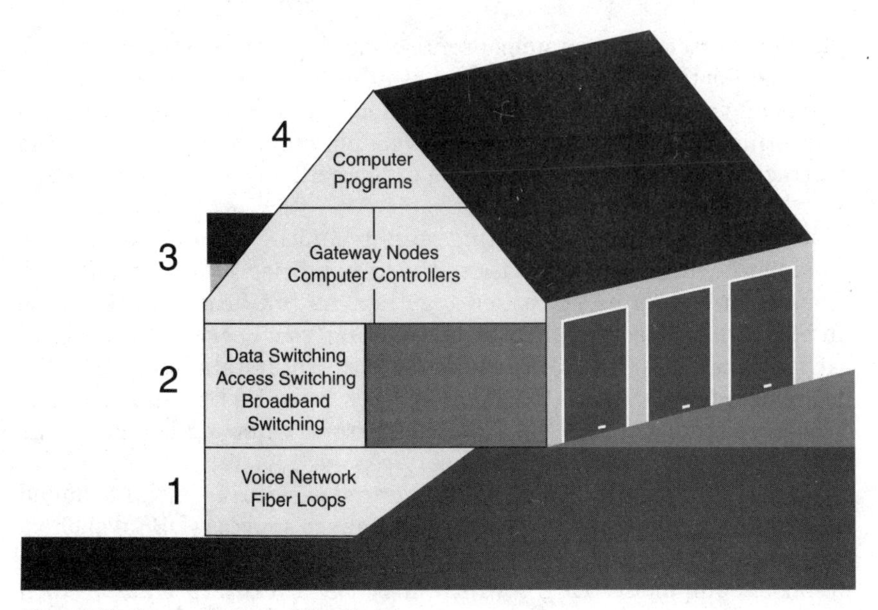

Figure 5-20. Communication building blocks.

specialized community of interest, offerings such as a new information centrex or enhanced services from platforms in the nonregulated marketplace. Therefore, on top of this transport infrastructure, service centers can exist as feature-rich nodes or as gateways to databases. In

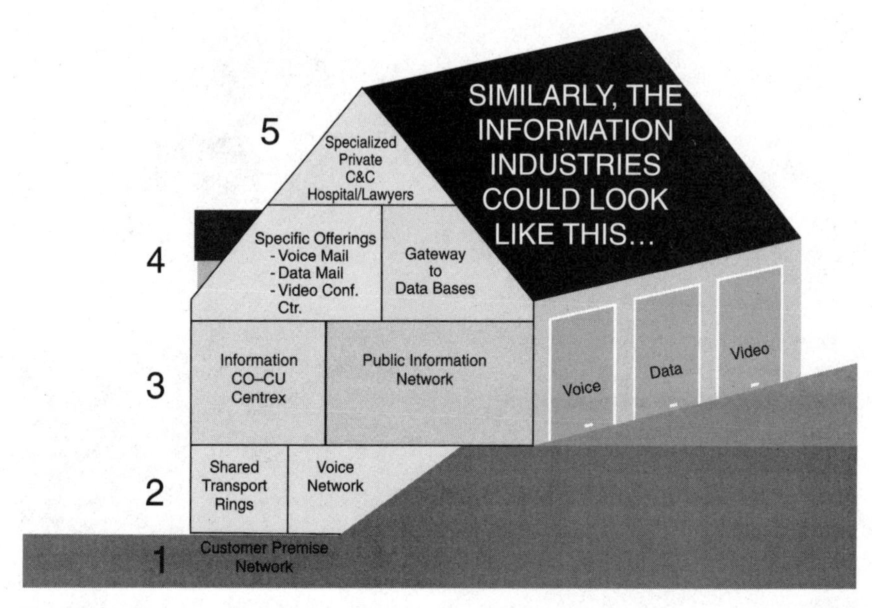

Figure 5-21. Layers on layers.

addition, internal customer-premise application systems can be accessed for their unique services. (See Fig. 5-22.)

To achieve this, we must bring technical possibilities together with marketing opportunities via knowledgeable management, just as the contractor enabled the tradespeople to correctly meet the homeowner's (user's) needs. (See Fig. 5-23.)

Not only must the marketing and technical capabilities of one provider be integrated together; the capabilities of many different types of providers must be tied together. As the future marketplace unfolds, we will see applications requiring voice services from local exchange carriers (LECs) as well as from the interstate (long distance) interexchange carriers (IECs), or simply inter-LATA interexchange carriers such as AT&T or Intercom. From value added networks (VANs), we may need additional services such as Telenet's data-handling capabilities. Also, there can be shared services from value added resellers' (VARs) microwave links, or database sources (DBSs) such as LegalNet. There will be customer services from internal customer premise equipment (CPE), which may be a PBX (private branch exchange), local area network (LAN), or simply a cluster controller or more complex internal broadband services switching center. (See Fig. 5-24.) The Telecommunications Act now opens all arenas to everyone.

As we leave the multilayered building analogy to focus on the movement and management of information, we cannot help but note the similarities and cost savings of layering networks upon each other to achieve the desired infrastructure, as well as layering services on top

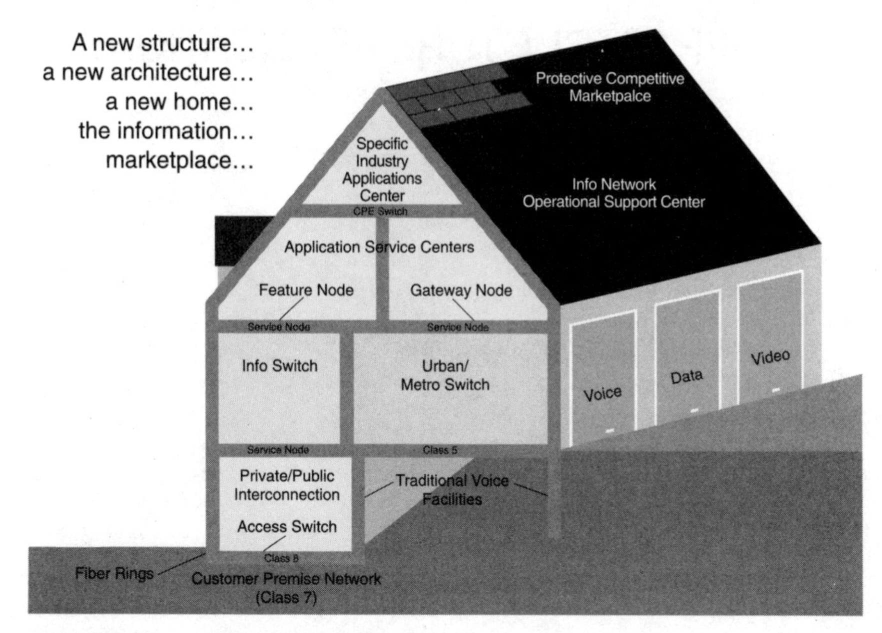

Figure 5-22. A new information structure.

of this infrastructure. Pursuit of this thinking results in a layered networks' layer approach, such that each layer physically provides a logical piece of the customer's solution. (See Fig. 5-25.)

Layered networks' layered services

As we begin layering services within the local community, several immediate separations come to mind that are quite distinct, from either a private or public point of view. Within the local community

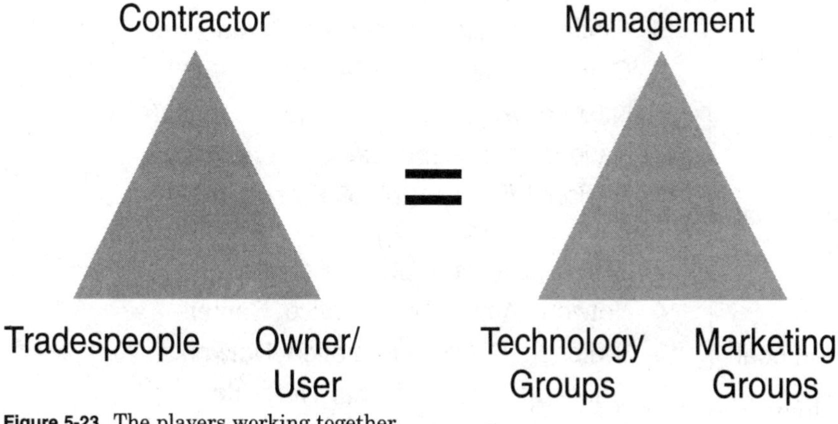

Figure 5-23. The players working together.

THE INFRASTRUCTURE APPLICATIONS

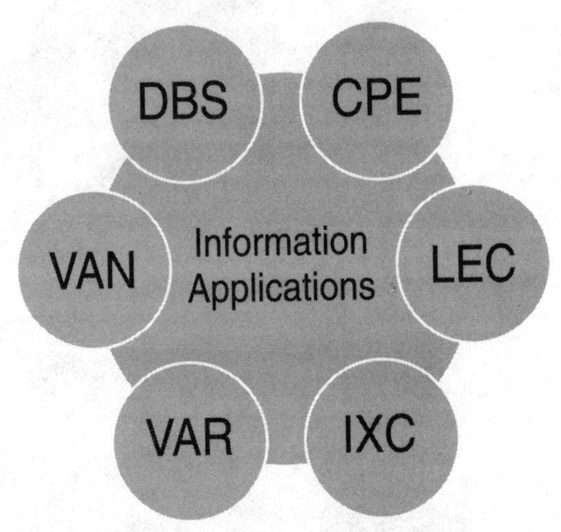

Figure 5-24. Multiple providers for infrastructure applications.

there are five layers of service offerings. Each service adds information-transport or processing value to the communication. Without such a layered approach, each private venture would need to stand alone with little sharing, and maximum cost for added value. This would limit applications to only the closed user groups who could afford to pay for them; but this also limits closed user groups from the opportunity of paying only for what they need when they need it. In many cases, not all the users within the private group want or will ever use all the services for which they must fully pay. This is similar to joining a private country club for golf and tennis, and then getting a $10,000 assessment for a new pool, even though you may not use it. (See Fig. 5-26.)

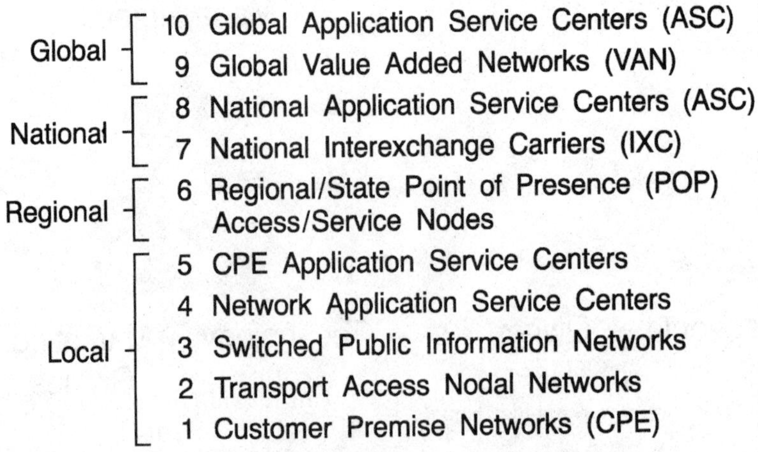

Figure 5-25. Global layered networks' layered sevices model.

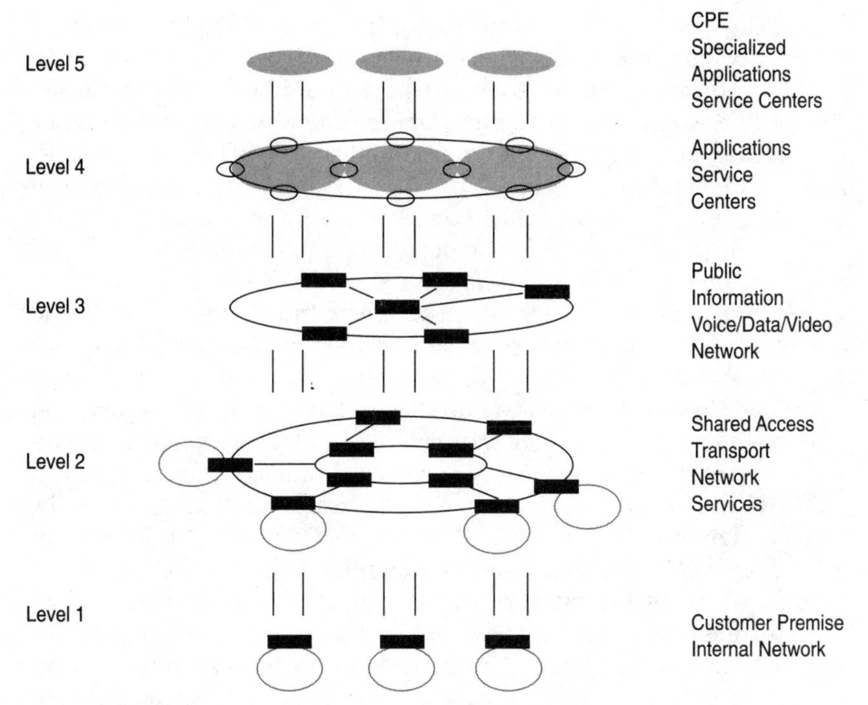

Figure 5-26. Local layered networks' layered services.

Local layers

Layer one: customer premise network services. This is basically ground level in the home analogy. In terms of information networks, "layer one" is the customer's basic internal network, be this a pair of wires to a phone, a multiplexor, a concentrator, a PAX, a key system, a PBX, a cluster controller, a local area network (LAN), a bridge, a router, a gateway, a microwave transceiver, a VSAT, a modem-data set, or a cellular or personal phone.

Behind this array of signal receivers and generators is a fascinating, increasingly complex group of terminals and computer systems. Besides the basic phone, there are now voice-message recording systems; expanding credit-card systems such as data pay phones, point-of-sale systems, and remote teller machines; electronic typewriters; printers; storage devices; personal computers; front-end processors; minicomputers; microcomputers; mainframes; special-purpose computers; sensors; environmental controllers; facsimile machines; and image generator-receivers such as x-rays, CAD-CAM systems, workstations, picturephones (9.6 Kbps, 64 Kbps, 1.5 Mbps, 6.2 Mbps, and 45 Mbps), video conference centers, high-definition TV, advanced TV or enhanced TV. Again, there will be the terminal side of cellular and personal phones, as well as internal alarms and remote control-sensing systems and internal base-satellite communications networks.

All in all, the customer premise equipment (CPE) constitutes an exceedingly complex group of voice, data, text, image, picturephone and video offerings that for the most part would be greatly enhanced if they could communicate to any remote or distant port. These entities together, both terminal and internal network I/O devices, constitute the lowest entry level of the traditional telephone five-level network. Since access to other new and traditional network layers will now be achievable from the customer's premise, these internal private switching systems will be referred to in this analysis as Class 7 systems to denote that a whole host of new internal addressing and route-control features need to become a part of the overall network design.

Layer two: transport access network services. These transport and access services are provided by private-to-public networking nodes located closer to the user. They are situated in such a manner that a CPE LAN can access a node, transport across to another access node, and exit to another LAN using SMDS or FDDI 50- to 100-Mbps connectionless MAN transport-type capability. Alternative transport exists in which nodes interconnect via survivable rings that ensure that multiple paths exist in the event of a cable cut. Similarly, multiple remote nodes "home" onto multiple network base switches, therefore survivability is assured. Here, traditional dial-tone delivery occurs, as well as frame relay interfaces, ATM cell switches, connection-oriented and connectionless data movement, dynamic bandwidth allocation, channel switching, STM circuit switching, various protocol conversions, and broadcasts to submultiplexed nodal points (called by some "sub-6 nodes"). The main nodes can be called Class 6 transport access nodes.

Layer three: public information voice-data-video network. The traditional Class 5 central office (CO) switching node is where call processing, traffic metering, line-to-trunk switching, maintenance testing, data-circuit switching, data-packet switching, ISDN interfacing, and broadband switching occur. The Class 5 central office also handles system 7 signal transfer points (STP), network interconnects for ONA offerings, and operational administration. Direct connections (trunking) exist between central offices (fives) or to a tandem system, which provides direct links to local exchanges, as well as access to the world. This direct access can also be accomplished on the rings between level two nodal switching (Class 6) offices, thereby reducing the need for tandem systems in the future.

Layer four: application service centers. These feature-rich service systems that exist above the network are usually in the nonregulated environment, where value is added to the call. Examples are facsimile or E-mail store-and-forward systems, voice message (regulated or unregulated) storage systems, gateways to various videotex databases, 800/900 databases, etc. These centers can be accessed directly from any layer in the network. They are provided to both open or closed user groups for any particular community of interest.

Layer five: CPE application service centers. Here, internal customer premise equipment (CPE) voice/data/image or video-handling takes place. Specialized processing of information, access to files, video presentation, and data manipulation complete the list of services for customers at layer one who wish to access some private data file someplace within the local area. Alternatively, they reach a public file on a shared basis, usually at some access cost.

For example, a doctor within a medical office wishes to access all the medical files pertaining to her patient. A request is made from a layer-one CPE workstation within a LAN. Access is made to the world via a layer-two transport node. The call is routed to the layer-three switching node, which recognizes the medical gateway address and routes the request to a layer-four medical gateway. The medical gateway reviews the request for the patient's records and initiates a sequence of calls to all the hospitals and archive files in the local area. These requests are routed by layer-three and layer-two systems to CPE application service centers within hospitals, etc., where searches are made and information is returned to the medical gateway ASC. Here, patient records are collected, sequenced, and transferred back to the doctor's workstation. Hence, all layers of the local network, both private and public, assist in the call, each providing its own level of participation, each providing its appropriate layer of service to complete the application. (See Fig. 5-27.)

Extended layers. As we consider the total solution, we might need to cross LATA boundaries to complete the call to some regional, national, or global location. Indeed, a firm today does cross many boundaries; this requires several (or many) providers to assist in the movement and management of the information (IM&M). To achieve this, the

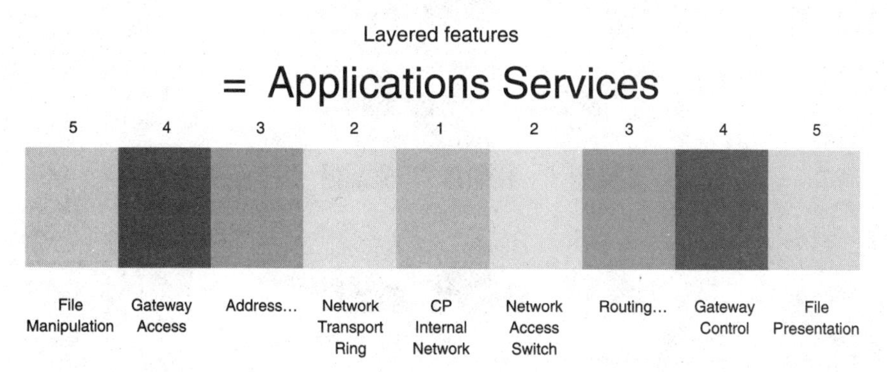

Figure 5-27. Layered services.

LNLS nodes must be extended to show the full range of services that can be provided by several additional layers of networks. With this objective, the model is extended to encompass regional, national, and global offerings in layers six to ten. (See Fig. 5-28.)

Layer six: point-of-presence access service node. The interexchange carrier (IXC) or value added network (VAN) provider locates their local, state, or regional point of presence (POP) at layer six. POPs are usually connected to a Class 5 central office or local tandem system in layer three, or can be accessed directly from the layer-two transport rings (Class 6). Or they are directly accessible via leased lines or trunks from the customer premise switches (Class 7), as well as by direct satellite, microwave, or radio links. In any event, this is the first-level switching node of the IXCs or VANs. From this, access to a national service node (layer eight) can occur, where services such as the following can be provided: changing routes dynamically, combining different types of services over common facilities, or using software definition systems. In reality, at the physical location of the carrier, there may be

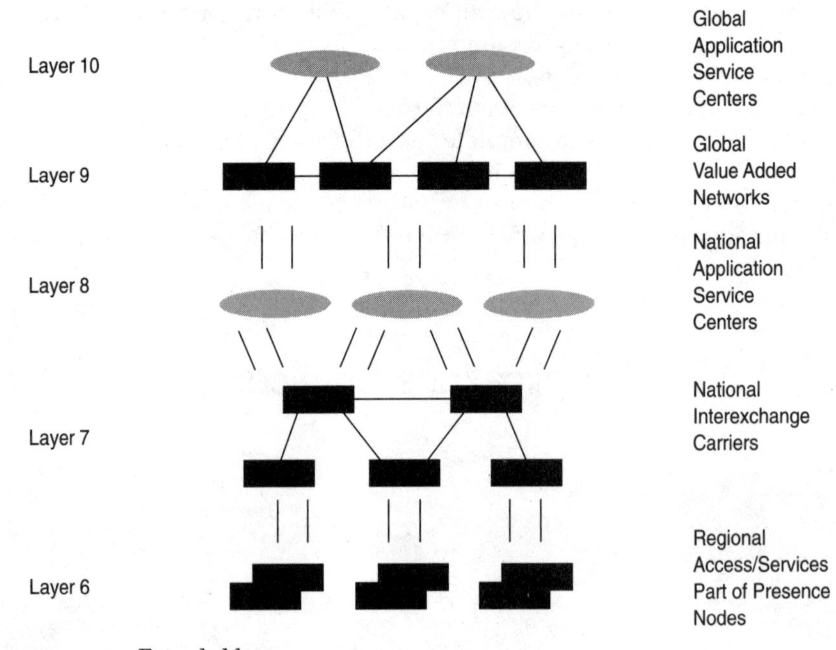

Figure 5-28. Extended layers.

a combined layer-six switch and a higher-layer service node. AT&T uses many of their Class 4 systems as their access switch node (POP) and software-defined service node, which are logically layer six and eight, respectively, in this multiservice layer model.

Layer seven: national interexchange carrier network. From the local-state-regional access node (layer six), information is carried by an IXC network across the RBOCs to other local-state-regional access nodes (layer six) via a hierarchy of switching nodes. Traditionally, the Class 1, 2, and 3 switches interconnect the Class 4s. However, as new fiber is deployed and local satellites are launched, more and more direct-connection (trunks or links) capacity exists between the traditional first-, second-, third-, and fourth-class switching nodes. Therefore, just as new nodes are introduced closer to the customer using high-capacity fiber, now less nodes are needed on the upper class levels of the traditional switching hierarchy. In time, this national network will eliminate several national-level nodes, but will then introduce several global-level nodes, as noted earlier in Fig. 5-25.

Layer eight: national application service centers. There will be a series of service nodes across the country that provide unique services on the national layer, as well as for individual states and regions. These centers will be considered national in that their location is independent of services such as 800-information centers. Each of the interexchange carriers will have these service nodes located separately or coexisting at their point of presence; these nodes can be accessed from both layer-six and layer-seven access transport nodes (Class 4, 3, 2, 1).

Many new services will be provided above the LEC by the IXC and other providers. In time, IXCs will offer data and video service centers, as well as the traditional voice features, as these multimedia offerings become more and more accessible to the local users.

Layer nine: global value added networks. Value added networks (VANs) will continue to overlay on traditional carriers, both in the local-state-regional-national arena, as well as in the global arena. In time, the entering, storing and processing of information will be performed at the most economical location. Hence, there is the need for these global networks to interconnect countries with a host of challenging information services. Local, regional, and national value added networks will simply be extended to the global marketplace. Therefore, there will be a point of presence for the VANs at layer six; there will also be a parallel national value added network at layer seven, and then a final global VAN at layer nine. In the new century, as IXCs continue to go global (e.g., MCI and British Telecom) and global VANs become national-regional-state and local, there will be a blur of these differences in layers; but for the late 1990s, the challenge will be to construct these global VANs to interconnect the world and national IXC voice, data, and video networks to the regions and states.

Layer ten: global application services centers. As global networks interconnect the various countries into a "global village," the "global marketplace" will emerge. Hence, global application service centers will exist to provide many interservices between national application service centers. For example, there will be global databases denoting medical, financial, and legal information that bridge countries and nationalities. All in all, it will be an exciting and far-reaching world as these global application service centers make the world quite neighborly, indeed.

LNLS cubics

To better appreciate what is happening in the information marketplace, it's interesting to look at the layered networks' layered services model in terms of why we use various technology. Let's look at who is using what forms of media to deliver what type of information services to what users. Let's also examine which array of suppliers' products are being used in interconnected private and public networks, phased in time to globally extend the application. This requires a little thinking, doesn't it? (See Figs. 5-29 through 5-32.)

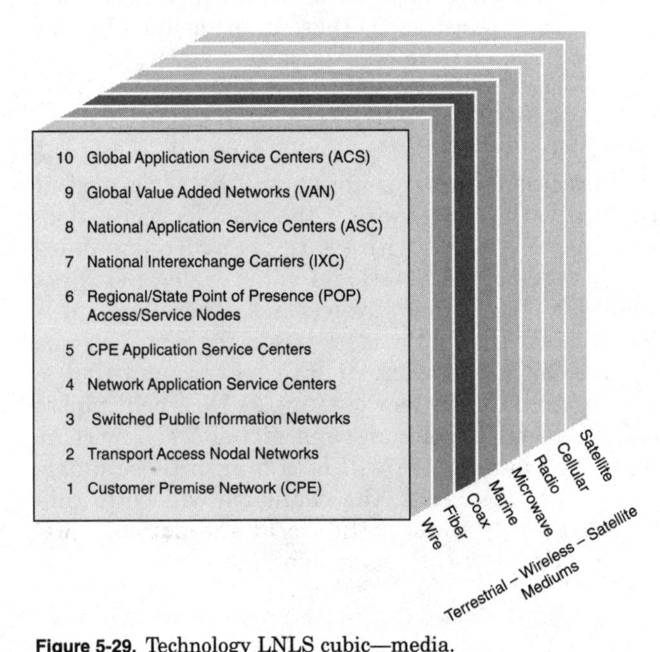

Figure 5-29. Technology LNLS cubic—media.

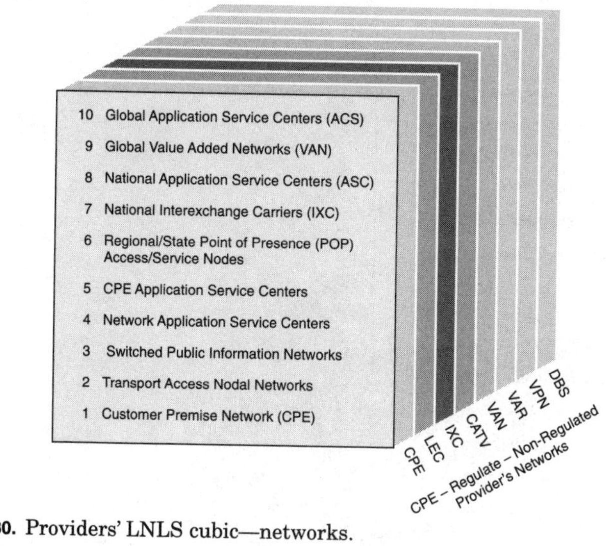

Figure 5-30. Providers' LNLS cubic—networks.

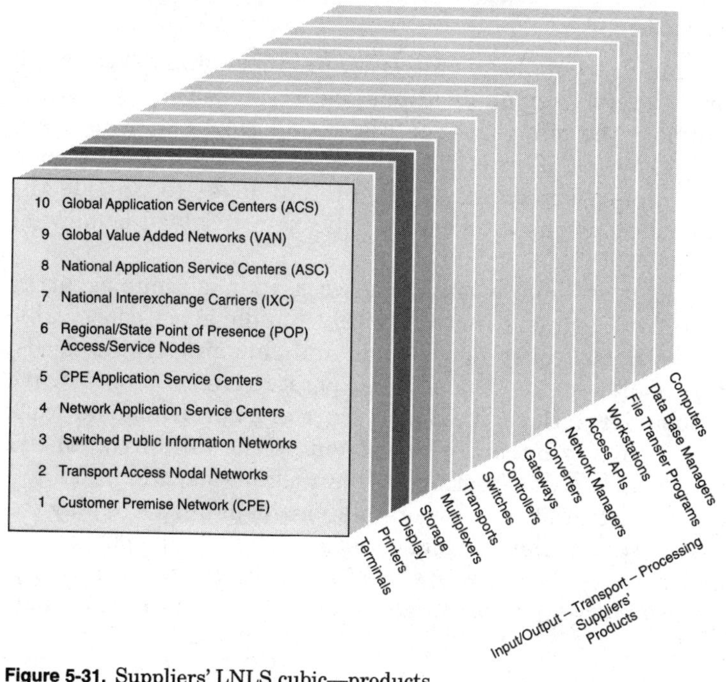

Figure 5-31. Suppliers' LNLS cubic—products.

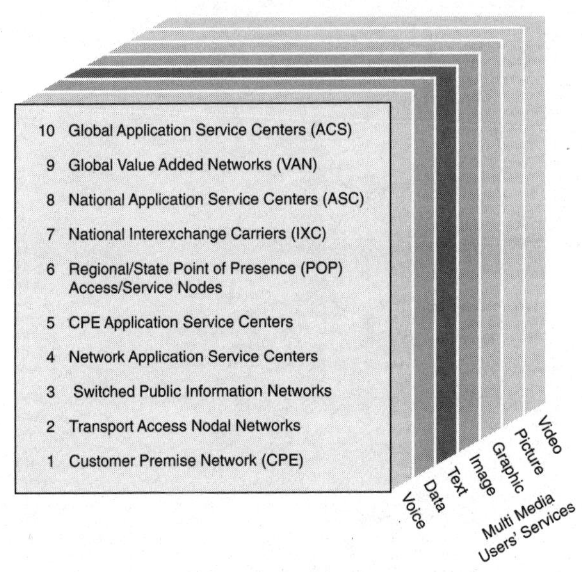

Figure 5-32. Users' LNLS cubic—services.

Application specific

We mix and match technology, networks, products, and services differently for different user applications, as we step from one to the other to complete the offering. (See Fig. 5-33.) However, who keeps the structure together? This effort will become more and more complex as each application requires this or that layer to provide the total end-to-end solution, with each application resolved somewhat differently.

It's much easier to offer a single network than a complex array of interconnected "seamless" networks that handle every conceivable call mix, with various "busy hour" loads and changing traffic needs. This is indeed a challenging task as we attempt to create a competitive arena on a global level. Here, the S&S (secure and survivable) network features are as equally important as providing the initial end-to-end connection. We cannot underestimate this task of operation, maintenance, and administration; as network management begins to play the major role in assuring that not just 99.99 percent, but 100 percent operational service applications are feasible. This will require considerable standards, testing, verification, and qualification programs, as

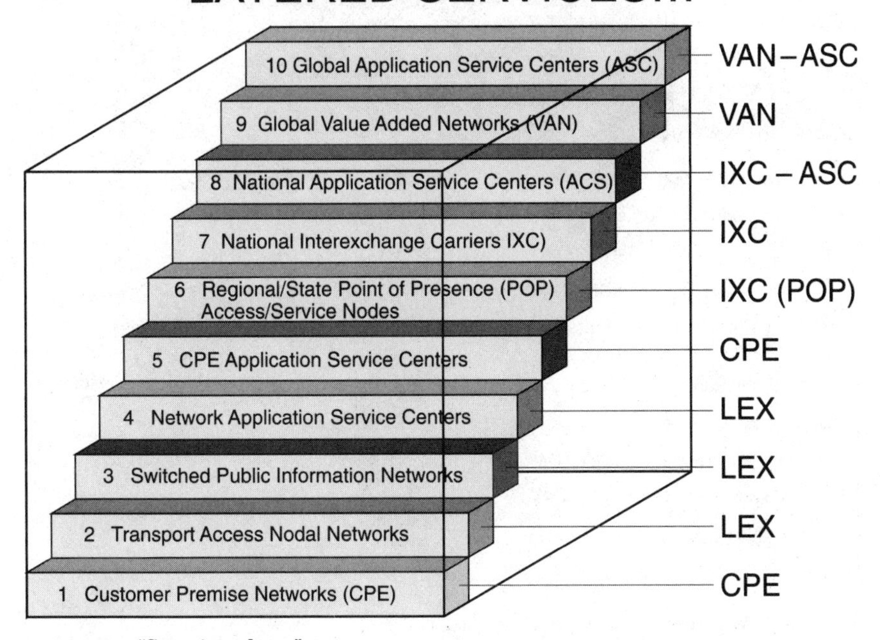

Figure 5-33. "Stepping along."

well as achievement of the "ability" objectives of maintainability, reliability, availability, portability, changeability, and growth-ability, as well as securability and survivability. (See Fig. 5-34.)

Service nodes

As we step back and review what has been said about new layers of networks closer to the user, including recognition that the customers now have their own initial switching networks, it should be noted that the traditional five-level network of Class 1 to 5 switching centers was separated at divestiture. The LECs retained the Class 5 and tandems, and the interexchange carriers (AT&T) kept the Class 4.

As access to the network is provided closer to the customer at new transport access nodes, it's essential that traffic can move directly from these nodes to alternative providers and alternative central office switches. Similarly, it's equally important that overlay service nodes or application service centers can be accessed from any layer in the network, depending on the type of service needed. Hence, the network expands on the customer side and shrinks on the national and regional high-end to take advantage of high-capacity transport systems.

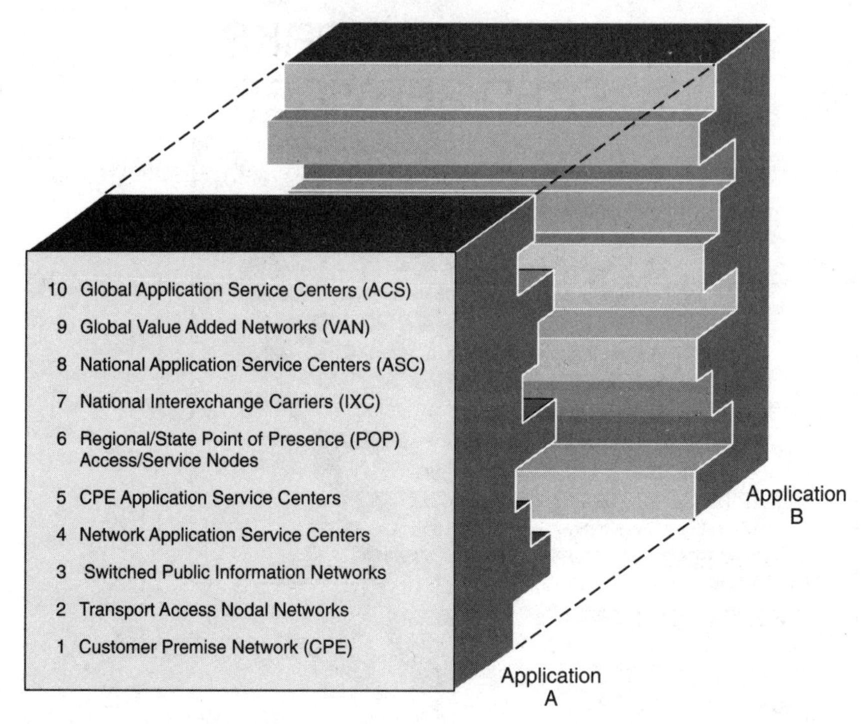

Figure 5-34. Who keeps the structure together?

The emerging new network infrastructure, as noted in Fig. 5-35, will include new switches at network access points and within customer premises called Class 6 and Class 7, respectively. It's essential that these systems, as well as traditional Class 5 and Class 4 (as well as LEC Tandem) systems access the multilevel service nodes that overlay or gateway new services to the users. These concepts will be discussed in detail in the network and technology analyses.

Summary

In reviewing this analysis, it's interesting to pause and review each layer of the model to mentally determine how a given service would be achieved by which layers. It's also interesting to note the importance of obtaining access to the customer. In the past, this has been called "the bottleneck," "the last mile," or "the local monopoly." As new super highway rings are put in place, with new access switching nodes that are indeed class-type offerings that enable limited front-end translations and routing to take place, the call could move off the public network directly to a layer-six point of presence (POP) of some other provider, then to the major local operating company (layer three). This would open up the access to the customer from an alternative carrier

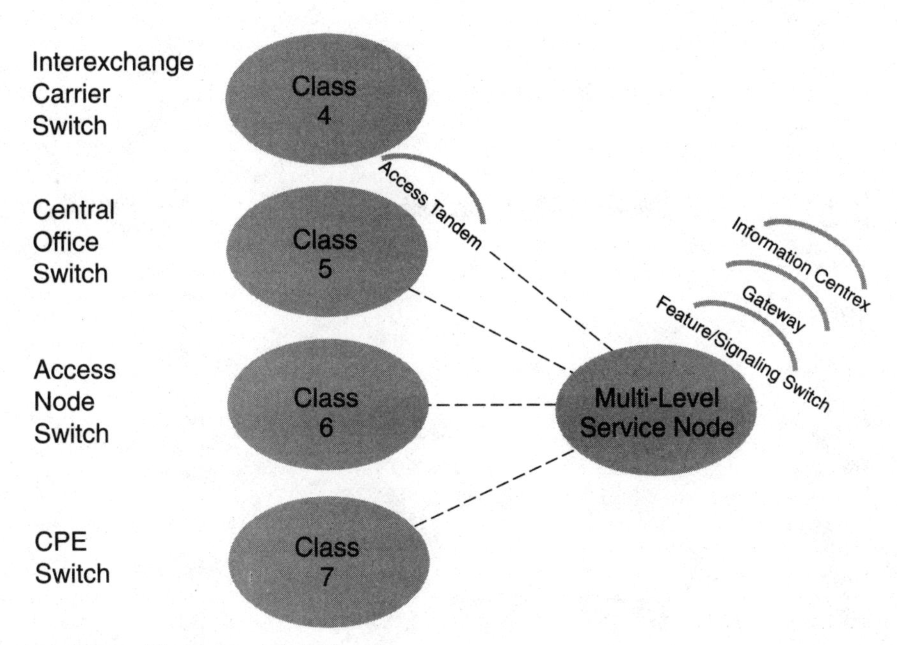

Figure 5-35. New network infrastructure.

location that is reasonably close to the user, using standards for transfer that ensure that competition can flourish.

From this layer-six switching node, access can be made directly to a national carrier (IXC) or global VAN. Alternatively, the new broadband public superswitches at Class 5 central offices, residing at layer three, will still have a lot to accomplish in order to achieve intra-LATA multimedia call connections. Service nodes called application service centers will provide a given service on a local network, both at layer four or internally within the customer's premises at layer five. Alternatively, an IXC or global VAN may wish to offer their services at layers eight and ten. These service centers will return the calls back to the network to be routed to their destination, or the service centers will be the completed destination of the call, depending on the customer application. (See Fig. 5-36.)

Perspective: The Structure

We, providers, suppliers, regulators and users, need to be able to visualize what is happening in terms of where we want to go and what we want to do to get there. The layered networks' layered services model helps fulfill this need, as it provides the framework upon which we can overlay the various networks, products, services, applications, and users models to formulate a bigger picture, the "broadview" of the information marketplace.

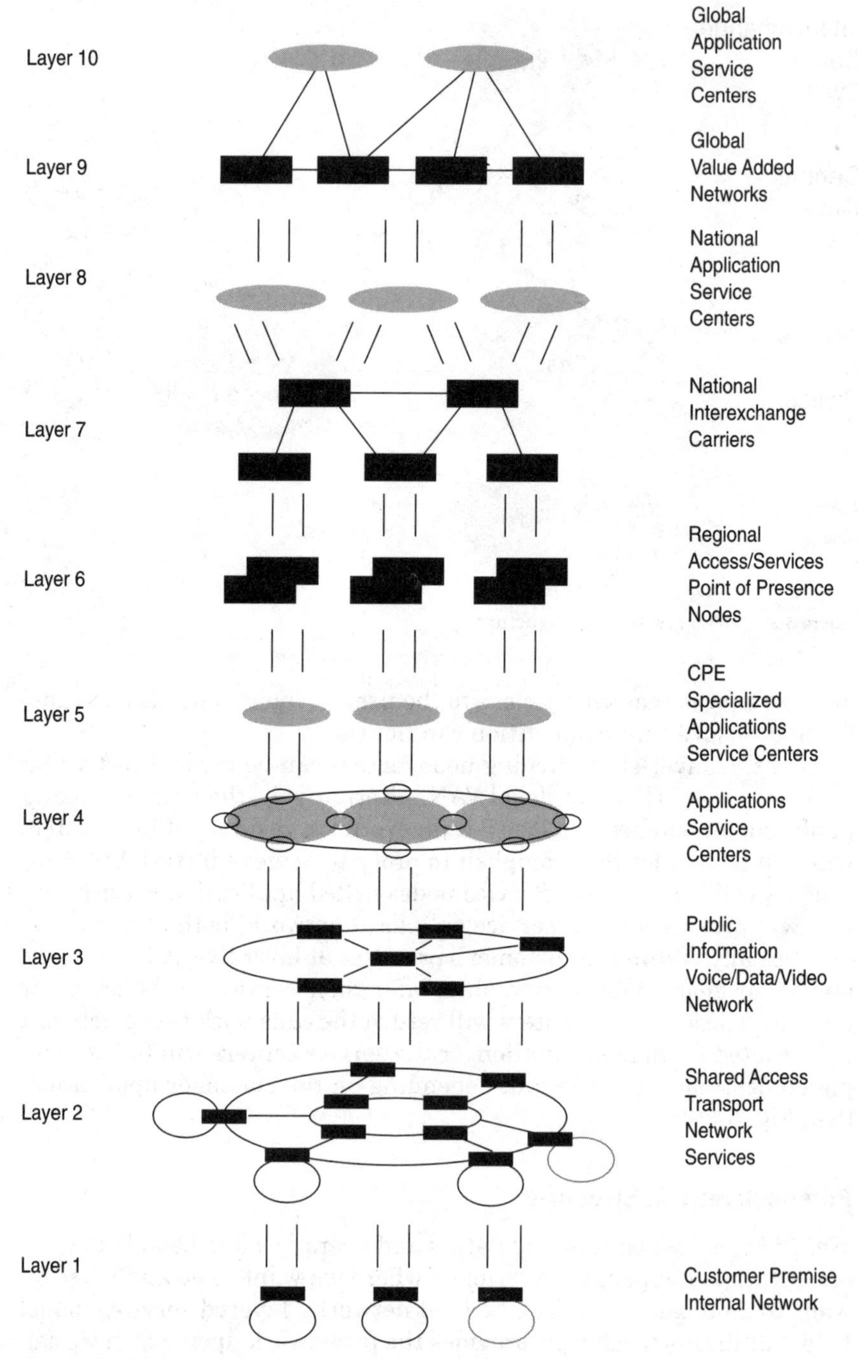

Figure 5-36. Global layered networks' layered services.

Telecommunications Management Planning Cubic

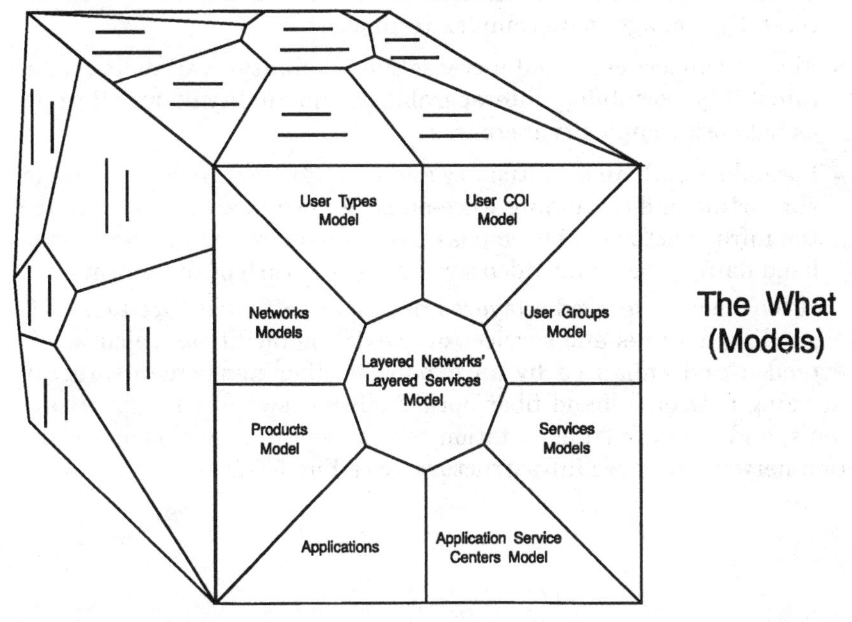

Figure 5-37. Global layered networks' layered services planning cubic.

This LNLS model will help us to better understand what is required to meet the following management, marketing, network, and financial needs:

- Establish a competitive base for future revenue opportunities well into the turn of the century.
- Be able to evolve new features in a timely manner from previous features, as user needs become more complex and extensive.
- Obtain a new family of voice and data revenue-producing services from existing copper-base plant.
- Obtain a new family of revenue-producing services from private-to-public internetworking opportunities.
- Obtain a new family of revenue-producing services for new data, image, graphic, and video users from new fiber facilities.
- Establish a multilayered but integrated information network infrastructure.

- Achieve this infrastructure for specific markets, as well as for general markets for more ubiquitous applications.
- Obtain an automated self-diagnosing, self-recovering, self-healing infrastructure that enables less sophisticated technicians to successfully manage more complex technology.
- Reduce support costs and increase the reliability, availability, maintainability, portability, interoperability, and quality of our offerings, as network complexity increases.
- Formulate both an evolutionary and revolutionary infrastructure for supporting and expanding voice-audio services, as well as extending the infrastructure to encompass narrowband, wideband, and broadband data, image, and video services by the turn of the century.

In conclusion, we need a layered networks' infrastructure to provide expanding features and services over existing facilities, which will be extended and enhanced by an expansive fiber deployment program utilizing new broadband fiber optic facilities, switching, support systems, and wireless transportation to achieve the new global information network services infrastructure. (See Fig. 5-37.)

Chapter

6

The Networks

Over the past forty years, emphasis has shifted from stand-alone products to systems to networks, as designers have expanded their world of "interconnect." The "systems approach" was developed in the 1960s to encourage designs that integrated both the computer and the terminal as a complete system. It was not until the late 1970s and early 1980s that the "network view" became more than simply leasing a transport facility to connect two distant systems. During the 1980s, private LANs blossomed, but it would be in the 1990s that "private networking" between LANs would take place in greater depth, as dissimilar systems become more "integrated" and "networked" together over dynamically changeable facilities. (See Figs. 6-1A and 6-1B.)

•Technology Shifts/Impacts
– Analog – Digital
– Main Frame – Distributed Processing
– Point to Point – Internetworking
– Batch Processing – Interprocessing
– Off Line – On Line
– Delayed – Real Time
– Bits – Gigabits
– Copper – Fiber
– Table Driven – Expert Systems
– Human/Machine – Machine/Human
– Lists – Graphs
– Voice – Voice/Data
– Manual – Automatic
– Transportation – Information
– Voice/Data – Voice/Video
– Separate Services – Interservices

Figure 6-1A. What's happening?

- Intelligent Networks
- Operational Support Systems
- Network Management Systems
- Open Network Architectures
- Gateway Systems
- CPE Systems
- Dynamic Bandwidth Management
- Customer Selected Services
- Customer Network Control
- Point to Point Private – Switch Private
- Class Platforms
- Signaling Networks
- Systemization – Office Automation
- Video Conference Centers
- High Definition TV
- High Speed Fax

Figure 6-1B. Where are we going?

As we pursue new technical and market opportunities, we might stop and ask where we are going. In so doing, we see that the systems engineer has given way to the network designer, who uses both private and public offerings to formulate the total "network solution" for his or her application. As market units from both the providers and suppliers address the various customers' needs, from residential through governmental, the question of type of supporting infrastructure, continually comes to mind. What is the new "information network infrastructure" that will support the needed new "information services"? (See Fig. 6-2.)

As we noted in the previous analysis, no longer will a single network provide a single solution for a single customer. We now live in a more global community that requires access to all information, anywhere,

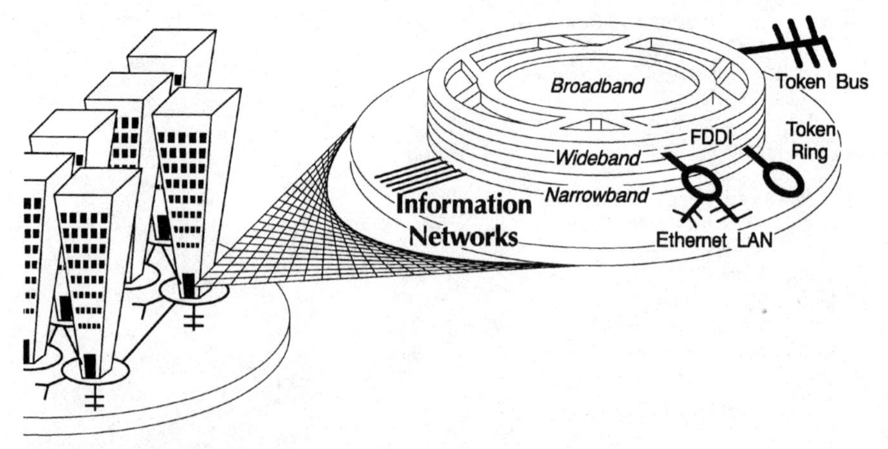

Figure 6-2. Information network's services infrastructure.

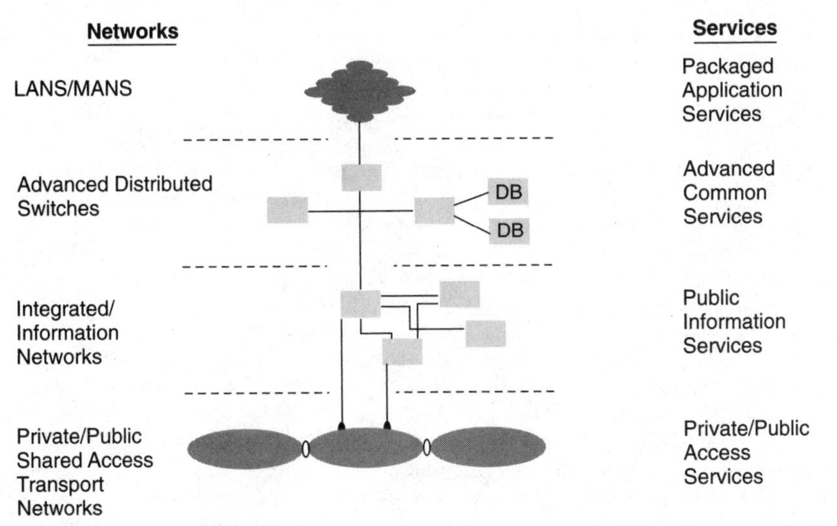

Figure 6-3. Layered network.

all the time, by anyone. No longer can the single product or system provide adequate access to all the needed solutions. There is an exploding need for private networks to access many different types of advanced and enhanced services via public networks in a layered network approach. (See Fig. 6-3.)

There will be a progression of networks over the next twenty years as ISDN is first overlayed on existing networks and then evolved, as noted in previous analyses.[1] In so doing, communications and computers (C&C) will become fully integrated over the 1990s using first narrowband and then wideband networks. Finally, wideband will give way to broadband, which will extend the range of transport and encompass previous offerings to support new service applications platforms, which will provide the full spectrum of new services for the Information Age. (See Fig. 6-4.)

Private and Public Internetworking

The information marketplace is irreversibly changed. By delaying their public data network offerings in the '70s, '80s, and '90s, the RBOCs have driven the MIS managers of large firms to create new networks that are quite autonomous from the telco's. Few suppliers' PBXs contained sufficient data switching capabilities to become competitive in the LAN environment. This lack of success encouraged considerable

---

[1]See Heldman, *ISDN and the Information Marketplace.*

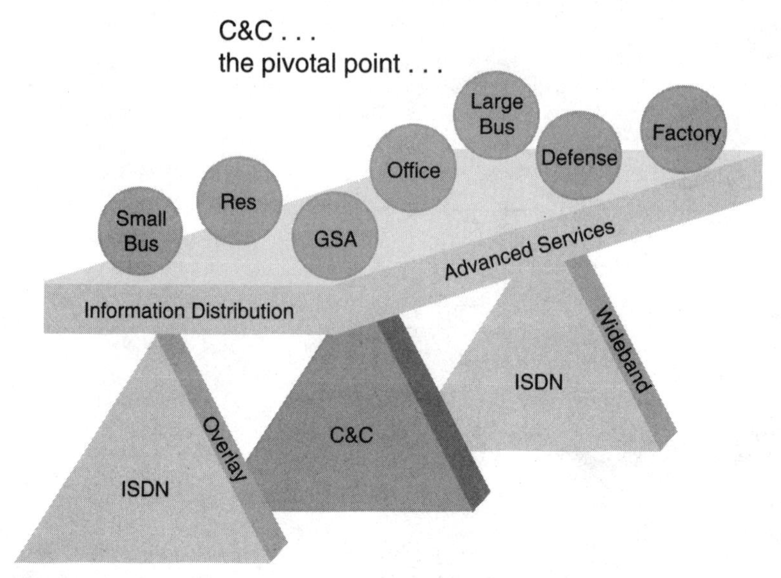

Figure 6-4. Information network's services applications.

growth of point-to-point direct connections via the LAN-type architectures, in which multiple terminals interfaced with each other directly using various addressing schemes. Users then expanded their networks to other types of distant structures by simply leasing more and more land lines, microwave circuits, or satellite links.

This has resulted in a separate "private networking" approach to interconnect LANs by gateways, bridges and routers. However, expansive growth has made this cumbersome, quite expensive and somewhat limited to closed user groups. In the new world of multiple systems interfacing together, there is a need for the more robust public networks to carry this traffic to and from the internal private networks. This realization has led to the layered network's layered services approach, which encourages a more realistic realization of Judge Green's competitive marketplace and Pete Huber's Geodesic Network. This is accomplished by "twelve networks" that will provide the underlying multimedia private/public structure that could be called the information network's infrastructure. (See Fig. 6-5.)

Pathways

There are many paths we can take to research the information marketplace, as we attempt to achieve a functional infrastructure that will enable the many new and different types of services to blossom. As seen in Fig. 6-5, this futuristic information highway enables the various market units' services (vehicles) to travel here and there.

Path one. Our first opportunity is to maximize the LECs' existing public narrowband network voice offerings by using new ISDN (access), SS7 (signaling), and IDN (Integrated Digital Networks) technologies to upgrade the current voice-based infrastructure to provide expanded offerings such as voice mail, calling-party ID, CLASS services, and selected call transfer, blockage, message routing, and priority override. In addition, we can achieve intelligent-network 800 and 900 database-access services, special audiotex services, and fast provisioning, number change, etc., support-type services over this expanded *public voice network.*

Path two. Next, LECs can establish a new family of narrowband data services over their existing copper facilities, using ISDN data interfaces to enable both the circuit and packet switched movement of data from terminal to computer, computer to computer, and terminal to terminal over the new *public data network.*

Path three. Using higher-speed technology, we can provide an exciting group of wideband services, including wide area networks that provide a variable number of 64-Kbps channels up to T1/T3 rates of 1.5/45 Mbps for data, imaging, and graphic display of information, for applications such as x-rays for the medical industry, stock forecasting for the securities industry, and CAD/CAM for manufacturing groups over the *public wideband network.*

Path four. In addition, to foster the movement of information between private networks over public networks, we can provide new high-speed

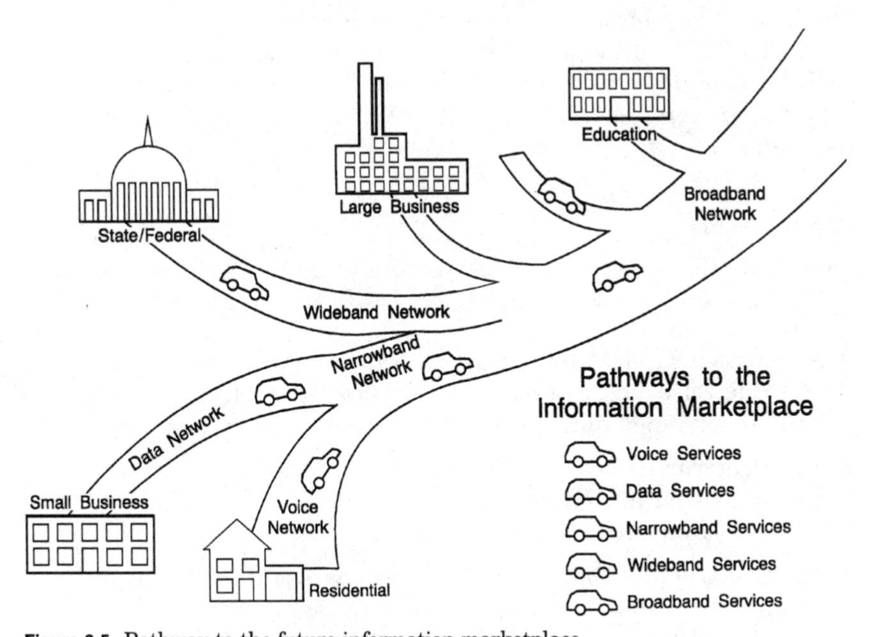

Figure 6-5. Pathway to the future information marketplace.

data transport services such as SMDS, dynamic bandwidth, bandwidth on demand, survivable transport, improved error rates, and access to multiple national carriers and international value added networks. This can be achieved by expanding copper-based wideband offerings to provide these services over new fiber-based *public broadband access networks*, using self-healing, survivable fiber deployment topologies with new switching centers located closer to the customer.

Path five. The 1990s also provide the opportunity to achieve new revolutionary broadband services such as high-resolution video services that enable visual imaging, desktop conferencing, high-quality graphic display, computer-to-computer data exchange, and video displays for education, media events, image archives, and entertainment-type video services. The *broadband public network* provides these services over a revolutionary new fiber-based voice, data, image, video switched network with automated support capabilities.

Path six. We can then establish extended, enhanced services in the nonregulated community via an overlay of information systems, such as info switches, which provide direct access to databases and advanced data-handling services or via gateway access to stand-alone application service centers for special services such as E-mail, and, access to internal CPE service centers for specialized databased services such as poison control centers, patient records, etc.

Path seven. Finally, to achieve "global telecommunications" we need to provide interconnected global value added networks from VANs or interexchange carriers using global gateway switches that channel traffic via the appropriate satellite, marine fiber, or terrestrial fiber networks. This will require direct local network access to national and international application service centers that then internetwork their information services via these "global VANS" for more global voice, data, and video applications.

Network needs

To achieve this marketplace, we require a multifaceted network consisting of several networks that meet the following needs:

- An expanded voice/audio network infrastructure that enables new versatile voice message services overlayed on existing capabilities to form an extended and enhanced *public voice network* for the 1990s.
- An infrastructure that fosters the economical movement of low- to medium-speed information over current copper facilities to formulate the *public data network* for the 1990s.
- An infrastructure that enables the internetworking of private facilities over unlimited, high-speed, secure and survivable (S&S) facilities on a demand-type usage basis to form the *public broadband-access transport network* for the 1990s.

- An expanded and highly functional support system infrastructure that enables timely provisioning, testing, maintenance and administration of the transport and delivery of the new voice, audio, data, image, graphic, text, and video information for the 1990s.
- A revolutionary network infrastructure to encourage the growth of many new high information-content services that require substantially higher bandwidth, faster movement, and lower error rates than obtained by evolutionary changes to the current network as it becomes a new *public broadband network* by the turn of the century.
- These networking needs require a layered networks' layered services infrastructure solution that supplies narrowband services over existing facilities, overlayed with conditioned wideband services and then augmented by a growing array of broadband services to formulate an expanding and encompassing fiber-based *broadband network services infrastructure.*

Twelve Networks

This new information network infrastructure will consist of twelve networks, as noted in Fig. 6-6. These networks range from the traditional voice only to the more global VANs. Each network will be discussed in depth to provide a clear understanding of both the logical and physical aspects of its structure. This information networks model can then later be used to help denote and differentiate why, where, and when various services could be or should be provided. In any event, the following analyses will attempt to answer the what and how. What is ISDN? What are the new information networks? How do we deploy them?

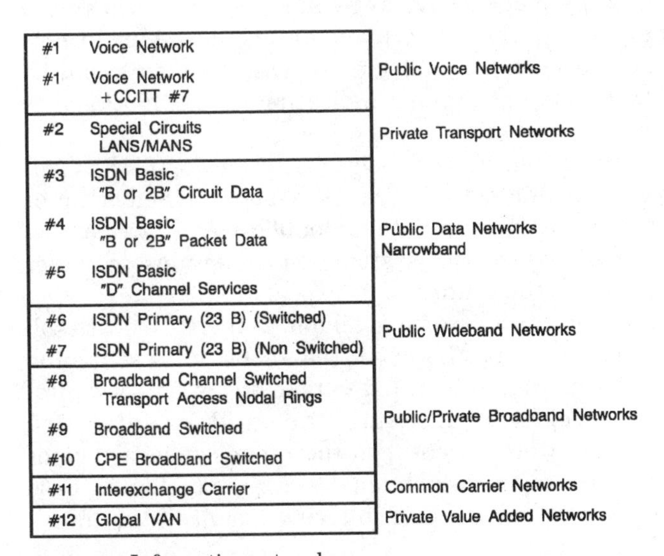

Figure 6-6. Information networks.

The public voice network

Let's talk fast;
it's a long-distance call.

Over the last decade, the public voice network has begun to take on a new form and structure. The traditional voice network was based on a pair of twisted voice-grade (3K bandwidth) lines from a black phone (maintaining an inventory of color phones was too expensive) to a Class 5 (central office) switching system location. Here, calls were processed by a long distance ten digit (area code + office code + subscriber code) or local seven digit (office + subscriber) translation that determined how to handle the call and where to route it. The wired-logic switching systems contained step-by-step, crossbar, or core-reed multistage matrixes. Later, path control across the matrixes became software controlled as it established a connection from port to port, thereby connecting the call from line to line or line to outgoing trunk. Considerable effort was spent automating maintenance and testing, and contacting and updating the large databases that provided individual line "class mark" or routing "translation tables" information on each subscriber and number group. Initial tariffs were based upon expensive "long lines" and cheaper local plant.

Transmission was initially separated from switching, where new central offices were established at limits of copper-pair transmission ranges of 12 to 18 thousand feet. There was a five-level switching hierarchy established for moving calls across the United States such that Class 5 local offices forwarded calls to Class 4 or directly to other Class 5s within an area. Or, tandem switches existed to handle high-volume urban trunk groups. At this location, code blockage could occur if some radio station disk jockey made a give-away announcement causing a rush of calls that could bury the network and bring it down due to abusive call-processing overload. The network switching center could be either path limited or attempt limited, indicating matrix blockage or computer processing limitations.

Private networks were constructed using feature group (A-F) tariff rates to move special traffic across town in a nonswitched manner from port to port. Similarly, long distance public traffic was multiplexed on higher-analog transport systems called groups or super groups, and moved across the country.

Digital transport capabilities became available in the 1970s with toll level (Class 4) digital switching. Later, in the 1980s, local switching and transport systems became integrated to take advantage of digital technology in new topology architectures called IDN, Integrated Digital Networks. These were based upon the remote-to-host-type or base-to-satellite-type concepts, which enabled a new digital base switch to be inserted in a central office in a rural county seat (town).

At the same time, the strategy was to take out the old mechanized central offices in all the small towns in the area with survivable digital remote switch units to form a digital cluster interconnected by digital T1 (1.5 Mbps) or T3 (45 Mbps) transport systems.

In this manner the earlier analog crossbar or analog stored program switching and transmission systems were being replaced by digital time-space-time switching systems based upon PCM-type pulse code modulation (PCM) technologies, where typically 24 voice channels time shared an incoming facility. They were then separated by a switch (in space) and switched to different outgoing facilities, which they again shared in time with other voice channels going in that direction.

However, it was not readily known by many BOC network planners in the early 1980s that the digital conversion was to also pave the way for integrated data and voice messages, since in PCM, voice was changed to digital numbers and data was easily represented as digital numbers. (See Fig. 6-7.)

In actuality, the original U.S. digital base-satellite system for the local environment noted earlier was initially called the #5 machine by AT&T and GTE (for example, #5ESS, #5GTD, etc.). (Other vendors such as ITT called their international digital machines other numbers such as System 12). There was, however, to be a new machine (called the #6 by the original GTE planners) for the urban environment, which would deliver integrated voice, data, and video features. It was yet to be defined in the 1990s. There would also be another new machine (called the #7) to provide new voice-data-video services to specialized networks or enhanced services to the existing network. This machine was to be called the Information Switch, the Business Serving Module, the Information Centrex, or the New Service Node.

As time passed, the digital transmission frequently had frame-timing synchronization problems as data came back to haunt the telcos. This loss of framing required new coding techniques to ensure that a continuous series of data 1s did not cause a loss of synchronization timing. This was reflected in new coding techniques for clear-channel transport to upgrade earlier T1 or T3 facilities to handle both voice and data.

After divestiture, the RCHs/RBOCs began an intensive search for more and more voice features to deliver to their customers, who had replaced their telco's black phones with their own multicolored touch-tone units. Customers had been given, before divestiture, a few custom calling features such as call waiting, call transfer, and abbreviated dialing, but now there was an opportunity to use out-of-band signaling systems, separate from the call path, to carry advanced information regarding the call.

Initially, CCITT Signaling System #7 had been chosen to replace earlier common channel signaling to help improve higher levels of network operational efficiency by determining if the called party is busy, and then preselecting call paths to avoid busy connections. This same system could be used to deliver calling-party identification to the called party. Also the local central office switches could use the SS7

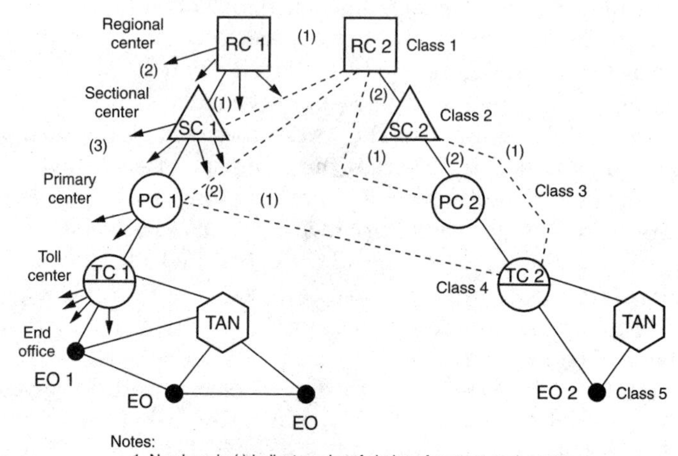

Notes:
1. Numbers in ( ) indicate order of choice of route at each center for calls originating at EO 1.
2. Arrows from a center indicate trunk groups to other lower rank centers that home on it. (Omitted in right chain.)
3. Dashed lines indicate high-usage groups.

(a) Choice of routes on assumed call.

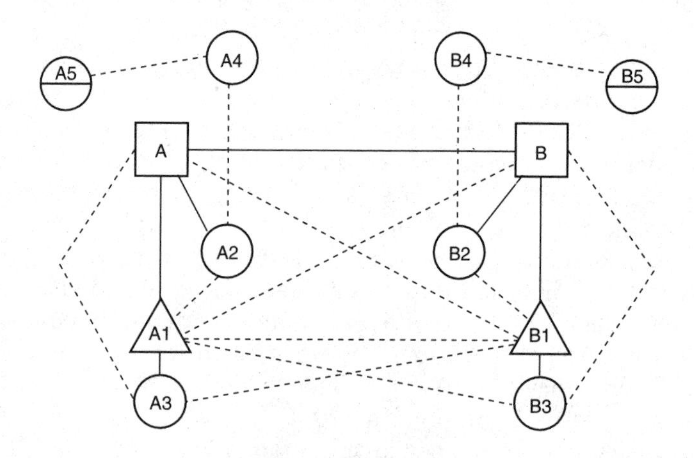

(b) Nationwide toll dialing
Typical intertoll trunk networks.

Key:

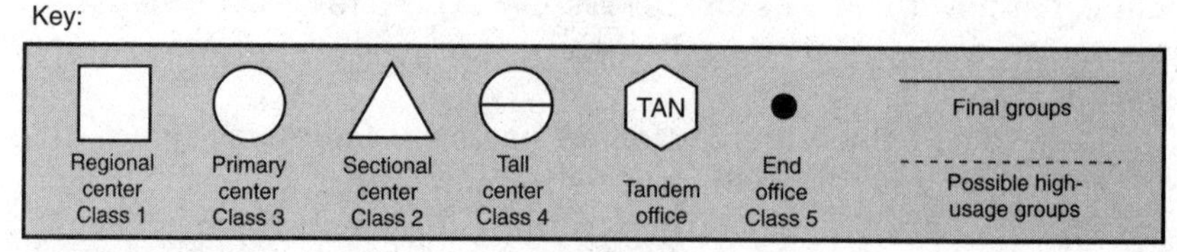

Figure 6-7. Public voice network. (Courtesy AT&T.)

signaling path to access an intelligent node above the network to provide special feature handling, such as database lookups or special translations such as 800 or 900 translations.

In this manner, the voice network began to address how to provide new voice services to the customer. Next, some suggested putting an adjacent processor next to the voice switch to provide additional features on an extended basis to the switching system. This would be done by branching from internal software states to the auxiliary processor, where RBOC or third-party software could enhance the call processing features. Or, the calls could be switched to autonomous service nodes, noted earlier as information switches that add value to the call. These systems can be selectively placed within a LATA in either the regulated or nonregulated side of the business.

In the 1980s, IDN was also the technology used to overlay digital switching into the urban domain by collocating these systems next to existing analog central office switches and extending digital capabilities via T1/T3 transmission systems to remote offices or suburban locations having business parks or shopping centers. These areas could then be served by IDN's digital remote switch units (RSUs), sometimes called optical remote modules (ORMs). This then became the plan for overlaying digital capabilities within the urban environment to provide extended digital voice services by connecting SS7 signaling to these systems.

Using this approach, the next step was to make these base and remote switch units ISDN compatible so that the D channel out-of-band signaling path was made available to each ISDN voice-data terminal. Hence, each terminal could be provided, via the ISDN "D" channel interface, the calling-party ID information from the SS7 signaling network. Therefore, a new family of voice services based upon knowing the caller ID could be extended to the customer. Other ISDN features were audio 7K quality stereo, the second voice line, ability to conference two calls together, easy transport of numbers within a location, and access to a new public data network.

This new services opportunity caused the supplier of the IDN switching system to overlay ISDN capabilities on their existing switch architecture, as the provider physically overlayed these IDN/ISDN systems into the urban environment next to existing analog systems. In this manner, the switching system initially designed for the rural environment was being used for the more complex urban community, extending its product life cycle.

Unfortunately, two other events occurred over the late 1980s and early 1990s. One was the realization that switched wideband urban services, using transport rings, would be a formidable market for the 1990s. The other was the delay in inserting host-remote switch units into the rural community, as some providers waited for new wideband-broadband technology to change out the rural plant. This delay was mainly due to the high costs for rural remote switch units that decreased the economic advantages of early change out to digital clusters. Fortunately, or unfortunately, depending on your perspective, state regulatory pressure continued to push for early change out to digital, but this regulatory pressure was only for digital voice, not data services.

The question of how existing voice switching systems can evolve to provide any new feature (voice, data, or video) becomes more complex as we began to understand what these new features and services were to be. This has led to the realization that there are several more networks. It's important to see what they are, and then determine how they can work together in order to see where one leaves off and another takes over, especially after the 1996 Telecommunications Act ruling.

Therefore, much of the voice network has remained visibly unchanged over the 1980s, except for IDN. However, the 1990s has seen more IDN changes, with the addition of these intelligent voice service nodes for database services above the network, new calling-party-ID related services (some served by adjacent processes), and new service node services that provide such services as voice mail–voice messaging. Hence, voice switches on an overlay basis will provide many new narrowband ISDN voice (and later data) features to the customer.

In addition, there will be more ISDN access to the new public data network, and the new wideband-broadband networks that will be discussed later. In this manner, the complexity of existing systems is increasing, requiring a clearer understanding of their ultimate purpose in order to cap their requirements by providing a family of new #6 or #7-type systems. As software programs become extremely large and extremely complex, software costs have become extremely high. Some believe that software expenditures for design, maintenance, and support are now 85 percent to 95 percent of the product life-cycle cost of these complex systems. (Suppliers may indeed elect to keep the same name as they add more and more capability to their existing systems in an effort to reuse existing software, but the net result will be a new machine.)

Parallel voice networks

In the past, parallel service networks were "piggy backed" on the existing network so that transmission systems could be shared but separate switching retained. In this manner, special-purpose, or closed-user-group networks provided separate voice-switched services to their large customers, such as a separate CCSA network for IBM. Here, smaller switches were collocated in the central office to switch private networks. Some providers used PBX switches to achieve these networks. From this demand came the need for shared networks for special customers, including smaller, local customer groups such as frame relay networks.

Internal CPE, small PBXs and key systems attempted to meet the needs of small business customers by serving 5 to 100 lines. Larger

PBXs of 100 to 1,000 or 1,000 to 4,000 lines served larger firms. Once these systems could be purchased by the customer, they became the new kid on the block, becoming extremely feature rich. Unfortunately, their maintenance and support became a challenging issue as the competition became so intense that prices were driven down and each vendor's market was shrunk to the point that few firms could reasonably compete for a long period of time. Many new vendors "bought into" the market by selling their products at unrealistically low prices to capture customers, so low that the overall support costs could not be absorbed. The classic story of too much success, too many miners looking for gold.

This drove customers, both large and small, to reconsider shared public network offerings, which led to a renewed interest in the 1990s in Centrex. (See Fig. 6-8.)

Intelligent network

As also noted by SIEMENS in their *Intelligent Network* book, Fig. 6-8 shows an overview of the joint study vision of the intelligent network (IN). The IN's main advantage is the ability to orchestrate exchange

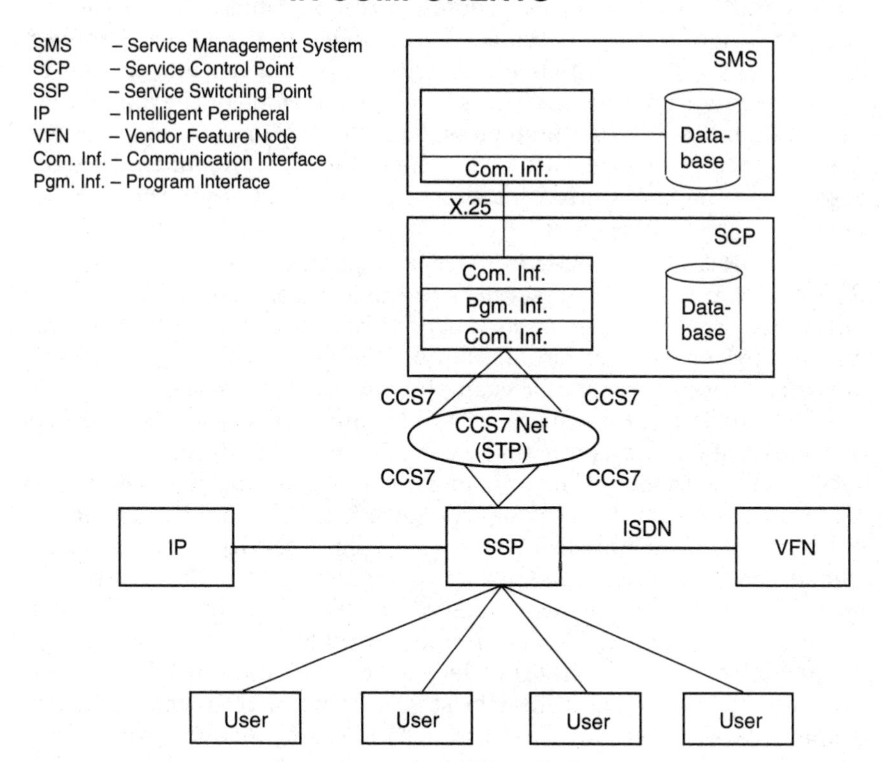

Figure 6-8. Intelligent networking. (Courtesy SIEMENS.)

service execution from a small set of intelligent network nodes known as service control points (SCPs). SCPs are connected to the network exchanges (known as service switching points (SSPs)) via a standardized interface CCS7. The CCS7 will facilitate a multivendor SCP and SSP marketplace, and the standardization of application interfaces allows a multivendor software marketplace for SCP applications (that is, the service control logic and its related data).

The service switching points detect when the SCP should handle a service. The SSP forwards a standardized CCS7 (TCAP) message containing relevant service information. Via the TCAP message, the service control logic in the SCP directs the SSPs to perform the individual functions that collectively constitute the service (such as connecting to a subscriber number or an announcement machine).

The IN's long-term goal is the ability to introduce new services or change existing services quickly, without having to adapt SSP software (only parameters or trigger updates). The adaptation will be confined to the SCP where parameters or stimuli are updated. This goal will be achieved in stages. This view is primarily concerned with the IN's initial form, known as IN/1. This does not require that services can be introduced without affecting SSP software. However, ideas and possibilities associated with the IN progression are considered as well.

Stage 1: IN/1. IN/1 requires updates in the SSP and SCP in order to support a new service. A typical IN/1 service is the Green Number Service (GNS), with which a subscriber can call a number free of charge. The SSP contains triggers (such as the value of the dialed digits) that tell the SSP to send a message to an SCP in order to get information about the destination to which the call should be routed. Migration from IN/1 to IN/2 implies significant changes in the SSPs to accommodate new services.

Stage 2: IN/2. Once IN/2 is in place, no updates need be made to the SSP's software when new services are introduced. The IN/2 triggers advise the SSP whether to let an SCP handle a service request or whether to complete execution locally. All SSPs and SCPs contain a set of basic service elements (for example, connect two lines, disconnect a line). The SCP also contains service-relevant data. These basic service elements are known as functional components (FCs), from which each service can be constructed. A customer could conceptualize a new service and the network operator, via the SMS/SCP, could construct it quite rapidly. Any successful and widely-used service may be downloaded (via the service logic) to, but transparent to, the SSPs (if this is more economical or provides a desired higher grade of service). This facilitates complete rapid service creation. Rapid service creation and user programmability will take place in the SCP and SMS. There will probably be one or more interim stages between IN/1 and IN/2, for example IN/1+, where the SSP provides increasing flexibility in accommodating rapid service creation.

Service management system (SMS). The SMS is owned by the network operator. It updates SCPs with new data or programs and collects statistics from the SCPs. The SMS also enables the service subscriber to control his own service parameters via a terminal linked to the SMS. (For example, the subscriber may define the day and time when an "800" number should be routed to a specific office.) This modification is filtered or validated by the network operator. The SMS is normally a commercial computer such as the IBM/ 370 or SIEMENS 7.5XX. The SMS may also provide a development environment for new services.

Service control point (SCP). The SCP is used when new IN services are introduced into the network and then activated. If a service is based on functional components (IN/1+ and IN/2), the FCs are executed with the help of a service logic interpreter (using an explanatory script). Some SCP services may require large amounts of data, which must reside on direct access storage devices such as disks. The service programs and the data are updated from the SMS. The SCP is a commercial computer or a modified switch. The critical factor is that the unit can access databases efficiently and reliably and provide a software platform for rapid service creation (through "user programmability" and "program portability").

Signaling transfer point (STP). The STP is part of the Common Channel Signaling Number Seven (CCS7) network. It switches CCS7 messages to different CCS7 nodes. The CCS7 is a standardized communication interface through which the goal of multivendor SCPs and SSPs can be achieved. The use of a stand-alone or integrated STP will depend on network-specific configurations. The STP is normally produced by traditional switch manufacturers.

Service switching point (SSP). The SSP serves as an access point for service users and executes heavily used services (as described for the SCP). In IN/1, no "user programmability" exists in the SSP. The SSP is produced by traditional switch manufacturers. The TCAP will be used to communicate between SSPs and SCPs.

Intelligent peripheral (IP). The IP provides enhanced services, controlled by an SCP or SSP. It's more economical for several users to share an IP because the capabilities in the IP are either not in the SSP or are too expensive to put in all SSPs. The following are examples of typical IP functions:

- Announcements.
- Speech synthesizing.
- Voice messaging.
- Speech recognition.
- Database information made accessible to the end user.

The IP is usually accessed from the SSP via a circuit or packet basis, such as ISDN.

Vendor feature node (VFN) or service provider (SP). The VFN is outside the network; it's owned and administered by a service subscriber. It can provide many services. The VFN provides the services mentioned for IPs, and can be connected via a CCS7 link. However, this type of connection includes filter logic, which prevents a VFN using CCS7 messages that could interfere with network operations.

Centrex

As part of the central office switch, CO-Centrex is an additional software package that enables customers to have separate private network capabilities from a privately shared portion of the machine. CU-Centrex provides similar services, but is from a separate software package on a separate switch, remote from the central office, near the customer premises. Besides many of the PBX-type direct-inward-dialing and line-control features that enable internal calling and fast station (phone) movement, several major new attractions for encouraging the retum to Centrex are the new Wide Area Centrex features with co-LAN-type data handling. Further return to the public network solution is occurring as ISDN Centrex features become readily available, enabling users to establish private data networks both internally and within a wide serving area by simply using the "B" channel data-handling capabilities of the ISDN Centrex switch. Similarly, key systems and more and more PBX features and functions can be handled by the new Information Centrexes in a move to be more competitive to the stand-alone systems. Finally, Centrex access to computer databases and number translation can occur so internal LANs can access the world via Centrex ports. All these features help make public shared services more attractive in the 1990s.

Therefore, as existing voice switching systems add IDN capabilities, intelligent network nodes and SS7 signaling, and, access to voice messaging application services centers, they will provide a new family of CLASS services and voice mail offerings. Similarly, as existing voice switching systems add ISDN data interface capabilities, the traditional public voice network will become better positioned as an integral part of the total "information network" solution for the customer, where these ISDN interfaces provide access to the new public data network.

The private transport networks

See Appendix B, which includes information on LANs, WANs, MANs, and VPNs.

The public data network

The misfortunes
hardest to bear
are those which
never happen.
JAMES RUSSELL LOWELL

In the late 1990s, a network planner dictated a paper on the ISDN "to be" user. It was not until he saw the final copy that the mistake was caught in communication. Or was it a mistake? Since ISDN and it's public data network were still not deployed by 1996, "to be" was actually meant to be "2B," meaning the ISDN user who has up to two 64-Kbps channels for data or a single voice and data "B" channel of 64-Kbps transport capabilities. Tests in the early 1990s demonstrated these ISDN capabilities for the medical industry by providing radiologists with the ability to move 128-Kbps data streams on switched "2B" facilities to more quickly display x-rays between remote doctors and hospitals. In this manner, switched data networks are making rapid inroads into the traditionally nonswitched point-to-point aspects of the marketplace.

Over the past several years, it has become readily apparent that the RBOCs have told a good story about new information networks (Bellcore even launched a new Information Networking Institute with Carnegie Mellon). But the reality was that few new data offerings had been provided that were beyond the scope of simply extending traditional voice-network base services. Even new gateway services to videotext databases, and facsimile over store and forward platforms, were based upon dial-up transport over the existing voice network or frame relay type traffic sensitive overlays.

Some had gone so far as to advise the senior decision makers within the RBOCs that narrowband ISDN was dead. They believed that no new data-handling networks were needed for low- to medium-speed data. Others continued to promote the expansion of private LANs instead of a new public data network and advised interconnection of these LANs using high-speed data networks such as switched megabit data services (SMDS) or traditional MANs and FDDI-type networks.

In this manner, the data world would only grow via private LAN-type networks, using leased special service circuits from the RBOCs, usually for large business applications. They believed that this strategy met the needs of the 1980s data users and would continue to meet their needs over the 1990s. So it goes. So it has been. So it will continue to be. So why establish a public data network? Or if so, what is it "to be" or not "to be"?

As we ask ourselves whether or not we want a new public data network infrastructure, a story comes to mind. Once power was lost due to a severe storm over a holiday weekend. As a result, customers were out of electricity for 12 hours from 5 a.m. to 5 p.m. The power company had told them that it could be out for 1 to 2 days. That was scary! It was interesting to note how dependent we (today's power customers/subscribers) have become on this infrastructure. As a result of having no electricity, we had no lights, no air conditioning, no TV, no telephone (system used electrical power for remote/base units), no food (couldn't open refrigerator or food would spoil), no water to drink (no electrical pump for the well), and no toilets (which require the water). All in all, the loss of the electrical infrastructure caused a loss in all the services that were layered on top of it to meet our human needs.

So what was life like before Thomas Edison wired the city of Philadelphia to prove the use of electricity to the world? How many new services like TV, air conditioning, and refrigerators were not known at the time of electricity's origination? How many new electrical devices exist today like irons, dryers, electrical drills, vacuum cleaners, etc. that only exist because the infrastructure is there to support their conception? Similarly, the future public data network will provide the supporting structure for many new market opportunities.

Some see change as threatening;
they pause and ask, "Why?"
Some see change as opportunity;
they embrace it and ask, "Why not?"

JOHN F. KENNEDY

Data-Market Opportunities

Data movement in military systems such as 465-L, AutoDin, etc., have demonstrated the need for a general-purpose data service network that provides numerous data-handling features to aid in the transport of data and the access to various databases throughout the world. The commercial and residential data users that could be, should be, would be, but aren't, have also adequately demonstrated their resourcefulness. They continue to perform functions manually that could have been automated, as they cope with the current limitations of the voice network. For example, credit check. How many times have you watched the person in the hardware store call someone on the phone and read a credit card number to them to verify your line of credit and log the sale. To fill the void of a public data network, enterprising businesses have noted that high-volume retail users can use a private data credit check network. Using leased lines from the telco, they directly access their computers to perform simple point-of-sale checks and debits. Others now provide automatic dial phones that dial credit check firms. Some simply dial up the remote database firm in the morning, then connect it to their credit-check box and leave the connection up for the entire day or month or year. Lost income from stolen credit cards or bad debts would help pay for a data network that enables small or large businesses' point-of-sale transactions to directly debit customer bank accounts, as the new SMART Cards or Debit Cards become available in the 1990s.

The sending and receiving of FAXes has always been a prime example of the commercial need for the types of clean, error-free systems that have been used by the military for the last 30 years. Sending and receiving FAXes especially need broadcast, delayed delivery, alternate routing, error detection, and correction techniques. I couldn't believe the excitement expressed by the secretaries the other day when they found a new service, provided by a limited voice network-based data FAX system, that could automatically send the message to several people by their simply providing a distribution list to the system.

Many new applications exist for the movement of data. As traffic congestion grows, you don't really want to drive to the bank to check your account balance or see if a check has cleared. A simple data call from home would be worth the cost of a reasonably priced packet of such information. Similarly, after the doctor writes the prescription and you drive over to the pharmacy and then wait a half-hour for the prescription to be filled or wait for the pharmacist to check back with the doctor, you would be willing to pay a small price to have the doctor send the prescription directly to the pharmacist before leaving the office, so you simply arrive later to pick up the prescription. Or how many times have you sat in the doctor's office, where they wanted you to fill out a form and verify your health plan before they would be willing to let you talk to the doctor? Wouldn't it be easier to simply provide them with a personal history access card to your health care information base? Then, they could simply send a data message to retrieve this information.

As noted, we have all seen point-of-sale activities being automated in the large stores to some extent as sale and inventory databases are accessed for each purchased item. But even today, fast-food chains have expressed the need for a public data network to enable their central control to access remote databases in each restaurant at 1 a.m. to determine how many items were sold. That would allow them to establish what should be loaded at 3 a.m. on the trucks from the central warehouse. (It's sad to note that Jack Lynch, financial V.P. of Burger King, requested that same service in the early 1960s.)

As firms become more dependent on other firms to provide all the pieces of their product (which are developed and built somewhere outside their local campus—now called *outsourcing*—either by another firm or another division) wouldn't it be nice to simply exchange information via data messages that are fast, error free, and inexpensive?

As we look at eighteen of our major industries and consider what tasks are performed by their people, we can overlay the particular type of data movement that enables each task to be completed. Then we are able to better identify who will be needing inquiry-response, data collection, data distribution, remote printing, and database access. Next, we can determine what type of network best meets their needs in terms of connect time, data transfer speed, error rate, security, transport delay, etc. We can also establish a range of values: low-, medium-, and high-speed data such as less than 64K, 64K–45 Mbps, or greater than 45 Mbps information movement. In this manner, we can determine specific data-user types for various real-world applications. In so doing, the need for a public data network would become quite apparent.

Applications

Applications are usually tied to a community of interest. Here we can observe who calls whom for what type of information. In this manner, a community of information exchange is established. This is called a community of interest. This is a network of people exchanging information by phone or personal visit. Many must access personal files, find memorandums, and send or receive various copies to and from the other persons. All this could be better supported by an "online," automated, interactive data exchange system on a real-time basis. The health care industry, the legal industry, and the protection industry, etc., are prime candidates for such a public data network. Or, if one is not available, they are also prime candidates for a separate switched private overlay data network by an alternate provider, who specializes in private virtual networks (PVNs), especially in the post- '96 era.

Data Users

It's important to differentiate both the uniqueness and the commonality of various data users. In general, you might consider that they indeed fall into five general groups. As noted in the preceding analysis on user groups, data-group III users will benefit immediately from the deployment of a public data network. As these users automate, they will generate much of the flow of low- to medium-speed digital data traffic. IEEE and ISDN interfaces will enable their CPE devices to interconnect, interchange, and internetwork information between the public and private arena. Behind these ISDN and IEEE interfaces there must be the supporting data networks to move the information to the appropriate application service centers, be they in the network, on customer premises, or on the World Wide Web.

Non-ISDN users can be connected to the public data network directly through twisted-pair interfaces to circuit or packet switches that eventually interface to the digital switched public network. So, let's first take a deeper look at what data is and then determine what narrowband ISDN (n-ISDN) has to offer for establishing a new public data network, before we are willing to say that narrowband ISDN is dead and we don't need a public data network.

What Is Data?

The Webster definition of data is: 1. Information, especially information organized for analysis or used as the basis for a decision. 2. Numerical information in a form suitable for processing by computer.

The computer user's description of data is that data is a sequence or string of alphanumeric characters that indicate a message or record of information. Traditionally, computers store data in memory systems in words containing one, two, or four bytes of information, where a byte consists of 8 bits representing a character such as the letter A. These bits are binary 1s or 0s, based upon EBCDIC, ASCII, or field data codes. Both characters and words contain parity bits (vertical and/or horizontal), which are used for detecting errors during transmission.

Errors usually occur in bursts during transmission over analog facilities. In some cases, these can be corrected using Hamming fire codes, without requiring retransmission of the message. Analog voice networks move information at 2,400, 4,800 and 9,600 bits per second at average error rates of one error per 10,000 bits per second. Digital facilities move voice and data at error rates of one error in ten million (10^7) bits per second. Voice is broken into 256 levels from high to low. Each level is represented by 8 bits. The voice conversation is sampled 8,000 times per second, so that $8 \times 8{,}000 = 64{,}000$ or 64 Kbps. Therefore, ISDN provides a digital interface to the user to transmit voice or data information in "2B" channels where B is 64 Kbps.

Today's "Data World"

Data is displayed on cathode ray tubes (CRTs) having 256 lines by 256 columns or 512 by 512. In contrast, television in America is 550 lines, or 650 lines in Europe. High-definition TV will be around 1,100 lines. Radiology images will be displayed on 1K by 1K, 2K by 2K, or 4K by 4K monochrome monitors. A video cross point can now denote up to 16 gray levels of the image. Similarly, printers have increased their speed and capabilities over the years, from the 60 or 100 words per minute Western Union teletype systems of the 1950s to the more advanced laser-type printers of the 1990s. Also, storage devices have developed from solid state flip flops that store 2 bits per chip at a cost of $55 in lots of a million, to those of today that economically store 4 million bits of information, or tomorrow's 40 million and 68 million VLSI arrays.

In the future, holograms will display hundreds of pages of information, depending on the perspective view.

Data is still transported through the telecommunications voice network via devices that modulate and demodulate bits of data on analog-type carrier signals using frequency shift keying (FSK), frequency modulation, amplitude modulation, phase modulation, or phase shift modulation. Digital techniques are based on generating digital pulses of data, using pulse code modulation (PCM) techniques to send the level numbers representing the sampled voice message.

Voice-grade data is limited by Shannon's Law for the voice band over the 4K and 7K bandwidth range of information transfer. Here, information causes change of state. Hence 9,600 using phase quadrature modulation techniques have been somewhat the upper limit for dial-up data on nonconditioned lines. This has been what the traditional switched voice network currently delivers, but some say only 4,800 bits per second is the norm at a reasonable error rate during busy hours for dial-up data calls.

Dial-Up Data

Over the past thirty years, data calls have been restricted to circuit switching delays and transport noise unique to the voice network. This has fostered end-to-end control of the data message via the originating and terminating terminal. To accomplish this, special transmitter-to-receiver protocols have been developed over the years that have a series of hellos and how-are-yous to ensure the orderly transmittal and reception of information. This results in a set of ACKs, RECs, and RTRANs codes between devices that have been established by international standards groups. However, in ISDN high-speed arrangements, these user-to-user (off net) protocols will actually severely limit throughput of devices. These devices could have taken advantage of the higher-speeds of greater error-free information movement under more sophisticated advance network error correction and control mechanisms.

Unfortunately, as data manipulation over the 1970s and 1980s moved from the accounting department mainframe computers to smaller but more powerful distributed online interconnected computers that automated every aspect of today's business, the voice network transport of data calls increased substantially, with different traffic patterns than traditional voice calls. For example, the calls from a credit-checking center are very short and so numerous that they could affect the call-attempt handling capability of voice circuit switching systems. By *circuit switching*, I mean the calls that were set up and held for the full duration by using physical circuit paths over the network, based upon origination and destination addresses. For this reason, new switching systems are providing packet switching to help move this load to other systems.

Some dial-up data users begin their work day, as noted earlier, by dialing and redialing a path through the network until they obtain a relatively error free circuit; then they leave it up during the duration of the work day, since they only have to pay a flat rate monthly fee, and some, in fact, leave it up indefinitely. Imagine future network management congestion if and when 30 percent of these types of customers will be doing this disservice to their fellow users.

Point-to-Point Data

Many users lease special service circuits from the telco in the form of lines and trunks and generate up to fully loaded trunk (36 hundred call seconds (ccs)) data traffic usage. They are usually on conditioned balanced facilities using data sets to send analog information up to 14.1 or 19.2 Kbps, or even 56 Kbps. Or, digital point-to-point facilities allow fractional T1 (multiples of 2, 4, 6, 8 channels of 64-Kbps data), T1 (24 channels up to 1.544 Mbits per second), or T3 at 45 Mbits per second (SYNTRAN) rates. More recently, the SONET OCX rates, as multiples of 51 Mbits per second, will become the new transport vehicles for data. The new trend is to switch data via private virtual networks (PVNs) at the T1 rate of 1.5 Mbps, if usage is 4 to 6 hours per day (or even less) in order to achieve wide area networking (WAN) advantages of point-to-multipoint without paying the cost of full-time usage to each destination using frame relay type technologies.

In the past, the telco has been able to make considerable revenue by tariffs that obtain considerable differential revenue from 2,400 bits per second to 56 Kbps. But, as SONET fiber capabilities dwarf these offerings, by-pass is encouraged as others take advantage of the capabilities of the fiber and the disparity in prices. AT&T has noted this in their reestablishment of new ACCUNET's Reserved Digital Services (ARDS), a software service for data users at more reasonable rates in AT&T's Software Defined Data Network (SDDN).

As telco marketing forces sell more and more T1 and T3, they need to realize they're removing the data opportunity from the switched public data network to private networks. They will lose potential revenues from the new services they could offer. As transport becomes cheaper and cheaper, and as fiber capabilities expand and expand, transport will simply become a cheap commodity, and the real revenue will come from switched services. In fact, some new strategies for MIS managers are to stop integrating voice and data, give voice back to the telephone company, lease their T1 frame relay trunks and build their own private data network, as a switched virtual private network (VPN).

Private Networks

In the past, books that describe LANs and how to interconnect LANs have outsold by 5 to 1 those books devoted to the future public network's new information services. There is now a proliferation of both private local area networks and new books that explain their different and varied protocols. These differences have caused a nightmare for the telecommunications managers of the large businesses, which may have as many as 10 to 20 different LANs of 1 Mbps to 10 Mbps, which need to be interconnected throughout the country.

Bridges, routers, brouters, and gateways have become the internetworking vehicles that enable LANs to interchange information. Distances between LANs are traversed over 50 to 100 Mbps by MANs or FDDI links that enable high-speed movement of information. Bellcore's SMDS is one form of a 50-Mbps MAN-type network designed to effectively handle bursty, high-speed data between corporate LANs.

Wide area networks (WANs) were formed to interconnect data terminals over a fifty or so mile area. Most WANs used the public network using dial-up 9,600 bits per second analog transport. Now the new WANs will use the voice-grade 64-Kbps digital switches. Some will use private digital switches to establish virtual networks at 1.5 Mbps as VPNs. At these rates, they provide switched point-to-multipoint services, where any terminal can reach any other terminal over circuit switch paths or frame relay circuits.

The private network has become very sophisticated as numerous and varied new transport protocols enable interconnection. The standards defined by IEEE for 802.3, .4, .5, and .6 are designed to move users' information from Ethernet to token bus to token ring to OSI interconnect compatibility. Hence, the layering of information interchange moves the transfer of information from physical connectivity standards to datalink-to-datalink, set up to network control.

As information is moved in packets with heads and tails embracing the data message, we see the opportunity for packet switching the information from node to node or from gateway to gateway. Here, we enter the world of connection-oriented data where logical and/or physical paths are established for a sequence of messages. This method requires less set-up time for subsequent transmittals. Connectionless data is independent of previous messages by requiring the routing and transfer-path information for each transmittal, but this requires more overhead for each segment of the message.

n-ISDN

As we know, ISDN basic rate provides a 2B-D interface where B is 64 Kbps and D is 16 Kbps. We also know that the T interface, from the network terminating device for the user is effectively a four-wire 192-Kbps interface over which 144-Kbps ($2 \times 64K + 16K$) information is moved up to 4,000 feet from the network terminating unit (NTU). It provides physical (level 1) or data link (level 2) interfaces of the OSI model. TAs (terminal adapters) can interface non-ISDN systems directly to the network terminating unit (NT1 or NT2).

The NT2 system connects directly to an NT1, which effectively provides a U interface back to the switch. Here, special coding (2B1Q) enables information to travel for 12 kilometers or so at rates up to 160 Mbps to and from the central office. At the NT1 device, level 3 signaling is achieved, allowing different protocols to communicate. The result is that n-ISDN provides three networks to move data, therefore the following networks are available:

1. "B" channel circuit-switched data network (64 Kbps).[2]
2. "B" channel packet-switched data network (64 Kbps).
3. "D" channel packetized switched data network (8 Kbps) + separate signaling (8 Kbps) for a total of 16 Kbps.

"B" channel circuit-switched data network

The "B" channel circuit-switched data network is an end-to-end digital 64-Kbps data transport network that allows the calling party to directly address the called party using current numbering systems to enable point-to-point-type data interchange over the switched facilities. The network does limited assistance to the call, other than establishing a path during the call duration. It does enable parallel voice communication so users can send and receive data simultaneously with the voice call. It serves as a 64K or 128K (in absence of voice) wide area network (WAN) that provides the set up and take down of data paths to any destination. This eliminates the need for separate conditional paths over point-to-point leased facilities to each and every destination from each and every originating location. It also enables any and all users from any and every remote location to call and access a firm's internal databases anywhere, anytime, without waiting for RBOC provisioning delays to set up private facilities.

On the other hand, the data call is limited and restricted to call-handling delays attributed to circuit switching the call. The call-handling features are limited, due to a lack of dynamic access to the information on the "B" channel during transport. In-line data handling can be accomplished by initially connecting specialized modems, security devices, or protocol converters across the transport path for the duration of the call by switching the call directly to specialized services provided by application service centers above the network (layer four of the five-layer local service model).

---

[2]As noted, "2B" doubles the throughput up to 128 Kbps.

"B" channel packet-switched data network

The "B" channel packet-switched data network provides the ability to packetize numerous users on a single 64-Kbps channel, and to packet switch them to remote destinations. It uses shared transport paths to increase economies of scale. Or, "B" channel packet features and services can be achieved by enabling the system to grab the data message during delivery. In this manner, error rate can be checked. If errors are detected by the network, message integrity can be guaranteed by retransmitting over alternate routing and by using specialized shared message-error detection and correction techniques. Similarly, network capabilities can now be expanded to include polling, broadcasting, delayed delivery, and specialized presentation features that can be dynamically provided to different segments of the message. Packetizing is readily adaptable to short messages of varying needs, while long messages of simple needs are best applied to circuit-switching or channel-switching networks.

"D" channel packetized switched data network

The "D" channel of the basic rate ISDN user provides a 16-Kbps interface between the terminal and the network. It's outside the movement of voice or data information. This out-of-band signaling and information exchange channel enables status indications and sensing telemetry information to be polled and/or sensed by the network to detect alarms, environmental-state changes, etc. The "D" channel also contains an 8-Kbps packetized interface to enable information to be sent to and from the terminal for such low-speed services as remote control of environmental control systems; energy management (to prevent brown outs); meter reading; home incarceration checks; and a whole new family of computer-to-computer, computer-to-terminal (via the network) exchanges of information that help clarify the progress of the "B" channel movement of information. These new "D" channel protocols will supersede earlier time-consuming data-exchange protocols that require the previously mentioned in-band acknowledgments, control characters, etc.

Hence, the public data network can consist of these three new data networks that augment and supersede dial-up voice-grade analog data to better foster the growth and expansion of the computer. Once the public data network is more ubiquitously available, the 18 major industries will begin to move more and more information between remote terminals and mainframes, causing a new wave of computer growth as we enable distributed processing to enter every facet of the market, creating and utilizing a new infrastructure for the information society.

Features and services

Why use a public data network instead of just dialing up over the voice network and sending modulated data over these analog facilities? Why not privately switch the calls within a CPE data switch to the various destinations over point-to-point leased line facilities, as a PVN? What services will a separate public data network deliver?

The answer to the last question is the answer to the first two questions. It's extremely important to recognize that the public data network as a public circuit data network (PCDN) or public packet data network (PPDN) must economically provide interconnectability, flexibility, reliability, and availability. If not, the universal dial-up mode will be integrated with the specialized private networks to provide limited but focused solutions to those who can afford them. Then there will not be a universal data network that is available anyplace, anytime, to anyone. So, to meet these challenging public network objectives, let's review a list of data features and network parameters that have been discussed in earlier works concerning features and services for the ISDN marketplace.

Reliability. The following are data features that enable the network to successfully and reliably move information from point A to point B.

Bit and byte interleaving. This ensures that burst noise affects several messages so that error correction and detection mechanisms can both detect and correct the errors.

Redirection of calls (alternative routing). In the event that a network node is down, such as Hinsdale, data calls are redirected to a different central office. In this case, local PVN point-to-point circuits lose their advantage to the public network switched calls if it can "home in" on multiple central offices.

Retry of network. The network will continue to try to deliver the information if the network segments are inoperative or congested over a short period of time.

Delayed delivery. The network will store information until the receiver is activated.

Duplex facility. This provides an alternate path to the receiver, such as switching information over self-healing paths (rings).

Low error rate. Different user classes have various error-rate tolerances (such as 10^{-7}, 10^{-9}, 10^{-11}). Therefore, they need to be able to select the most desirable transport. However, the default transport is still a significant level of magnitude greater than the current voice network of 1 in 10^4 error rate; for example, 1 in 10^7.

Interconnectability. Features are a must in today's data world, as LANs try to resolve their differences through bridges and gateways. Here a public data network with addressable users provides the following.

Protocol and code conversion. This enables unlike terminals and computers to talk to each other.

Barred access. This provides the ability to block unwanted users from accessing individual firms' private internal systems (on-net and off-net service protection).

Network-to-customer interface. These are specific interfaces that are provided to those who subscribe to a particular service. An example is PADS for X.25 interfaces, where PAD is a packet assembler/deassembler.

Remote terminal identification. This ensures that the terminating terminal device is active and accessible to the originating terminal network.

Speed format transforms. These are interconnectability transformations for dissimilar communication networks, as interchange carriers provide different facilities to move information at different rates.

Bit sequence independence. This provides the transfer mechanisms with independence from the content of the information and the form of transport.

Multiaddress call. This feature provides the ability to go from point to multipoint devices.

Packet switching, packet interleaving, and store and forward. These give the ability to integrate and switch packets of information from multiple servers in a store and forward mode of operation, thereby enabling a whole host of flexible new services based upon packetized messages.

There is also a basic set (family) of operational features such as:

- Three-attempt limit.
- Short set-up (less than 100 ms).
- Call back (automatic).
- Short clear-down (less than 10 ms).
- Manual/automatic calling.
- Manual/automatic answering.
- Direct call.
- High-grade service.

- Abbreviated address call.
- Multiple lines.

These are also reflected in extensive data network availability parameters such as:

- Transfer time.
- Overall grade of service.
- Overall quality of service.
- Call clear-down time.
- Call-request time.
- Format structure.
- Network synchronization.
- Route-selection modes.
- Speed conversion.
- Category of error rates.
- Type of network switching.
- Number plan.
- Usage recording.
- Transmission limits.
- Network signaling.

Application service centers

Once the network is ubiquitously available within selected areas (such as the Boston-New York-Washington corrider, Chicago, Dallas/Ft. Worth, Minneapolis /St. Paul, the greater Denver area, the Seattle/Tacoma, L.A./San Francisco area, Tampa/St. Petersburg, Miami) then the issue is no longer the network, but what services the network will offer to meet the applications. As noted in the Information F^3 Cubics, (in *Telecommunication Management Planning*) there are four levels of services over the basic transport. As the users obtain one level, they request and/or demand the next. Some insist that all levels must be available before they will move off of private networks to the public network. But most realize that the solution is really as indicated in the five-layer local network services model. It indicates that some services are best provided in the private arena and others in the public arena. The key is layering them together to provide the complete solution. Services will be provided by RBOCs or alternate providers of shared services at unique, specialized, or shared application service centers.

Application service centers enable the RBOC or other providers to deliver services such as:

- Polling—of remote sensors to provide records such as meter readings of water, electricity, and gas indicators.
- Broadcast—for data distribution of mail systems or closed user groups.
- Data service classes—a family of services for different types of data users. One type may be industry-specific, such as the medical industry radiologists who need to move x-rays at 128K rates, using both B channels and image storage at the destination. Others require specialized Apple-to-DOS format conversions or LAN Ethernet-to-Token Ring bridges. Some need specialized services requiring specialized handling, call redirection, or delayed delivery of calls; others desire priority override, etc.

As noted in the layered networks' layered services figures, numerous services can be integrated and layered on top of each other. This has been acknowledged as the "S" of ISDN. The ISDN services are differentiated by layers of the OSI model, where the first four are transport-related and the upper layers are computer-oriented. For example, the last layers contain standards such as X.400 File Management and X.500 Directory services.

These are being established to enable computers to interact with each other, to meet universal applications for data handling, processing, manipulation, and presentation. Other new security-type services are evolving into the commercial world. Banks are changing to more secure data networks as their users become aware of viruses that affect operation. These viruses can be created by someone outside the bank's domain. Customers are now requesting protection from access to their files. Hence, we have a new group of services such as:

- Access-attempt limits.
- Attempt-period limits.
- Encryption.
- Audit trails.
- Terminal verification.
- User verification.
- Access codes.
- Blockage (barred access).
- Criminal isolation.
- Criminal entrapment.

Finally, there will be a new level of terminal-to-network services for enabling terminals to transfer "D" channel information in parallel with "B" channel communication through the network and on to some distant receiver. This will enable a database to know what type of computer or terminal wishes to access it so that the right interfaces can be established. Similarly, zone control, priority overrides, network-to-network networking can enable hot-line-type access to selected families of data users, priority access to specialized databases, and internetworking of global or regional networks. All in all, it will be fascinating to see private-to-public-to-private internetworking of information at all five levels of the five-layer network services model.

Layered Data Networks' Layered Data Services

(See Fig. 6-9A.) As we consider the five-layer networks' services model, it's particularly interesting to note that each of the three data networks for ISDN basic rate can be overlayed against this model. At each layer, networks, products, and services can be identified, for example, for the "B" channel packet-switched network. (See Fig. 6-9B.)

As we look at the "D" channel services, it's quite evident that the applications service center at layer four could provide the platform for integrating, collecting, and distributing: CPE alarm, energy management, meter reading, and other polled information such as opinion polls.

Similarly, "B" channel circuit-switched data can easily be routed to the application service center that enhances their movement, such as the platform that enables E-mail and transactions and makes specialized routing translations to enable dissimilar systems "to be" interconnected around the world.

It's especially interesting to sit back and contemplate what layer provides what services for specific data applications such as the medical, legal, protection, manufacturing, government, etc., industries. From

Model Layers		Data Products	Services
5	CPE Application Service Centers	---	---
4	Application Service Centers	---	---
3	Switched Transport Services	---	---
2	Transport Access Node	---	---
1	CPE Network	---	---

Figure 6-9A. Layered data networks' layered data services.

Products	Services
5 CPE Applications Packet Service Center	- Specialized Packet Services (Hospital, etc.)
4 Application Packet Service Center	- Enhanced Packet Services (Polling, Global Network Interface, Delayed Delivery, Broadcast, Gateways, Data Bases)
3 Public Packet Switch	- Routings/Translations (Connection Oriented Services)
2 Packet Network Interface Node	- PADS (Connection-Less Services)
1 CPE Packet Switch	- Terminal Interface, Protocol Converter Cluster Control, Internal Switching, LAN Interface, PADS

Figure 6-9B. Products and services.

this type of thinking, products that are common across industries will be identified for each layer. Different layers may also provide services that are slightly more specific for each application. In this manner, one system does not attempt to do everything for everyone. We need to foster a network in which several systems will provide a piece of the problem's solution at the right location, at the right time, for the right application, at the right cost.

Some have attempted to promote only one general-purpose data switch for the LEC. However, an earlier analysis (entitled "Who's on First" in *Telecommunications Management Planning*) noted the need for a switch in the regulated portion of the telco that enables local community routing and interconnection to inter-LATA data carriers, national and international data networks (IXCs), and value added networks (VANs). The analysis also noted the need for special data-switched services in the nonregulated entity. Today the Telecommunications Act of 1996 has opened this entire arena to multiple players to together establish not only the transport network but also the enhanced transport services such as encryption, delay delivery, and bridging.

As we began in the early 1990s the next phase of the information game, in which Judge Harold Greene recommended data gateways (application service centers) that accessed databases for videotex, we saw the need for more enhanced ASC services that played with the information to a greater extent. For example, Greene's subsequent ruling on voice mail for storage and access may be applied to data mail and delayed x-ray movement or even patient archives stored in hospital gateways (ASCs). On July 25, 1991, Judge Harold Greene reluctantly relented in reversing a key part of the historic 1982 agreement that broke up "Ma Bell." In 1987, Greene ruled that the "baby bells" can only transmit information, not generate or own it. In April 1990, the Appeals Panel told Greene to reconsider the information services restrictions on the RBOCs. On July 25, 1991, he lifted the ban, but took the unusual step of staying his order until higher courts could review his decision and protestors could file appeals. RBOCs requested reconsideration of the delay; newspapers and publishing houses demanded limited Bell relief or congressional rulings to retain the status quo. Congressional leaders noted that removal of the ban would still require playing-field restrictions to be resolved over the 1990s. The subsequent October 1991 appeals court ruling allowed the RBOCs to provide information services without restrictions, causing further congressional pressure to reimpose some limitations. Therefore, new information handling services may no longer require extended ASC functions to enable full "content" manipulation and processing. Yet to be resolved was a 1990 ruling concerning single point of interconnection (SPOI). As a result of this ruling, RBOCs were required to have a signal transfer point (STP) in any LATA where inter-LATA call set up is being provided. Unfortunately, STPs are quite expensive, and some of the small LATAs fail economic deployment analyses.

This would eventually cause a dearth of advance services platforms in the more rural LATAs. (It's interesting to note that many of the independent telephone companies have gone to overlay platforms that provide services for large areas, some as large as the entire state of Iowa.) Though it's very difficult to predict how legal and regulatory decisions will evolve in the future, the 1996 Telecommunications Act opened up the game to enable both the RBOCs and their competitors to offer a full range of services anywhere, anytime. The subsequent FCC ruling supported this, which was supposedly what the RBOCs wanted. But when the prices for alternative providers to obtain access service was deemed too "cheap," the RBOCs appealed the FCC ruling, supported by both Appeals Court and Supreme Court rulings. Thus, the game has been left to be defined and redefined throughout the 1990s, as the new players jockey for the most competitive positions and the RBOCs delay and delay—as the users wait and wait for a much fuller spectrum of user-friendly offerings in the information marketplace.

In actuality, the "collection" data switches that are located in the local community are really layer-one CPE equipment. They will be providing a large amount of protocol conversions, code conversions and non-ISDN to ISDN internetworking functions for large business as nonregulated private data switches. In time, these systems should also be provided on a publicly shared basis for the small business users, who are not large enough to own or require a separate system for themselves.

Info switch

Here data can be routed to a new "Information Centrex"-type system. In previous analyses, this has been called the "info switch." A unique aspect of the system complex is that front-end systems are located on the residence or business and interact with the central control system. Some have called this a levelless system, since it can be accessed from all three of the lower layers of the five-layer model. Others locate the central control system at level four as an extended data switch / data services application service center platform. In any case, there will be several types of data switches deployed in the local arena to provide the full spectrum of offerings. To do this, some can be provided from the CPE side of the business, others from the network or even the VAN side. To solve all of the customer needs, both sets of services will be required.

The global view

If the more global layered network service model is considered, the three transport layers expand to include the national and global levels of two-tier national and global networks/VANs. In this manner, the five-layer local model expands to ten layers as international data switches are overlayed on the local network to complete the international network. Here, national VANs and international gateways play their role to move data across RBOCs and PTTs.

Again, specialized gateways services are available, as international applications service centers, for specialized lookups and translations. In this model, national and global ASCs are located on layers eight and ten as the two-tier global switches become layers six, and seven (or nine) for the IXCs or VANs. (See Fig. 6-10.)

Hence, it's important to keep in mind what services are best provided by the local application service center and which are best provided by the national or global application service center. In any event, data will be a "global game," as global info corps become predominate and as international boundaries soften. (For example, the 1991 break up of the USSR; after 1992, the European Community; and in 1997 Hong Kong/China.)

The public wideband network

A new phenomenon has occurred as a result of the Hinsdale fire and the New York City fiber cut and system outage. Private network managers have found that their leased trunks, which make up their private

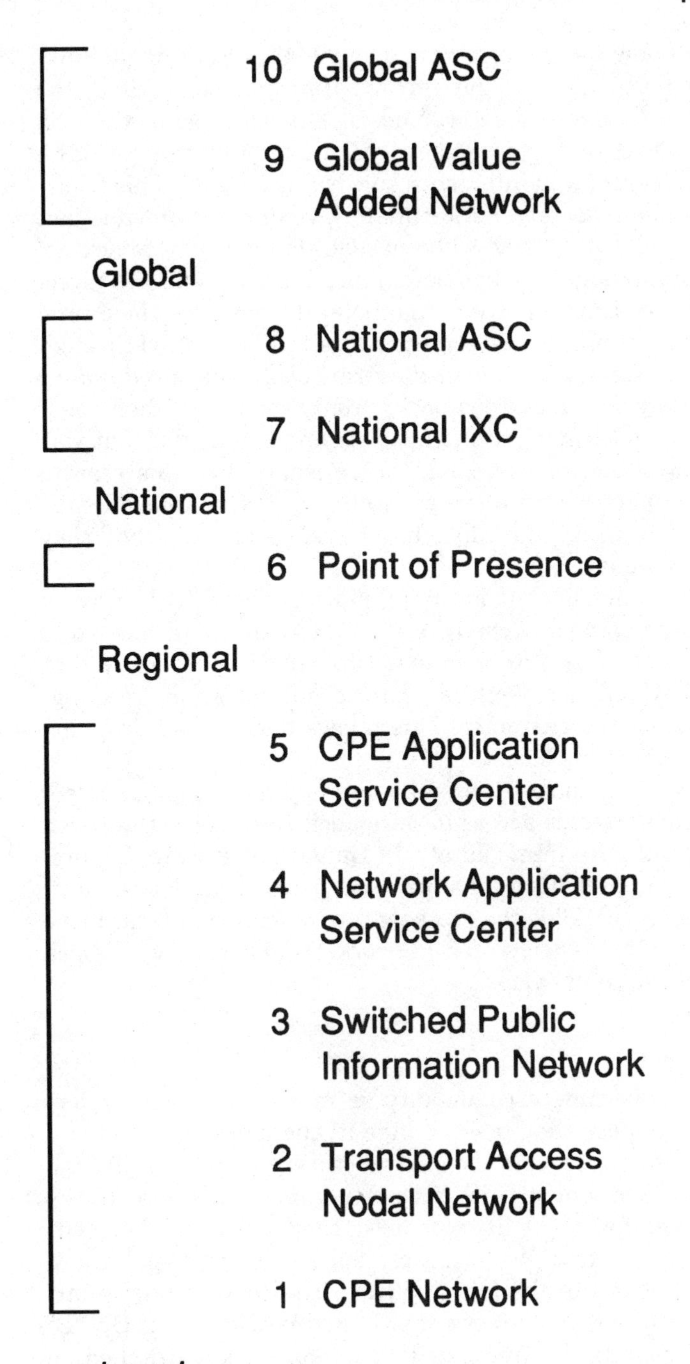

Figure 6-10. Global layered networks' layered services model.

networks, can go down and stay down. The 99.99 percent availability may be achieved for the year, but at one time, for some time, the .01 percent outage (3,168 seconds) can occur. This also means that the network can go down for one second every hour for 132 straight days. As a result, there is renewed interest in survivable circuits. This translates into a redundant facility. Unfortunately, redundancy means just that—redundant, idle, only used when needed—an expensive back up!

To provide redundancy, fiber has been routed in two separate directions. (hopefully not from the same manhole). In one city alone nine hundred fiber pairs supported 90 companies, with fiber routed through different facilities. However, none of the firms could share redundant fiber, nor could they obtain extra capacity from each other when needed; they could only use their redundant fiber and capacity if that was the direction they themselves wished to transport. For this reason, additional extra fibers were put in place during installation in each path (redundant as well). Initially, they were never used, but they were there for future use.

The second phenomenon that is taking place is the shift to private wide area networks using switched T1 and T3 circuits. This has led to digital cross connects that automate main distribution frames so that formerly manually patched facilities can be automatically reconfigured at the whim of the customer. This allows adding and dropping routes on a programmable basis.

As private network managers become more and more sophisticated, they are requesting dialable access to these facilities, where they wish to have them reconfigure immediately. In time, they wish to dynamically add and drop bandwidth (channels) on a real-time basis; thus, the trend in the early 1990s was to separate voice from data, move data on to programmable wide area networks, and return voice back to the telco public voice networks.

Wide area networks

Transport capacity became a commodity as more and more carriers deployed fiber. The more they tried to hide it, the more it bulged out. Depending on the distance from the central office, the price tariff for a limited number (*X*) of single DSO circuits became equivalent to the full T1 rate. Hence, if a T1 facility was used, then 24*X* of the DSO circuits were left over as "free" usage. As a result, T1 point-to-point networks became the focus of the late 1980s. In the early 1990s, smart managers also found the advantages of switched WANs. The ISDN "B" data channels were initially circuit switched, later packet switched, so that switched DSO (64 Kbps) wide area networks were being constructed from narrowband ISDN. Next, fractional T1 enabled users to purchase *N* number of the 24 channels so that users could select a variable number of 64-Kbps channels over a shared facility. This led to reconsideration of full T1 versus fractional T1, as new digital cross connect systems switched the T1's DSO units to any location.

Workstations, serving as front ends to computer mainframes, began to require more and more transport capacity. As a result, T3 transport point-to-point began to be a viable short-distance option, but not a long-distance choice, as T3 rates between major cities could be as high as $50,000 to $200,000 per month. This then led to 0/1/3 digital cross connects that provided DSO (64 Kbps) to DS1 (1.5 Mbps) to DS3 (45 Mbps) dynamic channel switching. Time slot interchange systems ensured that information was sequenced properly in sequential and parallel channels so that it arrived in the right order. Hence, the new world of wideband switched networks, serving as dynamically changeable wide area networks for MIS managers, was indeed a sought-after solution. Wide area networks were previously constructed over voice-grade facilities, usually switched at 9,600 bits per second maximum. Switched n-ISDN basic rate WANs can exist at 64 Kbps (DSO) or even "2B" (128 Kbps). Next, ISDN WANs will be 1.5 Mbps or even 45 Mbps. Wideband has never been "caped" as it also evolves from 64 Kbps to 1.5 Mbps via interum frame relay type offerings. In the 1990s, it will continue up to 45 Mbps in increments of DSO and/or DS1 units.

Primary rate ISDN

As more and more services take advantage of computer processing, higher and higher bandwidth limits are shrinking. Hence, more and more services will fit into the copper plant up to 1.5 Mbps. Primary rate ISDN provides 23 "B" channels and one 64-Kbps "D" channel. Multiple primary rate circuits can share one "D" channel so that the full 24 "B" channels are available. Hence, *N* number of primary rate ISDN circuits are rapidly becoming an alternative to T1 circuits, as we move to the world of fractional ISDN. Here, *N* number of primary rate channels are available on either a point-to-point or a switched basis, depending on the application. As primary rate interfaces become standard, and users require more and more variable capacity, *N* number of 64-Kbps "B" channels can be accessible from the traditional copper plant to new switching nodes that combine these channels into fiber T3-based facilities. Similarly, high-definition subscriber loop (HDSL) data can be sent over T1 facilities for longer distances requiring less repeaters. Hence, customers will be able to switch fractional-T1 (f-T1), T1, primary rate ISDN, and fractional-ISDN (f-ISDN) at *N* number of DSO channels up to 1.5 Mbps or multiples of 1.5 Mbps. As time progresses, this will continue into the T3 world, as *N* number of DSOs or *N* number of DS1s (T1s) are switched up to T3 45 Mbps (DS3), perhaps using SYNTRAN transport standards, and SONET (OC-1) at 51 Mbps.

In any event, as switching nodes move closer to the user, the only bottleneck will be the ability to use more than one copper pair or to have the unlimited bandwidth of the fiber directly into the home or business. With this in mind, the 1990s will be the first phase of broadband as users obtain higher and higher dynamic transport of wideband rates at more and more economical prices. This will extend the life of copper plant and provide the needed revenue to deploy the fiber. However, wideband will be a niche market for the 1990-2010 time frame, as the world waits for the future broadband switches. It should be noted that for network nodal switch capabilities for many CAD/ CAM/CAE, T1 picturephone, and video conference centers, wideband applications are the natural first phase of broadband development. They could be achieved as part of the new broadband fiber deployment plans.

Ring switches

As survivability becomes the key objective of the 1990s, fiber transport rings will be constructed at switching nodes closer to the customer. This enables shorter copper loop plant or fiber from these nodes to the home or business. Survivable rings can go in either direction in the event of a fiber cut between nodes. At these nodes will be ring switches that can initially perform digital cross connect-type functions to circuit switch a variable number of 64-Kbps or 1.5-Mbps channels as noted above. In time, other broadband functions will be placed at this node to make it a front-end switch (Class 6) to the future new broadband switch at the Class 5 central office, which is sometimes called the next generation switch, metro switch, or urban switch.

Result. Hence, wideband switching should be the "bridge switch" to move the leased lines' and trunks' traffic from the central office out to the new switching nodes (Class 6 locations) on the new fiber ring topology plant. Later these wideband switches will be replaced by future remote broadband switches as both wideband and broadband functions are switched along with narrowband ISDN at these new Class 6 nodes that are located closer to the customer. In addition, during the interim, wideband switching matrixes can be added to the existing Class 5 central office switches as separate fabrics to meet the overlay digital switch needs of a less heavily dispersed group of customers, who have immediate need of wideband switched services.

Hence, there will be a switched primary rate ISDN (PRI-ISDN), T1, f-T1 or fractional ISDN (*N* number of DSO channels) network and a point-to-point primary rate network/T1 network that provides switched and nonswitched wideband transport for wideband services such as x-rays, image transfer, T1 picturephone, etc. In time, these wideband network nodes will be expanded to include rates up to T3 services; later, they will be expanded to higher rates as broadband extends on to super broadband gigabit rates.[3] (See Fig. 6-11.)

The public broadband network

As transmission speeds increase to 2.4 and 4.8 Gbps, with microdot switches of 155 Mbps and clusters of highspeed computers, we are at the entrance of a new world, a new age, the Information Era, the information millennium. It's a time to embrace new ideas and explore new ways of delivering new services using new technology. No one can hold a cross section of cable, showing 600 pairs of copper wires in one hand, and in the other hold a single fiber that can carry all the information provided by the copper wires, and not realize that something big is about to happen with telecommunications.

This then requires different thinking, expanded thinking, big thinking to change our current mode of operation and begin taking the fullest advantage of the fiber. No longer can we say that transport is expensive and we need to charge extensively for it. No longer must people talk fast

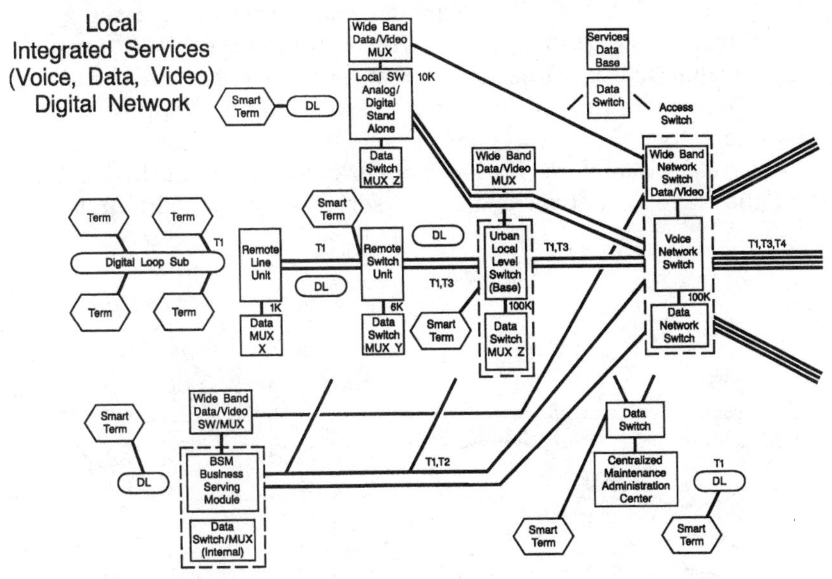

Figure 6-11. Narrowband/wideband local networks.

---

[3]As indicated in Fig. 6-11, data is added to voice switches to form integrated ISDN voice and data networks. Wideband capabilities are overlaid on this architecture to enable dynamic bandwidth management, LAN internetworking, and POP access from remote access nodes (Class 6), or as adjunct matrixes from digital Class 5s. In time, access switches will become superbroadband switches (as many as 5 or 6 per large metro area, or 2 or 3 per large rural sector), while some Class 5s become downgraded to Class 6s, as rings on rings are developed.

or hang up quickly to save large long distance charges. No longer do we need to deploy plant in the same traditional manner with everything homing in on central offices in the middle of our major cities.

Just as the shift from analog to digital changed the way we deployed switch centers, so will the shift to broadband. In the IDN network, remote switch units homing in on a base replaced remote central offices. Now is the time to continue the distributed switching concept to its completion, as we fully differentiate where and how transport will integrate with switching into the network of networks. (See Figs. 6-12A and 6-12B.)

Network of networks

As we pursued survivability in the wideband networks, we saw the advantage of placing switching nodes closer to the customer and using new ring-type topologies to obtain survivability in the event of a cable cut in one direction. As we look at the rapid changeover from point-to-point leased facilities to switched wide area networks, we see a rapid exodus of private network managers, who will no longer continue to accept the traditional long lead time for trunk ordering, servicing, and restructuring. As a result, "fiber" will be the new vehicle to replace long lead time "leased trunking." In fact "switched fiber" will be a vehicle, but to be more precise "distributed switched fiber" will be the correct vehicle. (See Fig. 6-13.)

There will be a new "network of networks," as noted in the local layers of the "layered networks' layered services" model. The first layer

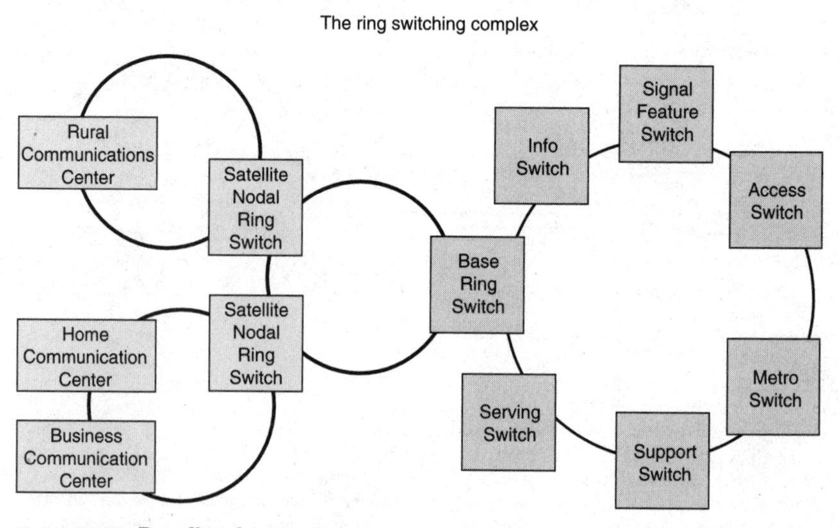

Figure 6-12A. Broadband networks.

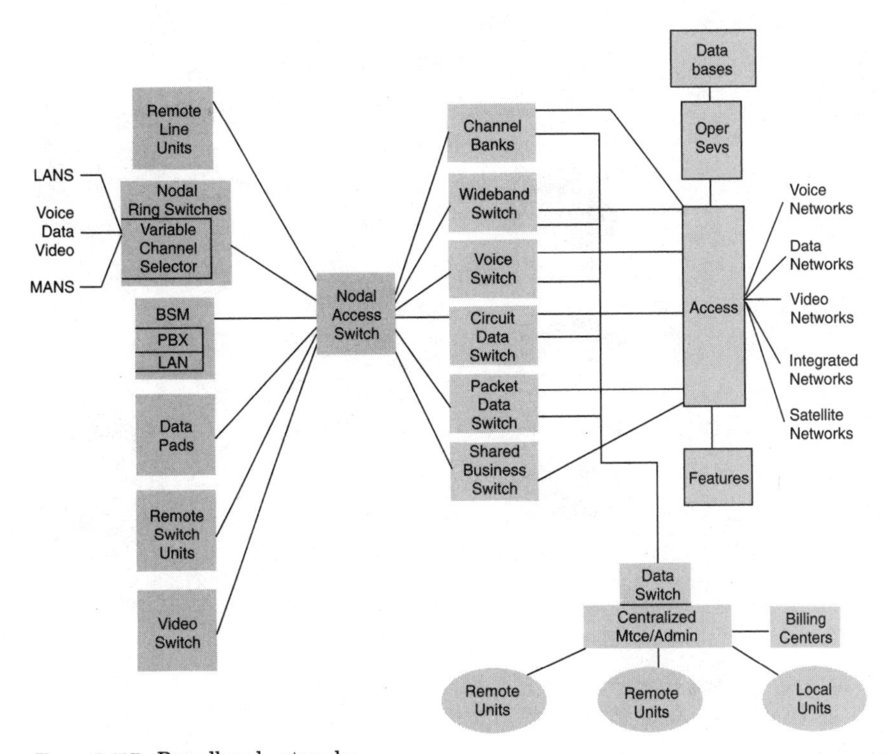

Figure 6-12B. Broadband networks.

will be on customer premises. Broadband switching will be a key tool in the intelligent buildings infrastructure. PBXs and LANs will eventually give way to fully distributed broadband switching nodes located throughout the customers' campus, be it university, manufacturing plant, office complex, or home. Just as fifth-generation PBXs have become more distributed, so will their broadband counterpart. As processors within the "campus" become interconnected to workstations to manipulate more and more graphics, images, and files throughout the complex, so the sophistication of their communication switches will increase, as these internal application service centers are accessed from the world. As internal workstations need to access the world through their switches, so will the world's interfaces to these switches require standardization, as these internal private networks become the first network of the network of networks. To demonstrate the need for clearly understanding all of the network tasks that these switching systems can and should perform, let's call them the new Class 7 as the traditional five-level network is expanded to include the customer, who now has a lot more than just a "black phone" hanging around.

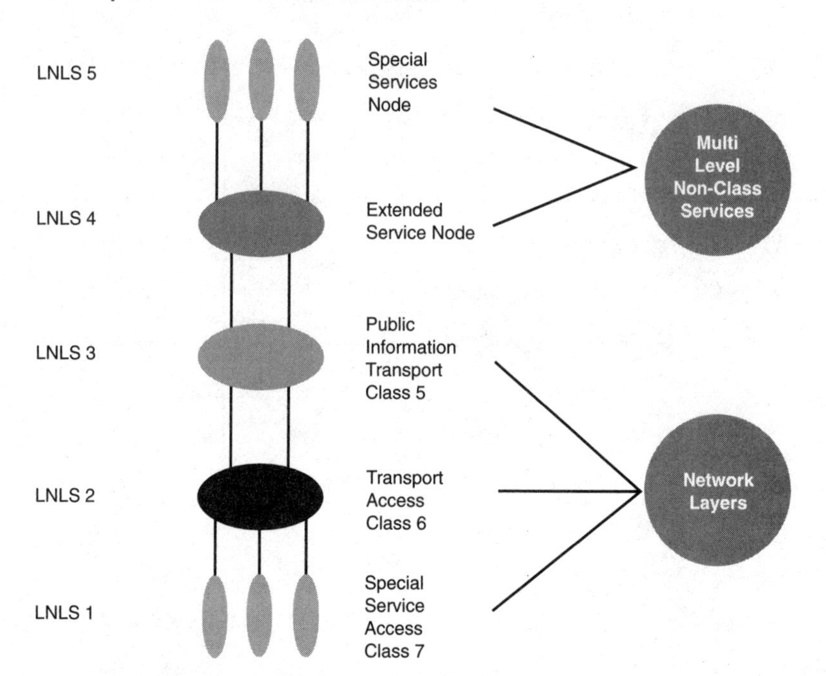

Figure 6-13. Local network layers.

CPE internal networks (Class 7)

The new customer-premise systems will be able to provide full information services for shopping centers, shared tenant services, universities, federal buildings, office complexes, hospitals, police control centers, banking institutions, and manufacturing plants. This will require a sophisticated online system that provides specialized transport, shared transport, and various computer offerings to its unique customers. On the network side, the CPE switch will act as a gateway to its internal services, just as more automated attendant services help facilitate the calling party to quickly reach the desired destination. Similarly, as the internal switch becomes the internal entrance and external exit interface to the network, there will be a growing amount of sophisticated network tasks that it will need to do. These include interfacing internal network management to local LEC network management systems; call billing and metering; specialized routing to POPs or specialized application service centers; LAN-to-SMDS interfacing; accessing commands to higher network layer databases over "D" channel signaling; termination testing; message encryption; audit trails; delayed delivery mechanization; multimedia image sequencing; packet assembly/disassembly; protocol and code conversions; and, of course, all the internal call switching and message control functions, to note just a few. (See Fig. 6-14.)

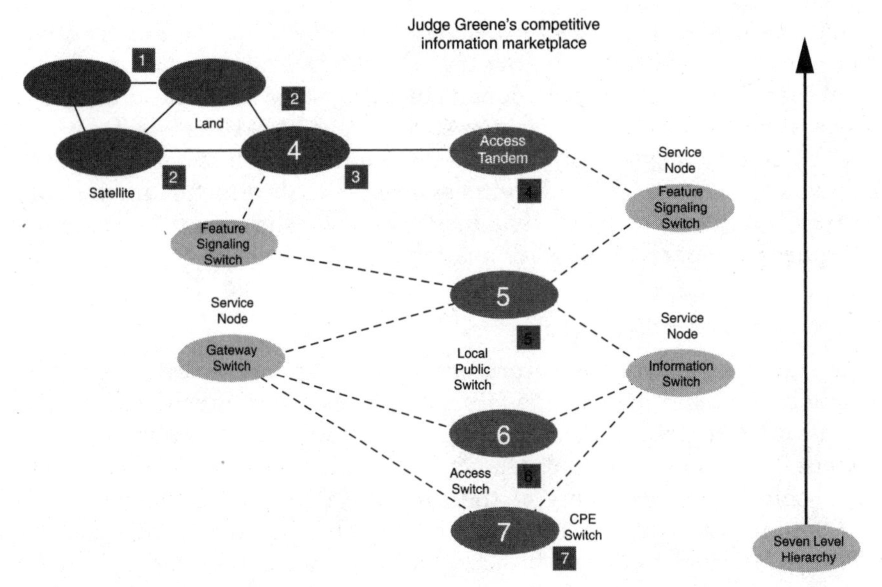

Figure 6-14. Seven-level hierarchy.

Transport-access nodal network (Class 6)

The broadband switch for the turn of the century will in fact be at least two physical, distributed switches. It will have a front-end stand-alone switch (Class 6) that will be located closer to the customer. This switch will control the huge gigabit rings that flow through high-density urban communities. These ring switches ensure survivability, as noted in the public wideband network analysis. They become the second phase of the earlier, more limited wideband network switches as they add broadband networking capabilities for frame relay interfaces, SMDS routing, ATM and STM switching, and special area and office code translations. These capabilities are added to the remote narrowband ISDN interfaces and digital cross connect capabilities offered by the wideband network. Similarly, network management, maintenance testing, customer-line administration, various protocol conversions, bridging, and routing will be performed at this node. It will also perform gatewaying of information to connectionless data transport interfaces such as FDDI over the rings between nodes.

In this manner, traffic can access a Class 6 node, travel along the ring to another Class 6 node and leave the network without need of further network features; calls can access POPs directly from these nodes, and access telco or private application service centers through information switches.

Much of the switched traffic will be handled by the public broadband network by moving the calls from these front-end nodes via survivable rings to the new broadband network superswitch centers. In the event that one center or routes to the center are down, then an alternative center can be addressed by the transport/access nodal switch through another Class 6. Hence, traffic can run along the ring to the next active system in the event of a disaster, such as the 1988 Hinsdale fire.

This ability to go directly to a POP or home on both primary and secondary central broadband switches from a multidirectional ring will greatly enhance call survivability as more and more information requires 100 percent network availability.

Broadband network (Class 5)

As the front end of the future broadband network evolves from the public wideband network nodal front-end switches, much work needs to be done to prepare the plant for full broadband information movement. Suppliers need to construct "superswitches" that will augment or begin to replace many of the overlay ISDN IDN switches of the 1980s and 1990s. A typical metro area may have twenty to fifty central offices. By 1999, many are projected to be digital as IDN digital switches are overlaid next to analog systems to cap growth, move business to digital facilities, provide ISDN access to customers desiring extended CLASS-type services, and enable ISDN access to the public data network. Wideband fabrics added to these systems supplemented by front-end wideband switches will complete the initial changes of the mid 1990s. Then in the late 1990s, selectively, superswitches can be inserted into the metro community as more and more broadband front-end switches are inserted into the network as ring nodes. By the new century, several superswitches will be bringing video services to business users both large and small, and to educational and state facilities.

Viewphone, videophone, picturephone, and high-definition TV become the leading candidates in order to ubiquitously offer broadband to the majority of the customers within the urban/suburban community. Over the first twenty years of the new millennium, the copper plant will be changed over to the fiber, replacing the current family of #5 systems (with traditional and overlay Class 5 offerings) with these new #6 systems, providing more integrated voice, data, and video Class 5 offerings. Perhaps ten or so superswitches (#6 ESS, #6 GTD) with full ATM/STM capabilities will now handle the entire metro area, along with thirty or so front-end broadband Class 6 switches.

Photonic switching, neurologic, expert systems, fuzzy logic, ATM, STM, SONET, SMDS, etc. technologies, along with sophisticated compression technologies and voice recognition algorithms, will form the bases for these new Class 5, Class 6, and Class 7 systems.

"Information please" (info switch)

As more and more services require access to more and more information, the info switches (layer four) that provide access directly to databases or to application service centers or service nodes will be the fourth new system of the broadband network. This system can exist in several forms. Some will use it as an intelligent node providing 800 voice, data, and video services on a call processing interrupt basis; others will switch calls to these service centers, either directly or through the information switches. Info switches exist parallel to the network and can be accessed from any layer. As noted earlier, these systems can function in the network, providing information Centrex-type services and access to common public databases. Or, they, with or without switching, will be in the new open arena to offer enhanced services. Here they will be supplied by many providers other than just the RBOC, similar to Japan's type 2 general VANs, some of which can be as simple as a small firm's database.

Once work on the call has been performed within the information switches, further work can be achieved at other locations by reinitiating the call at any layer, or by routing the call directly to any provider.

Broadband topology

Achieving the correct broadband network topology for deployment of fiber in the 1990s will be essential for later success. As superswitches are inserted and tied together by survivable rings, so are nodal switches.

In the configuration noted in Fig. 6-15, it's interesting to see how sub 6 nodes can extend fiber the short distance to the point of application. In time, sub 6s can also be connected by rings between sub 6s, and 6s can be networked to 6s for parallel survivability. It should be noted that Class 6 locations are easily determined, since many fit nicely into 2-4,000 line carrier serving area (CSA) hub locations. Sub 6s can be housed as pedestals at "B" boxes that are at the entrance to 100-home communities. Final pole-mounted remote optical units can passively split or collect fiber information for a small hub community of four or six homes. Power and cooling of these locations will be necessary environmental considerations of placement and type. We should also note that as transport capabilities increase there is less need for tandem switches, as direct Class 5 to Class 5 traffic is quite feasible on shared rings, and direct access to points of presence of IXCs and VANs.

A look at the future

As time continues and the phase over to broadband superswitches using supercomputers takes place, there will be a natural progression of services from the Class 5 to the Class 6 and from the Class 6 to the sub 6, and the Class 7. Initially, there will be considerable need for internetworking non-ISDN to ISDN and to dissimilar protocol interfaces. Similarly, front-end translators will be needed to route calls to new information centrexes, application service centers, or POPs. This distributed front-end work should continue to grow while the background supercomputer work will most likely change to more data storage,

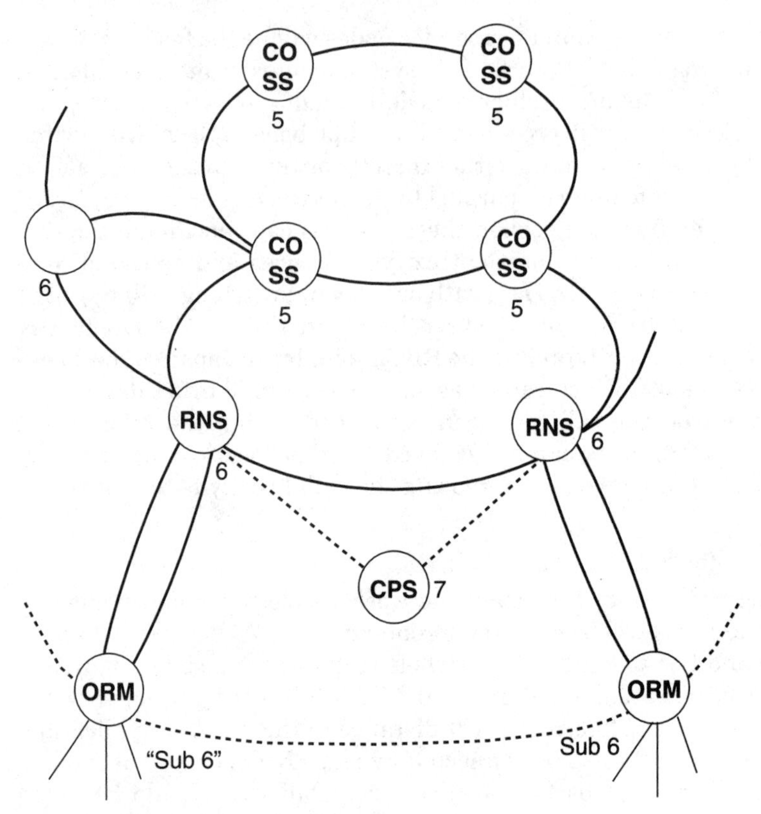

Figure 6-15. Example broadband topology.

manipulation and processing, as RBOC information services restrictions are lifted, thereby once again enabling various types of services within the regulated network. Or, the supercomputer work may shift more to network control as the need to interface with regional, national, and global networks increases in scope and complexity. In this manner, local tandems work and POP access are initially replaced by direct links from each superswitch. Then, network survivability, control of the Class 6s, and global internetworking functions become the growing concern of the superswitches as more and more work is downloaded to front-end systems offered by multiple players.

As both global and national networks become more and more broadband, then the world's population will indeed come closer together. (See Fig. 6-16.)

Interchange carriers networks (IXCs)

POP, POP! What is POPing? As noted earlier, each of the interchange carriers has a point of presence (POP) within the local exchange carrier's

(LEC's) local monopoly. For AT&T, their Class 4 switches have become their initial local POPs. Here, they provide access to their national network served by their Class 4, 3, 2, and 1 switch hierarchy. There, they provide national 800 services, operator services, switched voice call handling, switched 56-Kbps data-handling services, and direct trunking. Users can also lease lines or trunks to home directly onto this IXC node through LEC central offices, for various long distance data transfer services at T1, T3, and above rates. By the late 1990s their new Class 6 switches will be in service to collect customers throughout the business community, among other COIs.

Over time there will be an expanding army of POPs throughout the local community, providing access to a host of new higher-level services.

LEC operating companies can forward and receive traffic from the IXCs through their Class 5s or from their tandem switches. Call duration peg-count information can be noted at either of these locations. Some independent telephone carriers once restricted any IXC or local BOC LEC from directly trunking to their Class 5s, but required access

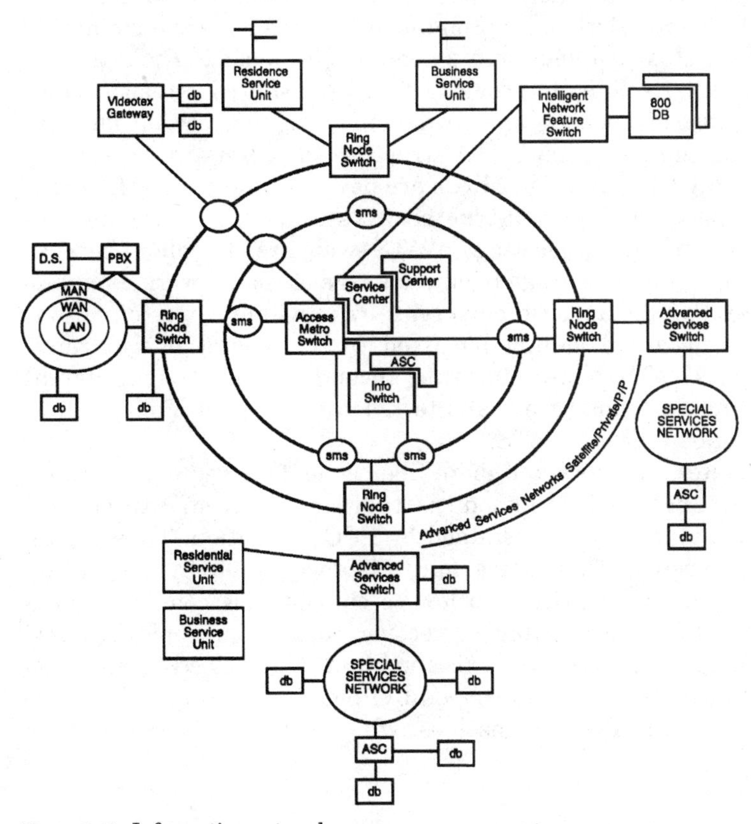

Figure 6-16. Information network.

through new area-wide tandem systems, some serving almost an entire region in an effort to provide additional services from this node. However, this is inconsistent with the need for providing access closer and closer to the customer, such as through the Class 6 at layer two of the LNLS nodal.

Therefore, the POP will exist within the local serving area at layer six. Its network will connect across the country (layers six and seven), and provide services from platforms either associated with the Class 4 or as separate stand-alone platforms. Both of these are logically noted here as application service centers at layer eight. These platforms provided software definition programs that enhance the movement of information to add value to the call as they compete with specialized VANs for the customer's long distance market. In Japan, they could be called a type 1 VAN, where they provide carrier services and information services.

The IXC's major concern is access to the customer. Some have developed VSAT-type satellite services, where 56-Kbps and above data customers access Class 4 systems via satellite. Or, some customers, rather than lease circuits (trunks) from the LEC, have opted to lease cheap conduit space and pull their own fiber, or use "dark fiber" where they provide their own electronics. Some use direct microwave circuits to access IXCs. This customer access need has been called the last mile, the bottleneck. As noted in previous analyses,[4] shared transport rings is one method of opening up the monopoly to competition but still maintaining the necessary LATA structure to preserve local service. US Sprint, MCI-BT, and other IXCs are now providing a new family of voice messaging and E-mail-type services and various volume discounts to encourage higher usage. WATS (wide area telephone service) was established to encourage business usage of the network, enabling economical long distance tiered service by providing zones of various tiers with unlimited calling for fixed (upper bound usage) rates. Similarly, INWATS allows flat-rate incoming calls from different zones, giving customers free calls to reach businesses. For example, the 800 numbers are INWATS.

Unfortunately, as more and more fiber is deployed across the country, transport becomes a commodity. Traditional price structures were initially being artificially sustained by FCC regulation. This was an attempt to keep a competitive long distance arena by preventing AT&T from dropping prices too low to cut out new competition, as these new players attempted to recover costs from deploying new plant to compete with AT&T's existing plant. Next, price caps formed an umbrella under which to (hopefully) blossom a competitive marketplace. Also, the IXCs themselves, who are naturally reluctant to

---

[4]See *Telecommunications Management Planning: ISDN Networks, Products and Services.*

give up today's revenues for tomorrow's growth, must reassess their transport point-to-point tariffs. But there's no denying that the fiber can economically deliver a tremendous number of new services such as picturephone, viewphone, and HDTV, as deployment costs reach the $1.00 a foot deployment target for unlimited bandwidth.

However, unless there is a shift to pricing data and video offerings on a service basis, instead of simply selling raw transport bandwidth, long-distance picturephone, viewphone, and videophone will never see the light of day. T1 and T3 transport rates for raw bandwidth make the service costs prohibitive. But this realization is occurring. Hopefully, by the turn of the century, or sooner, service pricing will enable $30 per month picturephone within the local LATA. Later, picturephone will be able to be expanded nationally across the nation for reasonable usage costs of $60 per month. Then, and only then, will we really be well within the formation of the information society.

Global VANs

In a way, VANs are somewhat like moving vans. Value added networks (VANs) are parallel transport networks that can go anywhere to provide services other than those provided by the traditional LECs or IXCs. In Japan, VANs come in all types: some own their own transport, some lease transports, some provide databases services and transport, and some only provide database services.[5]

VANs are local, national, and global. Telenet and Tymnet provide extensive data-handling national networks with various protocol and code conversions, enabling asynchronous local systems with separate timing to be synchronized to their national network and packaged within their internal protocols as packets of information that are stored and forwarded from end to end.

Of course, VANs that are initially intended for one form of traffic will attempt to adapt to whatever the market offers. One such firm was established to handle large data files. Soon, it found that the customers really only wanted to send small files for data transfers. Considerable redesign took place to save the business and reconstruct it around short low-speed data messages. Subsequently, with ISDN as the interface, combined voice and data was the next objective, as these firms pursued the economic opportunities of capturing both.

On the other hand, some countries, such as France, have found the opportunity for overlaying and interconnecting data networks across Europe, and providing access across the globe, through their data networks' gateways. Since data is not affected by the echo delays of multiple

---

[5]See the previous discussion on "Japan's VANs."

satellite hops, global data satellite networks are becoming increasingly popular. However, security is still a major consideration, as New York banks found out after building the monstrous, somewhat unsecure networks of the 1970s. Hence, the world of encryption, code locks, security passes, and protection algorithms have left the military and entered the commercial domain, with considerable resistance from internal intelligence agencies, who still want to break bank transfer codes to know what today's current group of bad guys are doing.

However, global VANs are here to stay. Book writers in the Caribbean send and receive information from their New York publishers; financial clearing houses in Singapore and Hong Kong do business before sending the transaction on to Japan. As more marine cables such as TAT-10 and TAT-11 are deployed by consortiums to connect the east coast of America to Europe, and the west coast to the far east, there will be an explosion of global data traffic to interconnect major cities of the world, as noted earlier in reviewing the customer's needs.

In parallel with these global activities will be local VANs. We have already seen end users move to selected cities' networks, such as Metropolitan Fiber in Chicago, or parallel area networks, where the New York City Port Authority Network captures a growing amount of the financial sector's data traffic from New York, Staten Island, and New Jersey. Once personal computers from the office are linked together to citywide databases, a new host of software programs will enable imagery and graphic display of information for commercial applications. These will then provide upward pressure (market pull) for cheaper long-distance national data transfer bandwidth. The pressure will then mount for global satellite bypass, because once airborne, why stop at state or national boundaries? Why not continue on to the most economical location that meets the need? For this reason, telecommunications can no longer be thought of in the local context, but must always be considered globally, as global telecommunications.

Urban telephone companies

With this in mind, local urban VANs will be a key deliverer of customers to the more global VANs. It's readily apparent that a new urban telephone company is in the making, whether it's Metropolitan Fiber or some other alternative network services type of offering. It appears that urban telephone companies have recognized the following possibilities and strategies:

1. Technology now offers economical megabit transport capabilities.
2. LANs on private facilities need to be interconnected.
3. Special service circuits provided by the telco for these LAN interconnections have to be established and are, in some cases, too

expensively tariffed. Some do not allow for today's growing and dynamic traffic that needs new variable bandwidth services.

4. Firms have now discovered that they want to move data and voice. The traditional LECs have not provided them with economical data networks. Therefore, if they can obtain transport from new providers, such as Metropolitan Fiber, they will then do their own data switching and data handling internally. In the future, local VANS will find the data-handling opportunity as a new growing market to provide many new data services.
5. Later, they will establish their own application service centers for specific communities of interest, such as the medical industry, as they use open network architecture and CEI to collocate equipment. If open systems architecture (OSA) is extended into internal LEC switch software, there will be further service competition, as VANs obtain access to calling and called party identity numbers. Via OSA, they will also obtain access to the internal database information of the individual user, thereby increasing the opportunity to deliver many additional new services.
6. Over the next twenty years, 40 percent of the LEC's point-to-point special services will become variable bandwidth (channel) switched; 40 percent will be data circuit and packet switched, and the remaining 20 percent will remain on point-to-point networks that transport both high-speed connection-oriented and connectionless data streams. This will happen whether the LEC switches it, it's switched by the customer on his or her local facilities, or the new local VAN switches it.

To counter VAN movement in the future, the RBOCs will offer large dynamic channel-switched transport rings that provide S&S (security and survivability) features, perhaps as a shared provider medium. But if not, they will need to take a more offensive position, not just a defensive position to plug the gaps, especially in the areas of switching bandwidth and data image networking. To do this, a new layer of transport movement (layer two) will be established, initially using new remote digital cross-connects. This enables the movement of the automated main distribution frame closer to the user.

The digital cross-connects can channel-switch information to secure survivable routes, providing new flexibility so that multiples of 64K to 155 megabits (or even up to 2.4 gigabits) can be shared and moved for the "large business" market in the private community under special, competitive pricing structures. As LAN interconnection grows with the explosion of T1, T3, and fractional T1 (f-T1) network management services, customers will continue to move to separate, overlay, private VAN networks to interconnect their LANs.

Summary

Hence, any firm that needs its business units to communicate directly from point A to point B across town, or to transport information to a remote location via a POP, is a likely candidate to move from the special services of the RBOC to a new local VAN. They will move from LEC special service offerings to these private, separate facilities. In the future, we will see a lot of CPE-type (Class 7) offices blossoming on the customer premise, accessing these new (Class 6) transport access nodal switches to get around the city. These new networks may be LECs or belong to someone else. New services will be provided by many different providers' application platforms. The LECs will provide the layer two as a new aspect of the transport business on a public offering, and the faster they do it, the better. Or, the RBOCs need to provide layer-two services at least initially on a private offering, to try to remove some of the pressure. As noted earlier, the other reason customers are moving to these networks will be to do their own data handling. Therefore, again, the sooner the LECs can get an overlay data network going, the better. Then later, with broadband switch services up and going as a public network offering, the local VANs will shift to providing more local databases and providing access to the global VANs through their gateways.

In any event, the need is there; the technology is there, and this translates to an urban VAN opportunity—so the local and global VAN competition will not go away.

Technology and strategies

(See Appendix B and Tom Bystrzycki's observations in Chapter 9, both in this book.)

Layered Networking Observations

As we conclude our assessment of the future in terms of each of the LECs, IXCs and VANs' potential networks, it's interesting to return to the layered network model to review what types of networks, services, and products would exist at the various layers. This can be readily developed by reviewing the following example of the local arena, noting the buildup of public services from the second to fourth layers, and the private CPE services in the fifth layer. (See Figs. 6-17 through 6-21.) This form of analysis can be performed for all layers, in terms of both services and networks. Figure 6-22 shows the full spectrum of offerings for the first five layers.

Information modeling

Considering the complexity and opportunity noted by the various models, such as the ten user groups of Chapter 3, the layered networks'

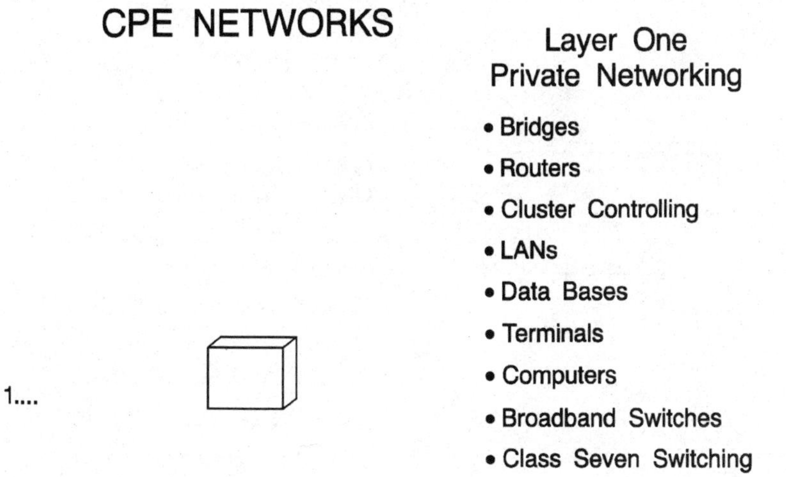

CPE NETWORKS

Layer One
Private Networking

- Bridges
- Routers
- Cluster Controlling
- LANs
- Data Bases
- Terminals
- Computers
- Broadband Switches
- Class Seven Switching

1....

Figure 6-17. CPE networks (layer one, private networking).

PRIVATE/PUBLIC
INTER-NETWORKING

Layer Two
Shared Access Transport

- ONA
- POP Access
- Private to Public Inter-Networking
- IEEE LAN/MAN Standard Interfaces
- Distributed Access Nodes Closer to the USER
- Bridges
- Frame Relay
- Fiber Rings
- FDDI – II
- SMDS
- Fiber to the Home
- Nodal Point Access Switches
- Class Six Switching

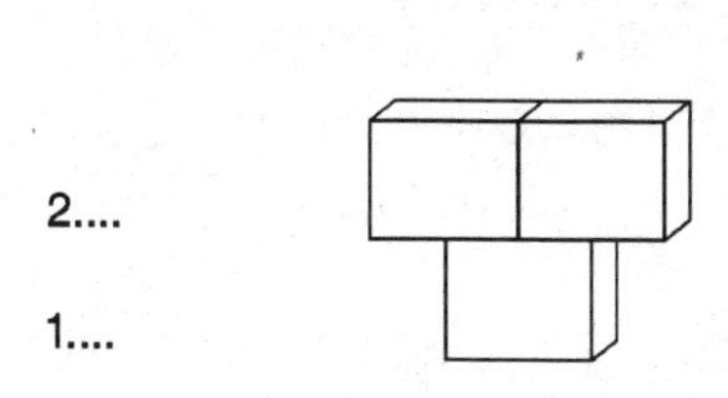

Figure 6-18. Private/public interrnetworking (layer two, shared access transport).

layered services model of Chapter 5, and now the twelve networks model, let's pause and take a moment to interrelate these models and form some general observations and conclusions.

Let's begin by reviewing just the local networks and services layers from the LEC point of view. (See Figs. 6-23A and 6-23B.) The first five or so user groups provide some interesting cross analysis as we tie their

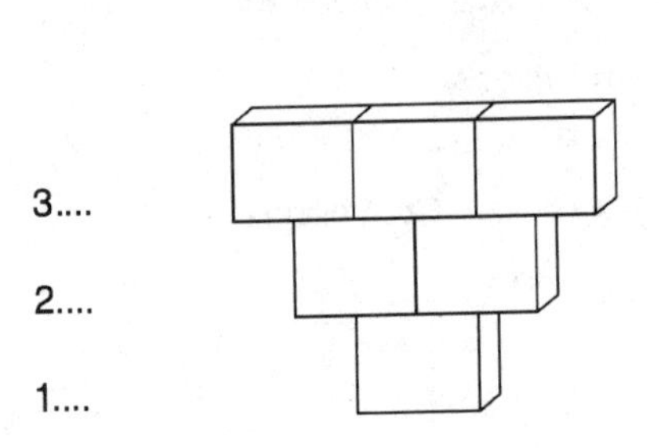

- Inter Office Network
- Data Switched Nets
- Inter-Exchange Access/ Information
- Routing/Addressing Control-Urban Switch
- Operational Support Centers
- Network Management Centers
- Access to Gateway Centers
- Signalling Networks
- Class Five Switching

Figure 6-19. Information infrastructure (layer three, public information network).

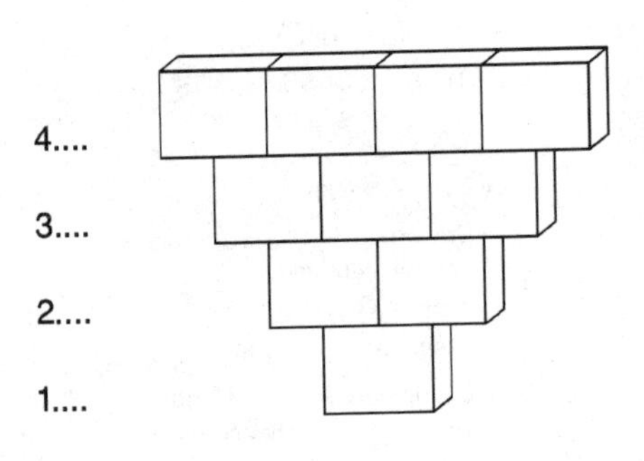

- Gateway Services
- 800 Services
- Advanced Data Services
- Top Layers of OSI Model
  - File Management
- Billing Services
- Operator Services (Voice/Data/Video)
- Feature Switch Services

Figure 6-20. Service information applications (layer four, shared advanced services).

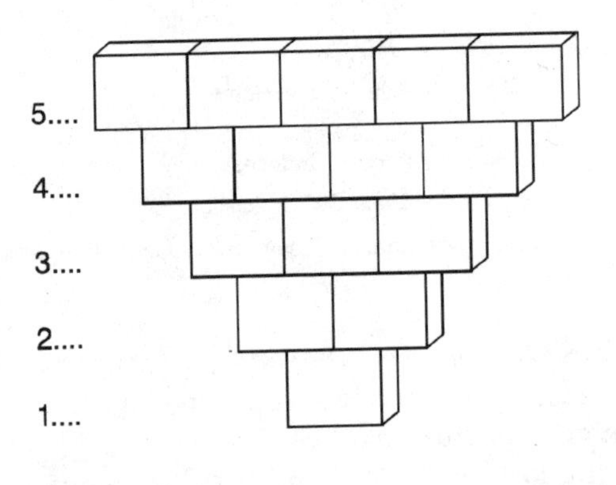

- Specialized Services
- Application Packages
  - Hospitals
  - Lawyers
  - Etc....
- Unique Gateways
- Private Network Management System
- Customized USER Services
- Specialized Programmable Info Switch

Figure 6-21. Service information applications (layer five, specialized services).

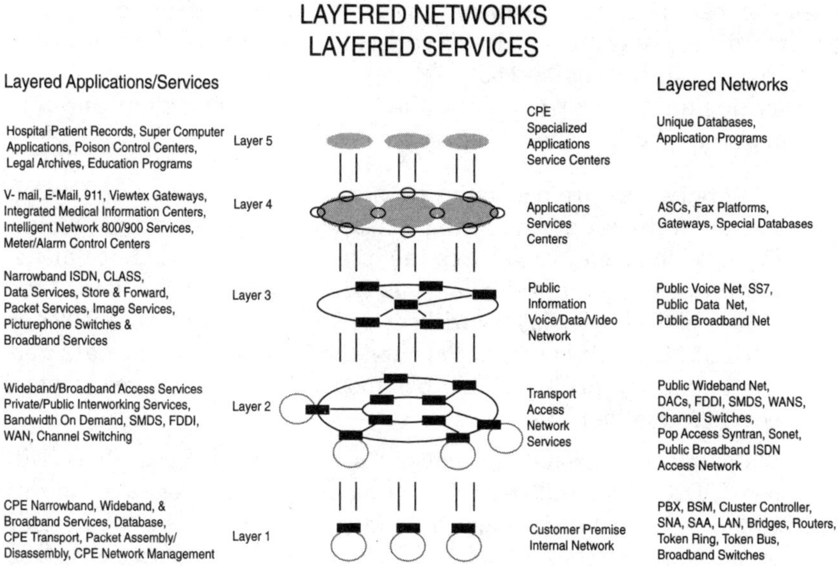

LAYERED NETWORKS
LAYERED SERVICES

Layered Applications/Services			Layered Networks
Hospital Patient Records, Super Computer Applications, Poison Control Centers, Legal Archives, Education Programs	Layer 5	CPE Specialized Applications Service Centers	Unique Databases, Application Programs
V- mail, E-Mail, 911, Viewtex Gateways, Integrated Medical Information Centers, Intelligent Network 800/900 Services, Meter/Alarm Control Centers	Layer 4	Applications Services Centers	ASCs, Fax Platforms, Gateways, Special Databases
Narrowband ISDN, CLASS, Data Services, Store & Forward, Packet Services, Image Services, Picturephone Switches & Broadband Services	Layer 3	Public Information Voice/Data/Video Network	Public Voice Net, SS7, Public Data Net, Public Broadband Net
Wideband/Broadband Access Services Private/Public Interworking Services, Bandwidth On Demand, SMDS, FDDI, WAN, Channel Switching	Layer 2	Transport Access Network Services	Public Wideband Net, DACs, FDDI, SMDS, WANS, Channel Switches, Pop Access Syntran, Sonet, Public Broadband ISDN Access Network
CPE Narrowband, Wideband, & Broadband Services, Database, CPE Transport, Packet Assembly/ Disassembly, CPE Network Management	Layer 1	Customer Premise Internal Network	PBX, BSM, Cluster Controller, SNA, SAA, LAN, Bridges, Routers, Token Ring, Token Bus, Broadband Switches

Figure 6-22. LNLS.

THE FIVE USER GROUPS — FIGURE 1

GROUP	USER	RATE	APPLICATION (*1)
Group 1	Dial Up Data Users	1.2 - 9.6 kb/s Voice Grade	Inquiry/Response
Group 2	LAN Data Users	1.2k - 10 mb/s Non-Switched	Data Collection Data Distribution
Group 3 (*3)	"2B" Data Users	(64k - 128k) b/s Switched Data Grade	Transaction Picturephone II (*2)
Group 4 (*4)	"VB" Data Users	Variable 64k 64k - 1.5m - 45mb/s	Data Manipulation Picturephone III Distributed Video Conf.
Group 5 (*5)	"OC"-N Data Users	Variable 50mb/s Super Channel OCI - OC12 50-155.-600.mb/s	Data Presentation Picturephone IV Videophone 1 HDTV

*1 Note each application will have each type of data collection, distribution, storage, manipulation, access, transaction and presentation offerings. These are only presented as representative examples.
*2 Picturephone/Videophone will have ever increasing resolution as bandwidth (data rates) increases.
*3 Group 3 users can be handled by overlay data networks using existing system with ISDN basic rate interface.
*4 Group 4 users will be selected; offering will be limited to specific user groups using Overlay Broadbound Switching and Digital Cross Connects.
*5 Ubiquitous offering of Viewphone over the entire area via - distributed megabit facilities to the Home/Bus - requires the New Broadband Urban Switch and B-ISDN interfaces.

Figure 6-23A. User groups' needs.

needs to networks. This can be further seen by cross relating user groups to networks to the first five local layers of the LNLS model. (See Figs 6-24A through 6-24C.)

From this analysis we can draw the following observations and conclusions from the LEC point of view.

1. Data networks are needed to support ISDN interfaces for providing ubiquitous data-handling features and access to data platforms such as gateways. (See the information cubic service model in *Future Telecommunications*, as well as the higher layers of the OSI model.) Lower layers will connect OSI standard interfaces to IEEE, SNA, Netview, etc.) Perhaps the local area public data network could be priced at $20 per month for 20 messages (1 page per message) per day.
2. Wideband and broadband networks are needed. First phase will use ISDN wideband (f-T1 to T3) modules, matrixes, and fabrics added to existing systems. Second phase will use new photonic broadband switches as networks expand over the entire city. One key universal service will be picturephone or video-feature, viewphone, videophone telepicture, personvision, or infovision. They should have several levels of resolution, depending on the offering and pricing of terminal. But dissimilar systems still need to

THE FIVE USER GROUPS — FIGURE 2

GROUP	TECHNOLOGY	ACCESS	TOPOLOGY	APPLICATION SERVICE CENTERS (*14)
Group 1	Current Public Network	POTS	POTS	FAX 911 800/900
Group 2	Current Private Network	LAN	Bus/Ring	Hospital Internal User Files
Group 3	Overlay Data Network (*6)	ISDN (*9)	Current/ Distributed (*12)	Protocol Conversion Data Base Access Extend 911+
Group 4	Overlay Broadband SW Matrix (*7)	Digital Cross Connect (*10)	Distributed Access	Video Education File Centers
Group 5	New Broadband Urban Switch (Class 5) (*8)	New Class 6 Switch (*11) from Class 7 Switches (*13)	Distributed Access	Dial A Movie Radiology Service Centers

*6 Switched Public Packet Data Network - Capable of handling 100,000 + Packets.
*7 Broadband Switching Matrix Fabrics/Modules (ATM or Circuit Switch) added to current system.
*8 New Broadband Urban Switch for voice/data/video using CKT-PKT-ATM structure.
*9 ISDN basic rate or multiple basic rates B channel/D channel (circuit and packet) interfacing.
*10 New Digital Cross Connects for DSO-DS1-DS3 variable channel control (Pre-Class 6) to multiple base units (central offices) via alternative networks.
*11 New transport access switches (Class 6) to interface to new urban/metro broadband switches (Class 5) using a new fiber distribution plan based upon a distributed plan based upon a distributed nodal topology architecture with access nodes closer to the user.
*12 New fiber distribution plan topology.
*13 CPE Access from internal network controllers or new Customer Premise Switches (Class 7).
*14 New family of Application Service Centers—shared (Layer 4) or specialized (Layer 5)—Layer refers to Five Layer Information Network Services Model.

Figure 6-23B. User groups' needs.

User Group	Network	Type	Networks Layer 1 (7)	Layer 2 (6)	Layer 3 (5)	Services Layer 4 (ASC)	Layer 5 (CPE ASC)
1	#1	Voice Network	Telephone & Dial up Data Modem	RSU RLU MDF	Voice Circuit Switching • Voice • Dial up Data	• Voice Mail • Message Mailbox • Voice FAX • FAX (Dial up) • Custom Calling	
1	#1	Voice Network +CCITT #7 (Needed for "D" Channel Services)			• CENTREX/CENTRON • STP/SCP • Calling Party Busy Check • ANI	• CLASS • Selected Call • Acceptance • Transfer • Rejection • Call ID • by Number • by Name • 911, 800, 900 • Weather/Time • InterLATA ID • Specialized Billing	
2	#2	Special Circuits LANS/MANS	LANS PBX's	DACS 1/3	• MDF • 56K Circuit Switch • Data Testers • Data Line Conditioning C1, C2	• T1 & T3 • Fractional T1 Network Mgmt • Point to Point 1200, 2400, 9600 19.2 K Data Transport • Special Handling • MANS, FDDI • Bridges, Routers	CPE ASC

Figure 6-24A. Twelve information networks chart.

interface, perhaps with shrunken screen sizes, as larger screen, better resolution systems communicate with smaller, less resolution systems. The other reason for deploying broadband is high-definition video for entertainment, education, business, and medical computer information exchange. Hence, broadband rates of 50, 155, and 600 Mbps will be needed for the different applications. Perhaps with nominal compression techniques, 155, Mbps will be standard, if not, 600 Mbps on the high end and maybe 6.3 Mbps on the low end, as compression techniques get better and better.

3. New topology plans are needed by LECs for competing with IXCs and VANs, and for deploying the fiber so that private and public networking can occur to obtain S&S (secure and survivable) routing with access nodes closer to the customer, using new Class 6-type, level-two nodal switches. Interim digital cross-connects can help, but they won't solve the problem.

4. Three new broadband network products and two levels of application service centers have been identified. Initial platforms for voice network dial-up data transaction needs are only a short-term transitional solution.

5. Several models have been provided to achieve the above networks based upon a layered network service hierarchy. Here, both the what and the how need to be addressed to not only encourage more global conceptual planning, but also to achieve realistic plans of action.

User Group	Network	Type	Networks Layer 1 (7)	Layer 2 (6)	Layer 3 (5)	Services Layer 4 (ASC)	Layer 5 (CPE ASC)
3	#3	ISDN Basic "B or 2B" Circuit Data	PCs Computers Printers X-Ray Picturephone III	Class 6 Data Ckt Switches	Data Circuit Switches 64K/128K 20 Data Transport Features	• Data Base Access • Gateway • Encription • Alt. Routing • FAX S/F • 10 Advance Data Features	
3B	#4	ISDN Basic "B or 2B" Packet Data	• PCs • FAX Systems • Printers	Packet Assembles Disassembles (PADS) Class 6 Data Pckt Switches	64K Data Packet Switches 20 Data Transport Features	• Data Base Access • Gateways • Polling • Sensing • Broadcast • Delayed Delivery • FAX S/F (RGT) • 16 Advanced Data Features	
3C	#5	ISDN Basic "D" Channel Services	• Sensors • Environmental Control System • Energy Mgmt • Terminal to Network Info • Testing • ID • Access Codes • Meter Reading	Sub Packet Switches Class 6 Node	8K Data Packet Switches 10 Data Transport Features • Terminal/Network Signaling • Terminal/ Terminal Signaling	• Singular DATA Sensing • Singular Data Collection • Alarm Monitoring • Remote Control • Billing for water, heating, gas, electrical services	
4A	#6	ISDN Primary (23 B) (Switched)	• PBXs • X-Ray Systems • Picturephone IV	MUX DACS 0/1	Switched Primary Rate	V(B)=1.5M Bts Data Transport Switched	

Figure 6-24B. Twelve information networks chart.

Network conclusions. We need to address both the what and the how and then challenge this thinking with the why, when and where. All in all, it's a time of great opportunity for the future, but we must enter the year 2000 with a clear understanding of the what and how.

We need to formulate:

1. A clear grouping of current and future data, image, and video users in terms of key attributes such as transport rates. For example, the ten user group model.
2. A clear understanding of the layers of networks needed to meet the needs of these user groups. For example, the ten-layer network service model.
3. A clear selection of the types of products needed to provide the data, image, and video transport services for the layering of network services. For example, data switches, digital cross-connects, broadband switch fabrics, new urban broadband switches (Class 5), new transport access switches (Class 6), new application service centers, and new customer data/broadband CPE switches (Class 7).
4. A clear separation of what local services will be provided by each layer of the ten-layer model is needed for both today's (non-ISDN) and tomorrow's (ISDN) world—using these new network switches (Class 5, 6, and 7) and new application service centers (specific (layer 5) and shared (layer 4)). We need to then perform the same analyses for the extended regional, national, and global layers.

5. Public networks should not simply deploy fiber to the home to provide an alternative network for CATV to transmit network programming or movie-on-demand-type offerings. It's time to reconsider "interactive communication." We still have the need for the complete interchange of all forms of information. This means the visual and the voice. It is time to readdress "Picturephone III" or "Videophone I." This service may be the key reason for deploying the totally new network more ubiquitously. If we only attempt to meet one direction (broadcast) or selected broadcast needs, public networks will make the same mistakes CATV made when they built their tree-type analog network to the home. The use of this network should also embrace the high-definition TV-type market of 6 to 600 Mbps, depending on compression technologies. We need to do it right the first time and let the new services' higher-capacity needs catch up with the technology. We can do this in a two-step process. First, with wideband-broadband fabrics for a limited market, and then with the new broadband switches for the more universal market.
6. Fiber should not just be dropped in the ground along existing routes. Public LEC providers need to recognize the need for new nodal switching points for S&S (security and survivability) transport. Here, a user can "home" on multiple base units (central offices) using alternative paths. There is a need for new network management and testing at these nodes as well, as LAN and CPE access ONA standards. These new (Class 6) access points should

User Group	Network	Type	Networks Layer 1 (7)	Networks Layer 2 (6)	Networks Layer 3 (5)	Services Layer 4 (ASC)	Services Layer 5 (CPE ASC)
4B	#7	ISDN Primary (23 B) (Non Switched)	Computers	• DACS (DSO/DS1) • SMDS	Non Switched Primary Rate		
5A	#8	Broadband Channel Switched Transport Access Node	• LANS • MANS • Broadband Switches • Remote Info Switch • Res Module • Bus Module	• Ring Switch (Channel Switch) • DACS • Local tandum switch replacement • Layer 2 Services	[Layer 2 Services • Special Transport services • MANS, FDDI • SMDS • Connection & Connectionless • Home on Multiple base units]	• Info Switch Common/Shared Services	
5B	9	Broadband Switched ATM	• One Way Video entertainment CPE • Imagery CPE • Broadband CPE Switches	Ring ATM Switch	Broadband Switch ATM 0 - 155 Magabit Data Transport Data/Image		• CPE Broadband Services • Video Files • Image Data Bases
5C	10	Broadband Switched STM	• Picturephone V • Two Way Video • One Way Video entertainment CPE • Broadband CPE Switches	Ring Channel Switch	Broad band Switch STM N(50+) Megabit Data Transport 50, 155, 600, 2.4, 4.8Gb/s...		

Figure 6-24C. Twelve information networks chart.

be part of a new fiber distribution plan that encompasses fiber-to-the-pedestal, fiber-to-the-home, and fiber-to-the-desk strategies. Public network LECs will need new ring transport and digital cross connect strategies to compete, support, and interface with firms, such as metropolitan fiber-type VANs. This will formulate the structural transport base for new urban broadband switches to handle the new types of integrated traffic for user groups 3, 4, and 5, noted in Chapter 3. To meet this challenge, we will need a family of completely new broadband switches at the turn of the century, as we enter the next millennium—the Information Age.

7. Product trials need to be performed for narrowband and wideband, and especially broadband services to determine users' responses to different technical possibilities. The output of broadband marketing trial analyses should be the requirements for the various types of data users groups. It should identify and establish both the market and the technical requirements for new network switching systems—both the transport access (ring) switch (Class 6) and the new broadband urban (metro) switch (Class 5). It should also identify the new topology for redeploying fiber. In addition, trials offer an opportunity to define the new data and broadband switches (Class 7s) for the business and residential customer premise market, which support the family of new workstations, video conference centers, and home communication centers.

Once the what and how is determined, we then need to address the where, when, and why, which are:

1. Overlay a data network for data group 3 users for the urban cities, rural business sections, and county seats during the 1990s.
2. Overlay a wideband channel switching network for urban MAN-type private networking and group 4 channel switching by the turn of the century.
3. Overlay a broadband services network to provide videophone for closed user groups, radiologists, x-ray networks, video conference centers—group 4 and 5 users.
4. Establish a fiber deployment plan based upon new distributed topologies for group 4 and group 5 users.
5. Formulate a transition plan from non-ISDN and ISDN universal services deployment that encompasses group 5 ubiquitous broadband offerings.
6. Implement an application service center plan for shared (layer 4) and specialized (layer 5) offerings across market units for both today and tomorrow's services.

7. Aggressively attempt to better achieve more complete and integrated standardization plans for both private and public offerings based upon IEEE 802, CCITT OSI, IBM SNA/SAA, Department of Defense Military Standards, Home of the Future Bus, Bellcore SONET/SDH, SMDS, Frame Relay, AIN/INA, T1 STDS Study Group, Internet TCP/IP, and ISDN narrowband-broadband standards, etc.
8. Use integrated supplier-provider, preprogram, preproject, preproduct planning processes to move ideas from concepts to requirements to product definition before commitments from both parties.
9. Establish deployment plans to interconnect local offerings to the global marketplace to meet the needs of the growing number of global customers.
10. Perform benefits analyses to determine if providers and suppliers do indeed obtain the desired benefits from their offerings. This will enable their products and services to be expanded and modified to better meet the needs of their customers and obtain new revenues. (See Fig. 6-25.)

Perspective: The Network

From an overall perspective and with an eye to the future, we can draw the following conclusions:

- Customer-premise equipment will become total switching complexes that require direct interfaces to all of the IEC, IXC, and VANs networks.

LAYERED NETWORKS' ...
LAYERED SERVICES ...

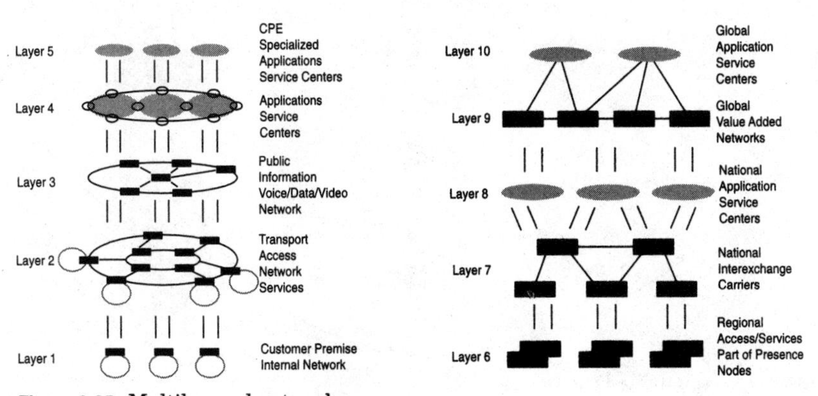

Figure 6-25. Multilayered networks.

- Public wideband-broadband ring networks with access nodes closer to the customer are needed in the 1990s. When deployed, they will open up the local monopoly by providing access to IXCs and VANs from these ring nodal switches. This should make possible compliance to the 1996 Act to enable IECs to directly provide information services, since many of the competitive bottleneck access issues are resolved.
- Data, image, and video transfer will require new public data networks, wideband networks and broadband networks that will become more and more integrated together throughout the turn of the century.
- Appliance service centers (service nodes) will provide new information services above the network and within the customer premises, as they become an integrated part of the information network.
- IXCs and VANs will obtain more and more access to customers, as their application service centers provide more and more information services.
- Global VANs will become a key economic tool of third-world countries, who will use them to provide lower-skill data entry functions. Or, various countries will use them to specialize in specific information-handling areas or provide global services to a specific market's communities of interest.
- Indeed, wherever we go, information networks will follow—here, there, and everywhere. (See Fig. 6-26.)

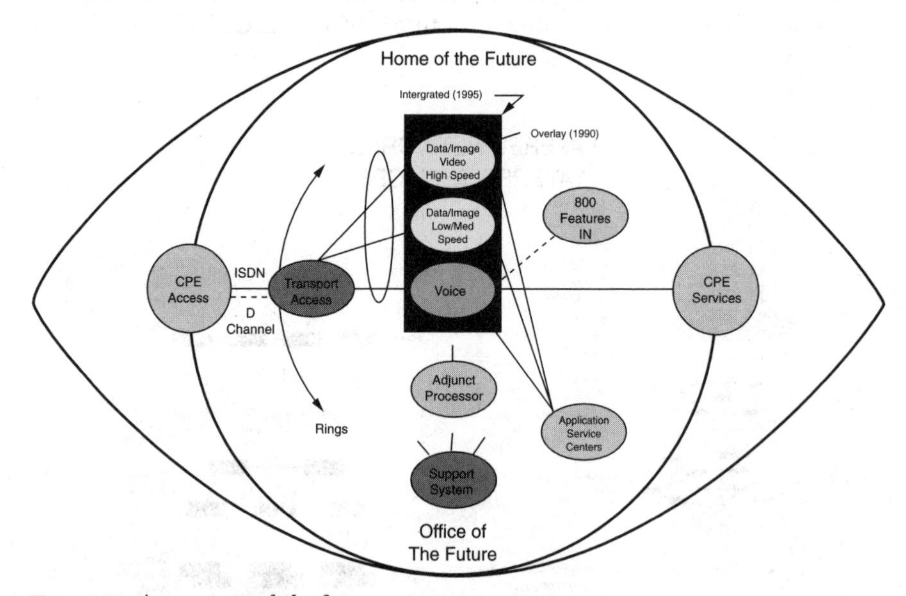

Figure 6-26. An eye toward the future.

Part

4

The Competitive Information Game

Before you can score,
You must first have a goal!

Chapter

7

The Challenges and Strategies of the New Information Millennium

Today, our society is becoming more and more complex as industry becomes more global and more interconnected. Hence, we find communications becoming more and more essential to help us manage our time more successfully and become more productive. It is indeed a time of more—the needs to communicate have literally exploded! Growth is increasing exponentially as we enable not only the ear to hear, but the eye to see and the mind to think. Computers have left the accounting arena, in the form of personal computers, to enter every aspect of our lives. We see them in the checkout aisles of our grocery store, or when we purchase parts for our automobiles. Computers monitor patients in the intensive care units of our hospitals and aid in the instruction of our children.

In the 1990s, applications will expand and interconnect, first vertically within each industry, and then horizontally across industries. In the past this has taken place privately via LANs, MANs, and WANs (local area networks, metropolitan area networks, and wide area networks) using point-to-point transport in the form of buses or rings. However, many of these data users have indicated their need for an alternative new switched public information network that is more versatile, standard, ubiquitous, and cost-effective to interconnect their disparate systems. Thus, they will no longer need to develop and maintain their own complex networks using leased lines, microwave, or satellite medians to internetwork their separate networks. Unless, of course, they want to do it for some reason other than convenience, performance, or cost, such as privacy and security. However, even these issues need to be further addressed by the new public data networks.

As noted, in the early 1920s, New York City had many separate telephone companies that did not interconnect. This required businesses to have numerous separate phones or message-intercept operators in order to talk to their customers. Seventy years later, we do not need to regress to this form of inhibiting communication. As we enter the next millennium, we will be entering the Information Era, with its many challenges and rewards. What we do over the next twenty years, more specifically the next ten, will be critical to our ability to successfully (or not so successfully) firmly position the United States as a major participant in the global information marketplace.

In the past, for their voice networks, the BOCs stressed quality, integrity, flexibility, and financial responsiveness to stockholders. The marketplace has subsequently become extremely complex with the entrance of new data and video opportunities. Hence, it is now essential to translate the original voice network's noble objectives into viable information networks, products, and services that will make the United States an acknowledged leader in the global information industry. In reviewing customer needs and noting the rapid advances in technology, our planning challenge of the late 1990s is to turn today's technical possibilities into tomorrow's products. To do this, both public and private providers and suppliers need to have a clearer, more focused vision of where they want to go, and an agreed understanding on how they are to get there. There is a need to establish the correct infrastructure to enable U.S. firms to successfully participate in this exciting information arena.

To help focus this vision and obtain this supporting infrastructure with its new revenue-producing services, twelve strategic challenges (goals) are identified that need to be achieved by the end of this century, the end of this millennium, in order to be successful in the next. (See Fig. 7-1.)

Our First Challenge

Providers and suppliers need to establish ISDN access interfaces to numerous information services. ISDN has been called the Integrated Services Digital Network. From a marketing perspective, it should have been named the Information Services Delivery Network. It provides seven levels of interconnect standards over the customer-network interface to enable the movement of voice, data, image, and video information to and from users' terminals, be they telephones, televisions, computers, environmental control systems, or workstations.

Initially, ISDN interfaces will be used as a vehicle to access the new digital public data networks, both low and medium speed, over its narrowband ISDN interfaces of 64,000 bits per second (basic rate) up to 1.5 million bits per second (primary rate). Later in the 2000s, multiples of 50 megabits per second of data, up to ten or so gigabits, will be

Figure 7-1. The game.

the norm. These superfast, high-speed transport interfaces will be provided by broadband ISDN.

Therefore, we need to aggressively provide these initial low-to-medium-speed ISDN access interfaces to every data user within the United States over the 1990s. This access must be priced to encourage usage and growth. Seven out of ten data users today will be satisfied by narrowband BRI ISDN, and two out of the remaining three with wideband

PRI-ISDN; but as usage and maturity increase, all will desire more and more data manipulation and presentation capability, as well as transport transparency. To meet these expanding needs, broadband ISDN interfaces should be readily available by the year 2000. We have seen the cycle of computer power–computer usage–greater computer power–greater computer usage continue until current computer capabilities and usage are doubling every 18 months. Similarly, private communication network capabilities have grown from 1 million to 100 million bits per second. So it will be with new distributed information services. Their growth must be fostered by providing private networks with wideband/broadband conduit to publicly facilitate these new information services via the new frame relay/SMDS/FDDI/B-ISDN access interfaces.

Challenge Two

Public network providers need to continue their U.S. digital infrastructure upgrade program by utilizing current host/remote (or sometimes called base/satellite) distributed switching technology in the rural/suburban environment. Here, a digital host (base) unit can be placed at the county seat to upgrade at the same time 15 to 20 small towns in the vicinity using remote satellite units. Maintenance and support operations can be automated using these digital capabilities. Where appropriate, digital bases can be collocated next to analog systems in urban environments to enable business centers to access digital ISDN centrex-type services for new services such as voice mail.

Similarly, they can access digital data networks by using the n-ISDN data interfaces provided from these remote switch units. As part of this upgrade, new signaling systems such as International CCITT SS7 Common Channel Signaling and CCITT ISDN "D" Channel Signaling will not only speed the calls through the network, but also will enable numerous new services based on calling-party identification. These services include: selected call transfer, call screening, selected messages, selected intercept, delayed delivery or reroute, as well as new phones and terminals that inform the customer exactly who is calling by name or number. Alarm monitoring, energy management, and environmental control systems will also utilize the new "D" channel transport services. Terminals of the future will utilize these signaling systems to obtain new features and services from the network for yet unknown applications. Private networks will need standard interfaces to the public network infrastructure to obtain P&P (private/public) internetworking.

Challenge Three

Providers need to overlay a public data network using ISDN interfaces to provide ubiquitous, low (less than 64 Kbps) to medium (64 Kbps to

1.5 Mbps) speed transport in a shared data packet (public) or dedicated (virtual) circuit (private) switched mode of operation. Here we will see the transition from analog to digital for user interfaces to the data network. To stimulate early removal of analog devices from the dial-up voice network and all the delivery problems they incur, change is needed from flat rates to usage-message rates. To encourage this, RBOCs and VANs will need to provide the full spectrum of data transport services from error detection and correction with alternative routing to message broadcast and data polling. This will enable the successful entrance of new services such as datafax, electronic mail, digital music transfer, remote medical treatment, home incarceration, weather monitoring, financial services (debt-card, smart-card, remote checking), and inquiry/response centers (reservations and ticket purchases), as well as work at home.

Challenge Four

Providers and suppliers need to establish a public and private network fiber-deployment topology plan to extend fiber transport capabilities to home and business customers. Once this plan is in place, we can then begin implementation in the late 1990s to "rewire America" with fiber. The completion target for restructuring all major wire centers serving both small and large business communities, as well as selected residential areas, will most likely be the second decade of the new century (2015 is the proposed congressional target for broadband networks deployment). As part of this strategy, transport access switching centers should be positioned closer to the user to provide alternative paths (routes) for survivability (by using new digital cross-connect and nodal switching centers that provide network management, and remote transmission testing), as well as open network architecture/open systems interface points for enabling private-to-public internetworking.

This allows for new services for bridging and routing local area networks together over the public network, as well as for providing gateway access to private MANs, WANs, and POPs (points of presence of national network providers such as MCI, US Sprint, and AT&T). These interconnect business campuses together. In this manner, competitive access nodes are established closer to the customer to set the stage for future regulatory concurrence that will enable full "information networking" of "information services."

Challenge Five

Private and public network providers need to overlay broadband switching service capabilities upon the digital network. The first step is to perform trials using suppliers' technologies to establish the new service requirements needed in the next-generation systems that are destined for the late 1990s urban environment. Here the suppliers and providers must determine (from both a technical and market perspective) the correct level of integration, synchronization, and mixture of bursty variable bit rate (VBR) or long continuous bit rate (CBR) voice, data, and video information that needs to be transported for various types of customers. Specifically, numerous ways must be explored for providing video images for the different applications in order to identify the type(s) of video services that will become the most successful in the new era, so as to channel broadband services in the desired direction.

The trials must sift through the ever-changing array of opportunities: from still-frozen images, to slow frame, to talking heads, as well as to commercial TV, advanced TV, and high-definition TV—we will jump from 9,600 to 64K to 384K, and on to 1.5M, 6.3M, 45M, 135-150M and 600 megabits while using various compression and enhanced imaging transmission techniques. Similarly, we must understand and identify megabit switched data services, which enable images to be transferred from front-end processors to mainframes for large database updates, graphics, simulations, and CAD/CAM/CAE operations. The results of these trials will be better requirements specs that correctly reflect, at the time of open RFQs, the appropriate needs of the data customers as identified in the trials, which would be published and defended in industry conferences and forums.

In this two-step approach, suppliers and providers can first explore these new opportunities in trials and then provide these services to the general public in a realistic and timely fashion. As we initiate low-speed and medium-speed services and explore the higher-speed opportunities, we are in actuality establishing a new U.S. customer base in the 1990s that will subsequently be ready to use the greater capabilities of new broadband network services as they become available at the turn of the century.

Challenge Six

Both private and public providers need to overlay value-added application services on top of the basic public transport services. As we consider numerous application opportunities, we see the need for platforms that provide additional work on the communication. These centers can be shared across industries, as they deliver similar services for their different applications. Here both computer and operator-assisted services can be provided for both voice and data messages, as well as image and video. For example, these new application service centers can provide data encryption, specialized routing, code conversions, protocol conversions, delayed delivery, broadcast, enhanced 911, 800 services, polling, sensing, alarm mounting, number translations, transaction processing, data manipulation, image generation, video conferencing, picture generation, file search and retrieval, data storage, operator database searches, and gateway services for a whole spectrum of common applications.

On the other hand, there will be a need for specialized CPE services tailored for specific applications such as internal x-ray imaging programs for radiologists. Hence, there will be two levels of application centers that exist above the network. These service complexes need to be identified and programs launched with enough design lead-time to ensure that the full range of offerings are available in a timely manner. This will stimulate new markets and continue to meet the expanding needs of new information users.

Challenge Seven

Public network providers need to plan and establish new automated network provisioning and support systems services to enable rapid deployment of customer requests and dynamic operational maintenance and administrative control of the communications network. They need to establish a new communications support infrastructure that not only automates functions for network management in the traditional sense, but one that also ensures that the survivability objectives of the most stringent data requirements are achievable. Here, automated customer requests to provide dynamic bandwidth allocation, bandwidth on demand, alternate routing, and extensive error-rate management must be available for voice, data, and video messages. We cannot implement a complex interconnected private/public information network that cannot stand the test of security and survivability, as the world turns to communication instead of transportation.

These support-system complexes will begin to deliver many new operational and administrative services, as new private network services are layered on top of the public offering. In this manner, future layered networks' layered services are supported by an expanding infrastructure, creating a very competitive, feature-rich, interactive environment between user terminals and both private and public networks. This enables these networks to become less restrictive and more transparent to their customers. It will then become the supporting infrastructure that is essential for supporting new competitive layered services offered by numerous providers in this blossoming, competitive information marketplace.

Challenge Eight

We need to use a layered networks' layered services model. As providers layer services to meet specific applications, the users will be accessing both private and public network services, as well as shared and specialized application service centers. This layering of network services can be effectively visualized on a layered networks' layered services model as having identifiable entrances and exits that are clarified by standards as open network architecture access points. We need to establish this reference model. It could be viewed as one that goes from CPE internal networks (layer 1) to network transport-access systems close to the user for interconnecting private systems to public systems (layer 2); to the internal public network call processing, routing, and control systems (layer 3); to the application service centers, both shared and specialized (layer 4); and finally to the internal CPE processing systems (layer 5), which complete the work on the call. This type of structure enables work to be quickly localized to one level or another to clearly differentiate who does what, where, and how. This is essential if we are not to have overlapping products providing overlapping services. Similarly, it is economically prudent to resist having a proliferation of application service centers performing similar functions or duplicating work performed by lower layers of the network, as the layered model extends to higher-layer offerings (layers 6-10).

Challenge Nine

We need new pricing tariffs and regulatory actions to establish the communication information infrastructure to support a competitive voice, data, and video marketplace. We need a structure in which the public and private offerings can coexist and grow. We must first take the time to understand the complexities of establishing integrated layered networks' services that enable interconnection of private and public offerings.

As the information marketplace demands more and more internetworking, interprocessing, and interservices for both the large and small business user, the home, government, and educators, we see the need for more ubiquitous offerings for both urban and rural customers. Hence, balance and scope are necessary. Application solution services cannot be limited by artificial restraints that make its achievement too complex, too expensive, or not attainable. Here, as indicated in the "critical connections" report to Congress by its Congressional Office of Technology Assessment, a realistic understanding is needed in order for artificial boundaries to be erased. "We must obtain the correct infrastructure that ensures that a strategic assent of the United States is established. This is an ubiquitous public communications network that supports the complete spectrum of voice, data and video Information Services to all the people in the United States, not just a selected few."

Challenge Ten

We need to provide global information services. Our industries are not limited to the boundaries of the United States. The offices of U.S. industries exist throughout the world. As foreign mergers, partnerships, and acquisitions continue to occur, the boundaries and differences between foreign and U.S. firms are becoming a blur, as the marketplace becomes more and more global. The next millennium will be the era of economic expansion and interconnection. Hence, value added networks (VANs) will extend over the globe, tying regions, countries, and continents together. As shifts occur from heavy equipment production-line industrialization to more software, information-based industries, locations for the workplace can become quite remote. Hence, distance is no longer a limiting or restraining barrier. This global information marketplace will require our participation in the form of interactive information exchange and utilization. Offerings provided within one region will need to be interconnected and available throughout the global information society (e.g., the MCI/BT merger).

Challenge Eleven

We need to create a formidable, competitive workforce. We cannot achieve the previous ten challenges without having knowledgeable, dedicated people. We need to educate our U.S. workforce in order for them to participate in this exciting information marketplace. Customer requirements can no longer go unanswered due to resistance, ignorance, noninterest, or lack of understanding. We must embrace new technologies to offer new services. Broadband ISDN and narrowband ISDN technological possibilities need to be understood and applied to market opportunities. No longer does "Ma Bell" give her "baby bells" the selected product of time to operate. RBOCs and independents with their suppliers must do their own thinking and planning. This means that long-term, five-year-lead-time products need to be identified and launched in order to be available in time for the marketplace. To achieve this, providers and suppliers need to use a planning process that supports preprogram, preproject, preproduct participation by people from every aspect of the industry. The providers must formulate long-term relationships with the suppliers, working with their customers, the users, to identify, modify, and specify the desired features and services that meet the users' complete communication-information needs of today and tomorrow.

The Twelfth Challenge

We, both public and private providers and suppliers, need long-term and short-term financial goals. The information marketplace will indeed become more and more complex, expensive and competitive. The rewards will be great for the successful players. We need to establish a long-term relationship with our financial stockholders to enable us to establish the desired supportive infrastructure that will foster an ever-growing array of rewarding services. On the other hand, we must continue to meet the realistic expectations for today's return on investment objectives. We will need to turn to creative financial partnerships with a new breed of shareholders to achieve longer-term, higher-reward ventures in order to put in place both the shared and specialized layers of new information network services.

Here are our challenges! The result is to create a new, growing, multilayered network service structure that enables internetworking, interprocessing, and interservices from multiple providers using multiple products from multiple suppliers—to meet our many customers' expanding expectations and unique service applications. This is the challenge of the 1990s—the challenge of the information marketplace!

Strategies and Plays to Win the Information Game

So, what do we do? A more detailed look at how to achieve these challenges is needed in terms of specific strategies and plans for future marketing and technical endeavors. Therefore, the following overview describes a general program in terms of specific technologies providing specific services to specific users.

To correctly position the United States, as noted earlier, and achieve a leadership role in this period of internal change, we need to have a detailed, clear vision denoting our key strategies for being successful. The industry has had a lot of activity over the past four years, but many sources have indicated that no one has stepped forward with a clear, innovative vision of where they are going and how they wish to get there. We need a vision for others to follow (see Chapter 9).

A national plan of action is needed. It should be based upon the key strategies and denote how we should play the game from a technological, regulatory/nonregulatory, market, and management perspective. Together, these underlying strategies and the plan of action should help us introduce the right type of networks, products, and services at the right time to allow the transition period required to encourage user acceptance and growth. As a starting point, let's review a brief description of the following ten strategies. (See Table 7-1.)

ISDN services strategy

Currently, "ISDN" is not considered "IS" but only "IDN," where "IS" means "Integrated Services," and "IDN" means "Integrated Digital Networks." We must begin to realize that ISDN really means

Table 7-1 Ten strategies

Market Strategy (M)	Technical Strategy (T)
1. ISDN service strategy	–ISDN interface strategy
2. Data services strategy	–Data network overlay strategy
3. Information pricing strategy	–Data trans/processing strategy
4. Information access strategy	–Gateway/IN strategy
5. Shared services strategy	–Wideband transport strategy
6. Urban service strategy	–Metro switching strategy
7. Res/bus/gov applications strategy	–Information switching strategy
8. Service management strategy	–Network management strategy
9. Competitive/monopolistic strategy	–Private/public network strategy
10. Integrated network services strategy	–Layered networks strategy

"Integrated Networks' Integrated Services" (INIS). As we begin the long journey from simply providing ISDN access, we must develop a layered set of services that are supported by the OSI model to provide enhanced throughput, manipulation, and processing, as information requires more and more internetworking, interprocessing, and interservices. These services must track those being applied for the private industry, as LANs and MANs become interconnected to enable file transfer and manipulation. Therefore, ISDN will mean basic, primary access, as well as H0, H11, H3, and H4 broadband access, but it will also mean layered protocols and code conversions, encryption, security, and reliability-type transport services. Later, ISDN will come to mean file transfer and presentation mechanisms.

ISDN interface strategy

Note that data networks can be built without ISDN, now that 802 standards for LANs/MANs transport are defined without ISDN. But even the Defense Department now requires an ISDN interface strategy to provide a building-block growth of ISDN access to the public facilities. This access must go from the low 2B + D 64K interface to the 2.2-50 gigabit range. This strategy must be for the home as well as the business so that a home 2B + D digital channel is available to enable a complete digital transfer of a radiologist's information from the hospital to the home. We need an ISDN deployment strategy for fiber to the home as well as to the hospital, based upon providing medium-speed data rates to the home as well as wideband dial-up HDTV movies. Similarly, intelligent building and not so intelligent building in the business community needs to be interconnected, using baseband (less than 192 Kbps), wideband (144 Kbps-1.5 Mbps-45 Mbps), and broadband (50+ Mbps) facilities.

Data services strategy

This market-based strategy recognizes the current computer needs for generating, transmitting, processing, manipulating, and presenting information to the medical, legal, manufacturing, educational (home, school, and work), state, federal, and defense users. "Data communications" and "computer networking" are sometimes called "data C&C networking." More specifically, AT&T's thrust is called "data networking." Even Bellcore's new jargon is now "information networking." This has become a major endeavor of the 1990s, as sensing (home incarceration), polling (alarm monitoring), and database access and information exchange become switched rather than point-to-point. Hence, a family of data-handling services needs to be available, services such as protocol conversion, code conversion, network retry, delayed delivery, text storage, record access, file manipulation, etc., etc.

Data network overlay strategy

This is the data network overlay on the analog and digital voice networks. It enables a packet-circuit-switched data transport using the packet-circuit-handling capabilities of the switch suppliers to be deployed in a string-type manner across cities. It must be provided in an ubiquitous manner in order to enable small business to use the network. Here, Century 21, local drug stores, retail clothing stores, large banks, and hospitals share the common network. It is a "beginner network" strategy that will be enhanced by later strategies as a backbone level two and three network for data offerings.

Information pricing strategy

As noted in financial analyses on previous data packet pricing, 56K data pricing, T1 tariffs, etc., these services have not proven to be used by the many—only the few. Public providers must reassess their pricing structure. Even usage-sensitive pricing may not be the way to go as they attempt to monitor the "gigabit" throughput of information. Many studies have noted that fixed price levels with a range of throughput capabilities is probably the way to go, with limited or no access charges. This entry-level pricing strategy is designed to encourage growth and bring users to the more advanced services.

Data transmission pricing strategy

Today, 50 percent of a central office is point-to-point. As private networking encourages customers to leave point-to-point for switched facilities, LECs need to construct a transportation facility that has high-quality, high-availability objectives, especially once success takes place. This is key to the data network world; it cannot go down.

Similarly, its omnipresence is another essential aspect. As layers of protocols are available for file management, access, and transfer, the small business users will want to use the public network to interconnect their dissimilar systems. "Content" processing will be necessary (the seventh strategy) as providers begin to provide "packaged" offerings to the home, business, etc. This strategy is to achieve an ubiquitous high-quality data transport network that provides the full seven-layer OSI model internetworking capabilities. It is designed to handle tremendous growth in high-volume traffic. This will require a shift from strategy two, the beginner network, to a growth network in sync with cheap transport pricing for both low-speed kilobits and high-speed gigabits.

Information access strategy

As users internetwork information and access local and remote databases, they need a family of packaged data/image services that enable search, retrieval, manipulation, and presentation of information by interconnecting with gateways and intelligent nodes attached to the network. As more and more inquiry-response tasks of the eighteen major industries become electronic, from patient-record search to theater tickets and seating arrangements, transport mechanisms can be shared across markets as they are packaged into specific application offerings. Note: the set of transport services are lower-layer functions used by higher-layer service functions that are then packaged into specific offerings for user applications by many post-Act local players.

Gateway/intelligent network strategy

As we look at many of the features being provided from attachments to the network, where databases are available for 800 numbers, 900 offerings, shared features, etc., and gateways accessing databases for videotex offerings, we need to combine these strategies into a set of layered gateway offerings, as a building-block support structure. LECs do not need different mechanisms for intelligent nodes using signaling access, while videotex uses direct connection access.

Shared services strategy

As many users wish to access the customer over "the last mile," we need to construct a list of offerings that will be provided by LECs and ISPs to achieve energy management, alarm monitoring, CATV, HDTV, etc., to the home, or patient-information exchange, inventory control, library research, common-community information exchange, etc., to the business world. The result will be a list of common access/transport/manipulating/processing providers' services that could share wideband/broadband transport facilities to achieve closer and closer access to the user. As their services are interexchanged and internetworked, both the public and private sectors will benefit.

Wideband /broadband transport strategy

The access-local-loop facilities need to be restructured into multinode, variable-channel, higher-capacity, local-transport loops that enable the full range of services, from low-speed data to high-speed, high-definition TV available to both the home and small business communities. This has been described as a *rings on rings strategy*. It is key to the growth of the public information network and should be reflected in the local plant topology and deployment plans.

Urban service strategy

Here, voice, data, and video transport and processing services are integrated into a family of service offerings with initial availability in the urban business community. This arena has a great potential for internetworking and interprocessing of multiple provider services. Here the ISDN international routing/addressing occurs. Here the private-to-public addressing, translation, access, speed billing, record keeping, and file managements will occur to enable the higher levels of urban information exchange.

Metro switching strategy

The next-generation switch will be deployed as a central base to provide integrated voice, data, and video transport and control using fast packet-switching and photonic-switching capabilities. It will augment and replace the beginner start-up network of strategy one, which can be reapplied to the rural environment, time-phased for their delayed growth period. The metro switches planning and development will take 5-6 years in order to be available in the early 2000s. Current digital voice systems will take on additional switching matrices until the new technology is available. Some suppliers say their current systems will simply grow and change shape, slowly discarding the original. Others believe in a new superswitch with distributed switching structure. Whenever, a new switch will be needed for new broadband services.

Residential/business/governmental application strategy

Residential markets need to define specific data (information) network applications such as PC-to-PC, home-to-videotex, sensing (home incarceration), polling (patient monitoring, energy management), high-definition television, shopping at home, etc., to denote low-speed/high-speed facility strategies for the home. Institutions of business, education, and government must do the same now that security is a new attribute in addition to speed and quality. They have become new issues, especially for law enforcement, banks, defense, etc. These strategies should indicate what services will be available on the specialized networks, as well as the dependent strategies that require services from the other layers of the ten-layer information network service model. Finally, the interdependent service features must indicate cross-industry functionality. The results will show how the new centrex, info switch, or feature switch can function with shared and private nodal-point systems, as switch modules are remoted to locate on customers' premises.

Information switching strategy

The next-generation common information services switches will replace centrex and compete with private PBXs for shared information services on a public network. They will address the private shared-tenant, home, small business, shopping center, etc., environments. This switch is a key service switch strategy. It is needed as a first-generation and second-generation offering as both n-ISDN and B-ISDN for baseband and broadband offerings.

Service management strategy

Who will be "the keepers of the network" as we internetwork many new private and public networks, interconnecting many new services? Here, service management gives the customer, the private network manager, and the public provider access to network operations. It will be an essential network element as data transport networks are expanded and rerouted, depending on traffic needs of the user and time of day network operations. The set of network features and services that both private business and government wish the LECs to provide are contained in the lower layers of the INS information network services (LNLS) model. RBOCs will also use them on a private basis in their higher ASC layers. Note: the higher-layer services cannot exist without the lower network services. If these services are brought to the high level, they will be more expensive to provide ubiquitously.

Network management strategy

A multilayered network management strategy must provide access and control for not only the public network voice infrastructure, but also for data and video. Also, the ability to administer and control private networks will encourage the private networks to move from point-to-point to switched, shared facilities, especially as more private-to-private internetworking takes place through the public network. Hence, there is a need for a layered network management strategy between exchange, interexchange, special and private carriers. Without it, we are simply building a forthcoming disaster. It's as simple as that!

Competitive/monopolistic strategy

There may be a need for a shared local transport strategy to open up competitive "content" ventures by the LECs in the higher layers of the INS model. Then, the individual providers can package their more specific application-oriented services packages to meet the full spectrum of user needs without having to form expensive partnerships, which increase the costs and make the offerings less attractive.

Private/public network strategy

The internetworking of LANs, MANs, and WANs with the public networks is key to long-term success, as noted in the INS model. Here, a full range of standards need to be established by the LECs that lead the way in this arena, especially the RBOC who quickly leads, not delays, in opening up the local monopoly. Note: this lower layer (two) is needed even if the shared option is not elected.

Integrated network services strategy

As noted in the challenges, a multilayer service model needs to be developed to show how each of the above layers will use services from the lower layer. Here, nested functions are used by both the residential market and the business community. They will reside at the lower level to be shared across the higher level. For example, gateway mechanisms will exist that are used by both residential and business customers to access videotex services, but become part of each application package, having different prices. (See Figs. 6-13 and 6-14.)

Layered networks strategy

The ten-layer integrated networks' integrated services model enables shared, private, advanced, and specialized networks to be deployed in both ubiquitous and selected environments, enabling multiple providers to communicate more effectively throughout the information marketplace. The information network model denotes new switching transport systems that meet the needs of the advanced services and special services markets. Note how the info switch is used to provide residential/business nodal-point access to the advanced service market. Note also how the ring switch can be deployed to provide the base infrastructure for integrating private LANs, MANs, and WANs to the public network. Finally, note how communication service centers and network management centers can be deployed to meet the needs of all the layered networks. On this vehicle, the layers of information services can be delivered to user applications in both the private and public arena, across the residential/business/state/educational/federal information marketplace.

Players must carefully consider these strategies and apply them to their area of the marketplace. Where strategies do not exactly meet players' constraints or applications, they could modify, change, and enhance these strategies. However, it is essential to end up with a set of strategies that all players will see, understand, and implement. The strategies should clearly denote cross networking requirements and interdependence between market sectors and segments.

Perspective: The Challenges

If the LECs do not establish public data networks, then what? So what? We have seen the military and the educational firms build their 465-L and ARPANET networks of the 1960s. In the early 1970s data was ½ of 1 percent of the common carriers' business, so they elected to not add data-handling capabilities to their analog switching systems. Firms like DATRAN attempted to switch 4,800 bits-per-second commercial data. Digital occurred in the late 1970s. Many did not realize that the reason for going to digital (digitized voice) was to move voice and data down the same pipe and to switch data as well. By the 1980s, data had grown to 3 percent of the RBOC's business; still, its needs were ignored in the construction of T1 digital paths. It was known that data could arrive in a sequence of all fifteen 1s or 0s that would cause a slip in digital framing. It was not until millions of miles of T1 were deployed that special codes were considered to handle data on digital facilities. In the 1986-1987 time frame, clear-channel T1 became an issue.

Some say wait for the next wave of technology, broadband. This will not be widely deployed for the general public until the beginning of the new century and on through the 2010 timeframe. There will indeed be an opportunity for the major players in broadband in the late 1990s, but, again, this will be on a selected basis. This will be an overkill for many of the twenty-six data user types, who simply need to move small-rate to medium-rate information across the network today. In watching the history of computer usage growth, the early users did not generate tons of information. Usage begat usage. So it will be with the data world. We need first to interconnect the world to enable information exchange, multisystem database access, and multisystem problem solving. In this manner, we advance the computer evolution from large mainframes to PCs, to interconnected PCs, to PC to mainframe, to mainframe to mainframe, as we enter the world of distributed processing and proceed to intelligent processing and imaging.

So what will happen if the RBOC doesn't build the public data network? Well, data is only 3 percent of the business because it is going private. New alternative private networks will continue to be established.

Some switch manufacturers may move into the transport segment of the business, if the RBOCs don't use their more data-oriented products. If not, someone will. Within the late 1990s, the Information Age will blossom somewhere around the globe. Then, once successful there, it will be "cloned" elsewhere—especially here in the United States. This can occur in new urban information companies (local VANs) while the traditional telephone company remains "the phone company," perhaps with decreasing revenues.

What if, then what?

What if the LECs do go ahead and select specific areas for deploying ISDN data networks? What happens? The answer is quite simple. Without needing to deploy fiber, by simply extending the capabilities of existing twisted-pair wires, the local exchange carrier has now created a new second service with an entirely new expanding group of customers. The local exchange carrier does this instead of trying to take an existing customer base that is growing at only 4 to 6 percent for voice lines and selling them second lines or losing them to internal systems. There is a new bottomless pit of potential users. Once the infrastructure is established, then the RBOCs' market units will be able to overlay layer-four and layer-five application service centers, as the information game becomes one of services on top of services. As noted in a radiologists' ISDN trial, once the radiologists transported x-rays at 64 Kbps or 128 Kbps, the real issue became the moving of other entities besides their x-rays, such as patient files. Then it becomes an issue of storing x-rays and files, etc., etc.

Using the layered house model, once the garage is established, we will have a great deal of enjoyment designing and constructing the rooms on top of the garage and adapting them to our particular lifestyle and applications, be they bedrooms, bathrooms, offices, studies, lofts, or whatever.

Thus we are indeed at a crossroads in time. It's a timely opportunity to play the information game; the LECs didn't, for various reasons, in 1965 or 1995. If they don't step up to the plate now, perhaps the third time around they will be out. It may be time for another player to play the information game—in another place, in another time. (See Fig. 7-2.)

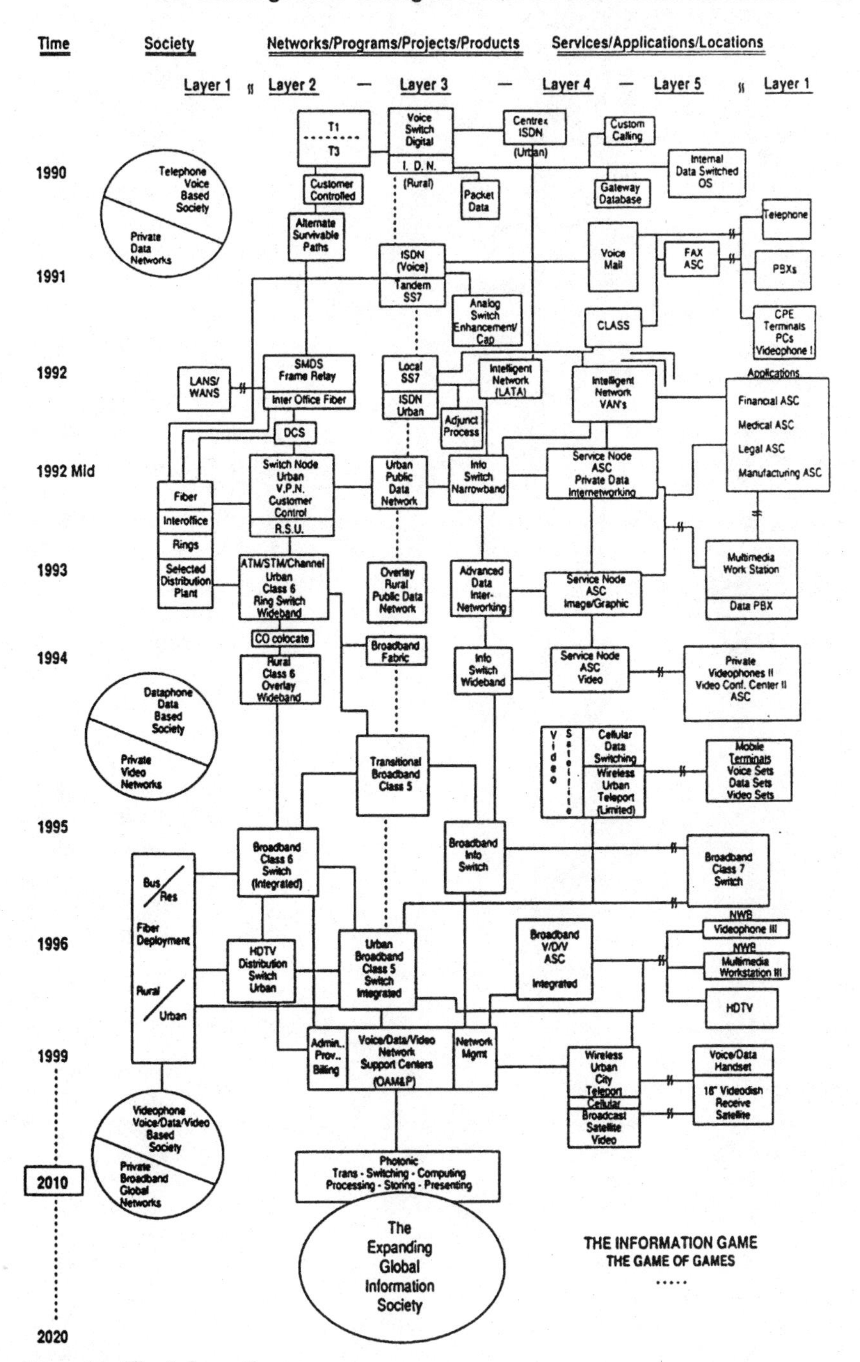

Figure 7-2. The information game arena.

Chapter

8

The Applications

In the ISDN market, everything works together. It's the chicken-and-the-egg puzzle; you need an Integrated Services Digital Network to provide the services, yet you must be able to prove the need for the services before you get an ISDN.

ISDN Market

Many people today are questioning ISDN (Integrated Services Digital Network). What is it? What will it cost? What services will it provide? When will it be available? Who will pay for it? Why do anything now? Why not wait and see what happens? These questions and attitudes are quite similar to those expressed when digital networks and services were first introduced in the 1975-1976 timeframe. There were those who resisted the change and hung on to analog technology until the mid-1980s. Many of those people have been left behind; they are no longer considered leaders or participants in the competitive arena. Others have embraced change and learned from each new transition to build a knowledge base to play the game. Remember the expression, "leapfrogging" technology.

Several large companies did not participate in the initial development of digital, only to find when they entered the new technology and services arena that they could neither perceive nor appreciate the subtleties and shifts of emphasis that the more seasoned players had developed. This made playing the game much harder and more risky. Even if we had been playing the game, to survive in this intense arena we need to pause and reflect on what's happening, where are we going, where we want to go and how we can get there. It's time to think. It's time to learn how to successfully survive the transition to the Information Age—providing, supplying and using the right services from the right products via the right network at the right time.

A service, no matter how attractive, will not become a major success unless it truly meets a human need in a manner that enables individual survival and success. With this in mind, and remembering Maslow's hierarchy, the AT&T Hawthorne experiment, and other social science research projects and experiments, let's pursue the ISDN information marketplace by first recognizing that there are certain basic human needs, and then noting the human operational needs that address the basic needs. Next, let's describe our current society's functional structures in which humans and machines participate. Finally, we can reference industry-related tasks for which information operations are performed. Recognizing that different individual attributes and characteristics encourage the "utilization of" or "resistance to" new information, personalities must be factored into the Information Age solution, since a full range of "people types" exist who are as unique as the "Archie Bunker" or the "Alex Keaton" type of television characters.

From this perspective, we can review what's happening with current telecommunication products and services and then look at where we want to be in terms of future information features and services. However, also realizing that many of these offerings are visualized separately as a technical possibility or a market opportunity, it's essential from a business strategy point of view to translate these possibilities/opportunities into recognizable customer application service packages. These consist of packaged features that are offered in specific geographical areas, priced to capture a specific percentage of a specific aspect of the information marketplace over a specific timeframe.

Therefore, there is an orderly progression as we shift from a technology-based industry view, where technology today can accomplish almost anything economically, to an industry view, having more of a market-based management emphasis, providing only networks and products that offer acceptable features and services that meet identified customer needs and aspirations.

When discussing ISDN, it's not sufficient to discuss network features and services solely in technical terms; rather, they must be restated in application terms to be more fully appreciated from a business perspective. For example, "home incarceration" is a new application for unloading prisons, an application that draws heavily on telecommunications. It enables minor offenses to be served without endangering the noncriminal offenders. Here, the offenders must remain in the homes without access to the world for a period of 30 days to one year. The participants pay for telecommunications links to a control system that constantly monitors the home. The service could be implemented by any form of technology; however, when technical solutions are discussed, they quickly boil down to sensing, polling, and notification. The incarcerated person wears an armband that generates an electronic signal to identify the person's whereabouts. These signals are detected and transmitted to a central location using a sensing/polling data network message interexchange. Therefore, simply saying we need to offer a data packet network that provides sensing and polling features is not enough to appreciate its potential in the marketplace. This can only be obtained by denoting specific applications.

Layered Services

A network layering model (Fig. 8-1) can be used to show the relationship of private and public networks. These layers can be viewed as having technology-based, market-based services. Layers two to five are noted in Fig. 8-2. Notice that the items in the first column for each of the four categories describe the service in technical terms, whereas those in the second column restate the service from the standpoint of the marketplace.

As systems and services are integrated, much of this interconnecting, internetworking, interprocessing, and interservicing can and will take place outside the public arena. However, the last mile to the customer remains an expensive venture for an individual service provider or interconnect firm, due to the relatively low use of any service other than television or a specialized pay-per-view retrieval system.

Similarly, regulatory pricing requirements that call for access charges for all terminals that interface to the public long distance carriers have inhibited the growth of shared facilities interconnecting

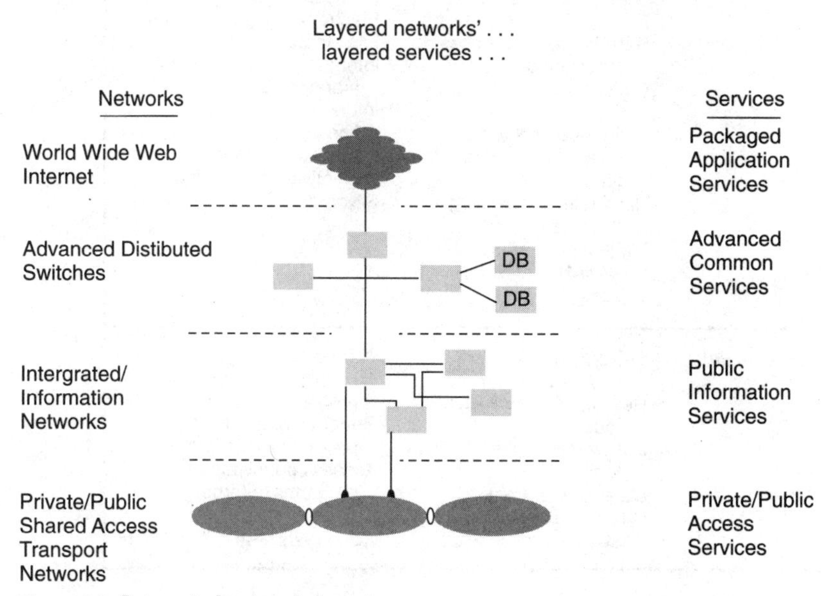

Figure 8-1. Integrated networks' services.

	Technology based	Market based
Shared transport services	Local loop access Shared tenant access LAN-to-LAN routing Private-to-private switched routing Private-to-interex-change access Multi-network interface Sensing, polling & en-cryption Alternate routing Access blockage	Alarm monitoring Shared tenant gateway Interexchange gateway Common data base ac-cess Multi-network menu Delayed delivery Nodal point message code conversion Nodal point processing program Billing
Public services	Basic POTS service Local data packet switching Local wideband video/ data switching Interface to local Data base gateways Interface to voice/data video interexchange carriers ONA interface diagnos-tics Network management	Centrex features Custom calling I, II, III CLASS features Data delayed delivery Error detection/correction Interchange access to voice/data video carriers Polling Billing
Advanced services	Gateway to special networks Gateways to data base Bridges to LANs, MANs Satellite nodal switches Cellular nodal switches Protocol/code conver-sion Inter-network routing Delayed delivery stor-age Distributed processing interfaces Multi-service interacting LAN/MAN/WAN inter-faces	Gateway menus Data base searches Image processing Electronic TV movie rental File storage/retrieval Inter-processing program management Inter-service transforma-tion Inter-system access Inter-private networking Billing
Specialized services	LANs, MANs, WANs service software Residential INFO switch node Business INFO switch node Data base storage (doctors, hospitals, lawyers, radiologists)	Source data bases Transport/storage retrieval programs Generation/processing/ presentation programs Access/control/reformat programs (Internet) Billing World Wide Web

Figure 8-2. Layered services chart.

private networks with the public network. There is still heated discussion concerning call charges applied to private, shared offerings that lease interconnecting facilities, demonstrating that state versus federal regulatory boundary conflicts will continue to be an issue.

Applications

As we consider the expanding set of technology-based and market-based information-handling features and services, we need to step back and consider them in terms of packaged offerings that often use networks supplied by specialized providers. The game (and it is a game) of positioning and playing together to achieve success has indeed become quite complex, as the offerings become more and more interdependent when we progress into the transition phase of the Information Age. We are moving from large, stand-alone mainframes or personal computers to distributed, interconnected processing, where computers and communication come together at every step: generation, transmission; storage, transmission; access, transmission; retrieval, transmission; processing, transmission; manipulation, transmission; and presentation processing.

Unfortunately, there is the American attitude of "let's get on with it," demonstrating our unique spirit of freedom, choice and action that we all know so well. However, without having a playing field that adequately supports our efforts, we have found over the last two decades that in reality we cannot "get on with it." The game has not been played. Services need networks that need products that provide interdependent services. Once a list of services is identified, an attempt to implement specific applications demonstrates the unavailability and nonubiquitous aspects of information handling over the last mile to the customers' premises.

This further demonstrates the need for understanding these potential offerings in terms of the interdependent arena in which they must exist. Only then can we establish the structure needed to support the large playing field on which these applications can develop. We can see this by simply noting a tentative list of interdependent information services applications that could develop over the late 1990s. (See Fig. 8-3.)

Conclusion

Each of these applications satisfies either a basic human need or an operational human need. From these flow charts you might conclude that we can't begin deployment technology until enough user applications are understood, both conceptually and quantitatively, to justify the network/product/service expenditure. This could then lead us to the chicken-and-the-egg situation, where quantitative figures of future

Law Update Service	- A service data base noting legal rulings/decisions that are indexed to various key issues.
Law Review Service	- A data base search firm that will perform not only a search for pertinent legal information, but provide an analysis as well, thereby adding value to content.
Medical History Service	- A search for all pertinent medical information on an individual by accessing numerous data bases throughout the country.
Radiologists Information Interexchange Service	- A network service solution employing a private network within a radiologist's office building which interconnects with other private radiologist networks throughout the city, using the public data transport network.
Patient Archive Retrieval Service	- A service that enables radiologists and other doctors to access the Medical History Service from home and office.
Automobile Collision Service	- A data base that tracks collisions on types of automobiles noting severity, cost to repair and medical damages.
Lawyer Information Service	- A service that enables lawyers to access the Law Update Service, Law Review Service, Automobile Collision Service and, through a secure gateway and with patient permission, the Patient Archive Retrieval Service.

Figure 8-3. Service applications.

endeavors can't be actually understood and justified until the future event actually has taken place. In the past, people asked, "Who would want to talk over the telephone? Who would want to use electricity over gas? Who would want to sit on top of an exploding engine?," etc., etc. Now it's, "Who would want to use such an information service?" These questions can't be answered easily. But many insights become apparent after this form of market application analysis is performed. In doing so, we see the high number of potential voice /data /video applications for the three forms of ISDN networks, once their number and type are projected into the world of narrowband ISDN, wideband ISDN, and broadband ISDN. As future applications take place, it will become quite interesting to see how the multilevel ten-layer network services model, consisting of shared, public, advanced, and special services capabilities, will support all of these future applications. Integrated services will begin to blossom as they become better time-phased with expanding

ISDN capabilities. Once the Telecom arena has become restructured to enable all the players to play a fully competitive information game, their interconnected computer usage will beget more and more usage, causing the information marketplace to grow and mature as we enter the global Information Age. (See Fig. 8-4.)

A review of potential applications in Fig. 8-3 demonstrates the power of information services networks as they build on top of each other across lines of business. These layers of networks within both the local public and private communities will use national networks to reach remote service gateways to access specialized database sources and search services, whereby:

- At home, a radiologist can review the x-rays of a patient in the emergency room.
- The process of looking for a house and closing can be completed in hours rather than weeks.
- A retired engineer in Florida can teach high school algebra to students in Boston.
- From their classroom, art students can view treasures like the National Gallery at their own pace.
- A person interested in buying a car can tap into the databases of all the local automobile dealers to locate a desired model at the lowest price.

Figure 8-4. Information world.

- Business people can communicate by a videophone rather than making a business trip.
- A bank can communicate with its branches and automated teller machines via the public switched network rather than private lines.
- From a college dorm room, a student can have instant and inexpensive access to the 15 million books, 37 million manuscripts and 11 million photographs stored by the Library of Congress or any other database in the world. (See Fig. 8-5.)

It's also interesting to see how transparent ISDN becomes in this analysis. One marketer once said, "I don't care what technology is used to achieve it." Unfortunately, there still exists an "I still don't know (ISDN)" attitude. This limits realization and understanding. Nothing of this interservice magnitude can be achieved without providing many layered services upon a layered network structure-and all must be supported by an ubiquitous public infrastructure. As traffic volume increases, a transition then may be made from narrowband ISDN to broadband ISDN using numerous C&C ISDN higher-level interprocessing protocols. All these internetworking, interprocessing, interservice aspects constitute the reason for having ISDN international standards.

As stated earlier, ISDN is actually a misnomer. It should have been called INIS, (Integrated Networks' Integrated Services), or even called LNLS (Layered Networks' Layered Services), which we will need as we formulate and position ISDN in the information marketplace. (See Fig. 8-6.)

$$\text{Industry Applications} = f\,(\text{Market Segments}\,(\text{Geographic Location}\,(\text{Community of Interest}\ \sum_{L=1}^{4} \text{Services}\,(\text{features}\,(\text{products}\,(\text{price}\,(L)))))))$$

$$\text{Layered Industry Applications} = \sum \text{Industry Applications}$$

Figure 8-5. Layer applications.

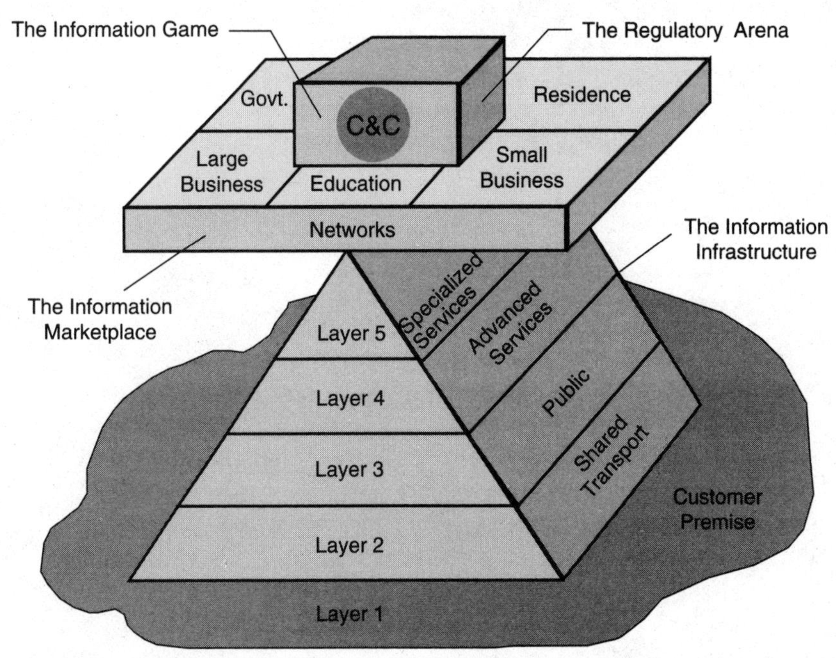

Figure 8-6. The information society.

Part

5

Building America's Future Information Society Today

A man needs a plan, he needs direction.
If he doesn't have this, he has nothing.
A man, like a ship at sea,
might change course many times
in getting to an eventual destination,
but he must always be going some where,
not simply drifting.

LOUIS L'AMOUR

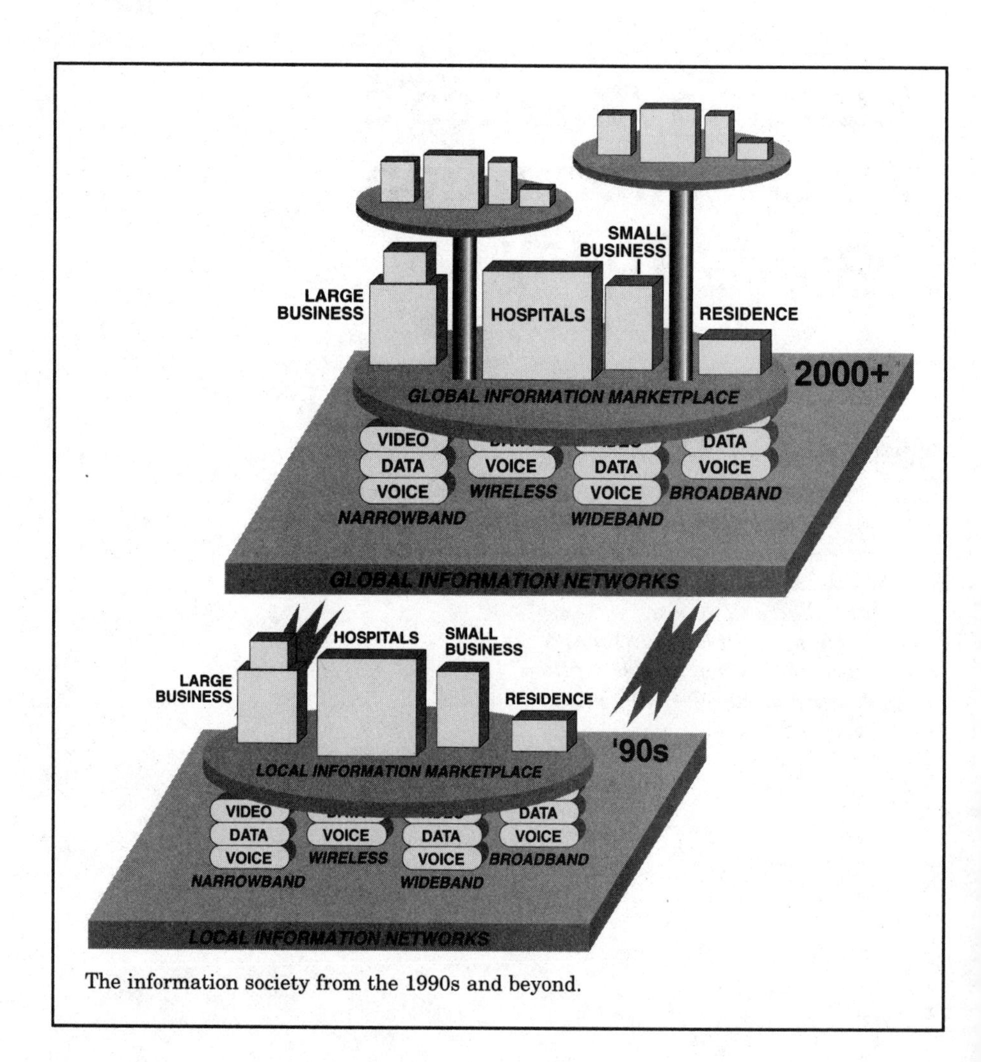

The information society from the 1990s and beyond.

Chapter

9

A Blueprint for Success: A Vision, Plan, and Plan of Action

Tom Bystrzycki

Foreword

Never before in the history of the 100-year-old telecommunications industry has it been positioned to experience such dramatic and rapid change. During the close of the twentieth century, the Telecommunications Act of 1996 has opened up the voice-based communication networks. Intent on fostering change, it will substantially restructure the communications industry to facilitate the transport of fully integrated voice, data, and video information. Highly complex communication networks will merge with highly sophisticated computer applications to formulate and establish the new, exciting information marketplace.

At this point in time, it is essential to pause and formulate a feasible, realistic vision for the future, as well as a reasonable, acceptable plan for the future to achieve this vision. It is equally mandatory to have a well-conceived, balanced plan of action to execute the plan for the future—to establish the proper information networks, with their many services, for today's world as well as tomorrow's. For society to grow and flourish over the next millennium, it is essential to have an economic, universally available, switched public data / video network, enabling "any-to-any" communications over shared facilities, providing access to distributed databases, promoting full interoperability among diverse systems, and facilitating the internetworking of private networks.

With this in mind, this guide will help minimize the current atmosphere of crisis and confusion, which has recently engulfed many providers, suppliers, and users during this intense period of voluminous, complex change. With this identified direction, as well as an appropriate, orderly program to achieve it, we can ensure that each of our endeavors becomes a building block for establishing a competitive telecommunications infrastructure, which will hopefully improve the quality of life for ourselves, our children, and our children's children.

TOM BYSTRZYCKI

INFORMATION TELECOMMUNICATIONS

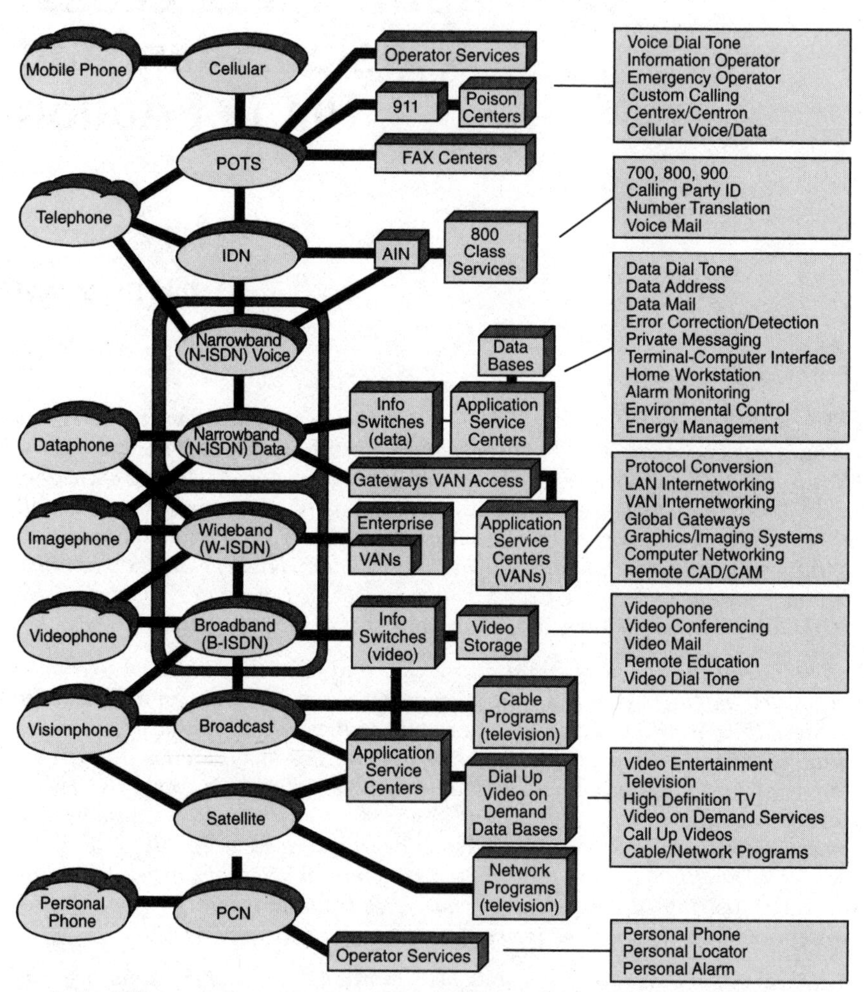

Figure 9-1. This overview portrays the narrowband, wideband, and broadband building blocks that form the centerpiece of the future information network infrastructure, supporting an ever-expanding array of versatile service platforms offering a plethora of new voice, data, and video services.

Vision of the Future

A well-founded vision—one that can last through the decades—must properly address real issues and encompass real needs. So it is with the future of telecommunications and its networks, products, and services. Fortunately, there are many exciting, new technologies in development or applied research from which to choose. However, with these increasingly complex technical possibilities come difficult choices requiring a new level of understanding to properly select the right technology for the right application. With this in mind, let's consider the new users' requirements of the forthcoming information marketplace, so we can offer realistic solutions to users' real needs in orderly, timely phases and stages of achievement.

Needs. It is now quite evident from both a market and network viewpoint that there is a need for a growing building-block support structure (Fig. 9-1) from which to offer various types of new, enhanced voice, data messaging, and interactive video services. This infrastructure (Fig. 9-2) must be robust in nature, ensuring security, privacy, and survivability; enabling interoperability of computers of differing capabilities; standardizing the interconnection of private and public networks from multiple providers; and supporting the need for ever-increasing throughput capacity to transport varying services for numerous applications. These will cover twenty or so functional areas, both distinct and interdependent, from health care to education to manufacturing to banking to retail/wholesale to home services.

Phases and stages. It is also quite evident that numerous market applications require technical solutions covering the entire transport range of narrowband (64 Kbps/128 Kbps), wideband (128 Kbps to 45 Mbps), and broadband (50+ Mbps) offerings. It is equally important to provide new offerings from existing copper plant and expand where economically feasible, on a timely basis, into a fully switched, fully interactive, fiber-based delivery system. Additionally, facilities can be augmented with passive broadcast and fully switched wireless cellular and private communications service (PCS) capabilities.

Hence, there is a need to time-phase these offerings for selective and universal availability, so they become staging platforms that adequately support many versatile services provided by information service providers (ISPs) and enhanced service providers (ESPs). These offerings will also provide access to the various value-added networks (VANs) and interexchange carriers (IXCs). Finally, to ensure continued successful offerings, especially as success begets increased usage, sufficient support systems must be in place to meet desired goals and expectations.

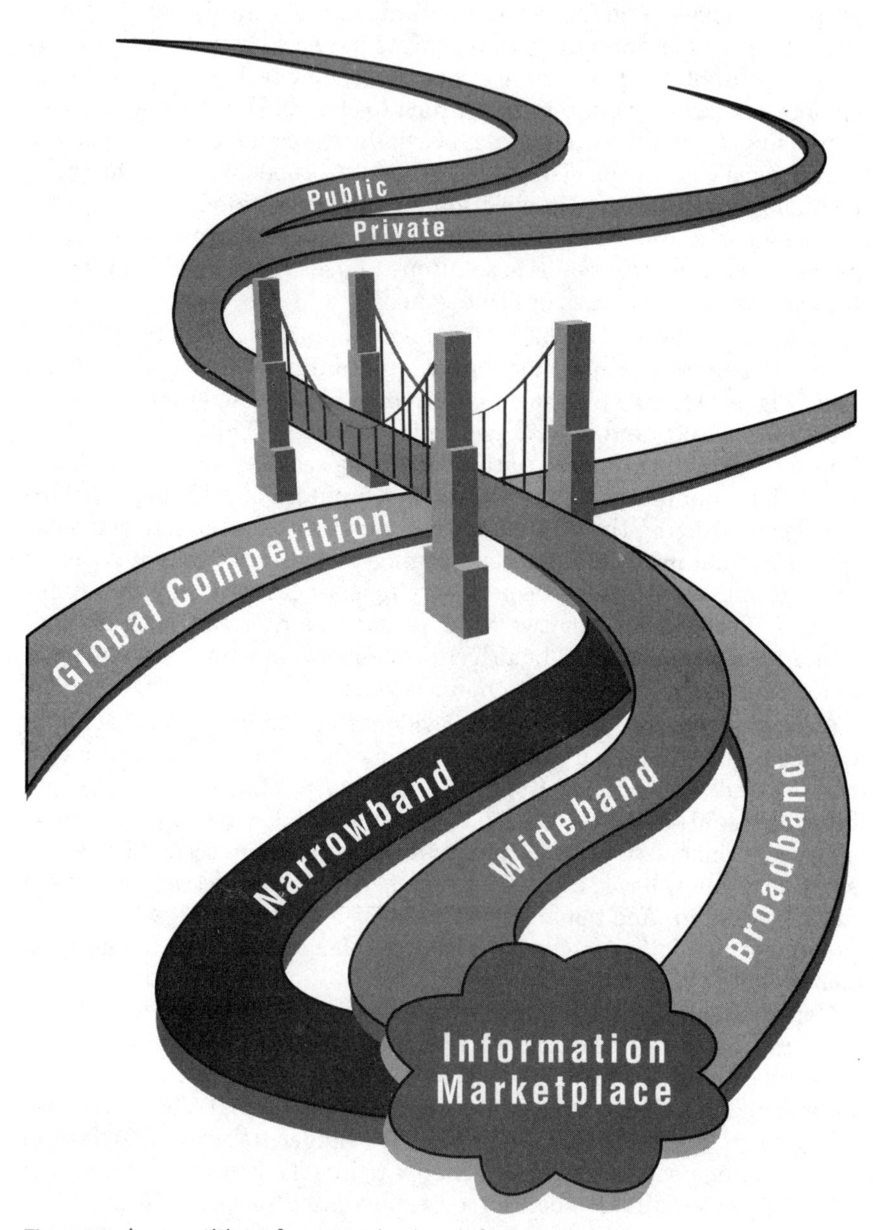

Figure 9-2. A competitive telecommunications infrastructure.

Narrowband

Establishing a public narrowband data network (Figs. 9-3 and 9-4) is the first step—an essential step—to take advantage of existing plant facilities to address 70 to 80 percent of today's information exchange application needs. Besides establishing a public digital voice network, it is interesting to see how several different types of narrowband data transport roads can be developed to meet the needs of differing transport vehicles. In pursuing the narrowband information highway, there are several distinct options with specific basic network feature packages. These offerings can be enhanced and expanded by advanced intelligent network platforms and applications service centers. Hence, with this step, telecommunications industry providers can successfully enable their customers to enter the global information marketplace.

Wideband

The wideband information network (Figs. 9-5 and 9-6) needs to be initially deployed as an overlay information offering for interactive voice, data, image, graphic, and video traffic—operating as fully switched multiples of 64,000-bps channels, up to T3 rates of 45 Mbps. These transport offerings will eventually become subrates of the SONET OC-1 (50 Mbps) broadband information network. The wideband network will initially facilitate existing LAN-to-LAN internetwork traffic but will primarily provide fully switched, interactive facilities between workstations, mainframes, and databases, as well as enable supercomputer-to-supercomputer interconnect traffic and interactive wideband videophone and videoconference-type communications. The offering will be a prelude to the forthcoming fully interactive, fully deployed broadband information network.

Broadband

The public broadband information network addresses a neverending, expanding family of customer needs (Fig. 9-7) as we move into the world of fully interactive voice, data, and video services. Here supercomputers internetwork among themselves and access distributed databases, as well as enable sophisticated workstations to perform complex analyses, presenting solutions in the form of video images and graphic displays. Videophone conversations and videoconferences will require fully switched, interactive, high-definition, high-resolution capabilities. Similarly, multimedia systems will require more and more information handling as these versatile tools are used in numerous applications throughout the marketplace. On these endeavors, broadcast capabilities will be overlaid to provide entertainment-type services. So, all in all, there is a need for an ever-expanding, interactive, robust, survivable, secure public broadband information network that facilitates the transition of today's industrial-revolution-based twentieth-century society into the information-based society of the twenty-first century.

Plan for the Future

There are multiple paths (Fig. 9-8) to pursue and tasks to perform to achieve the vision for the future as we venture forth along the following pathways to the future information marketplace:

Path one

Maximize existing narrowband network voice offerings by using new ISDN (access), SS7 (signaling), and IDN (integrated digital network) technologies to upgrade the current infrastructure to provide expanded offerings, such as voice mail; calling party ID; CLASS services; selected call: transfer, blockage, message routing, and priority override; intelligent network database access services (800 and 900); and special audiotext services; as well as fast provisioning, number change, and support-type services over the expanded . . .

. . .Public Narrowband Voice Network . . .

Path two

Establish a new family of narrowband data services over the existing copper facilities, using the ISDN data interface to enable both the circuit- and packet-switched movement of data from terminal to computer, computer to computer, and terminal to terminal over the new . . .

. . .Public Narrowband Data Network . . .

Path three

Foster the movement of wideband information between private networks over public networks by providing new, higher-speed data transport services such as frame relay, SMDS, dynamic bandwidth on demand, survivable transport, improved error rates, access to multiple national carriers, and international value-added networks. These transport services provide an exciting group of wideband services across areawide networks that provide a variable number of 64-Kbps channels up to T1/T3 rates of 1.5/45 Mbps for data, imaging, and graphic display of information. This will be achieved by expanding offerings to provide these services over new fiber-based networks with public broadband access nodes using self-healing, survivable fiber deployment topologies with new switching centers located closer to the customer, for applications such as x-rays for the medical industry, stock forecasting for the securities industry, CAD/CAM for manufacturing groups, and for WANs that internetwork LANs over the versatile, dynamic . . .

. . .Public Wideband Information Network . . .

Path four

Achieve new, revolutionary broadband services, such as high-resolution video services that enable visual imaging, videophone, desktop conferencing, high-quality graphic display, and computer-to-computer data exchange, as well as video displays for education, media events, and image archives services over a revolutionary, new, fiber-based voice, data, image, video switched network with automated support capabilities, called . . .

. . . The Public Broadband Information Network . . .

Path five

Overlay passive **broadcast offerings** on the fully switched, fully interactive broadband network, enabling entertainment, dial-a-video movies, shopping at home, games, and many other services.

Path six

Augment wireline with **wireless services** for cellular phone, personal communications services (PCS), wireless data, and wireless PBX applications, enhanced with access to AIN facilities.

Path seven

Establish **extended or enhanced services** in the nonregulated community via an overlay of information systems, such as info switches, that provide direct-access databases and advanced data-handling services or via gateway access to stand-alone application service centers for special services such as E-mail, as well as access to CPE service centers for specialized or unique data-based services such as poison control centers, patient records, and so on.

In conclusion

There are a whole host of new, exciting technical possibilities and market opportunities, but it is essential that they are provided on an overall basis (locally, regionally, nationally, and internationally) to ensure growth. These capabilities will enable firms to interexchange information locally and globally in a cost-effective manner, thereby reducing various energy-consuming modes of operation such as physical travel. This change achieves increased productivity and personal effectiveness not only in the business place, but also in the home. Here, personal issues such as quality of life, education, health, social interaction, and entertainment are enhanced as highway congestion reduces, queues are eliminated, rural homesteads blossom, etc. However, for these offerings to be successful, there is a need for an expanded and highly functional support-system infrastructure that enables timely provisioning, testing, maintenance, and administration of the transport and delivery of the new voice, audio, data, image, graphic, text, and video information for the 1990s, for the next century, for the next millennium—**the Telecommunications Information Millennium.**

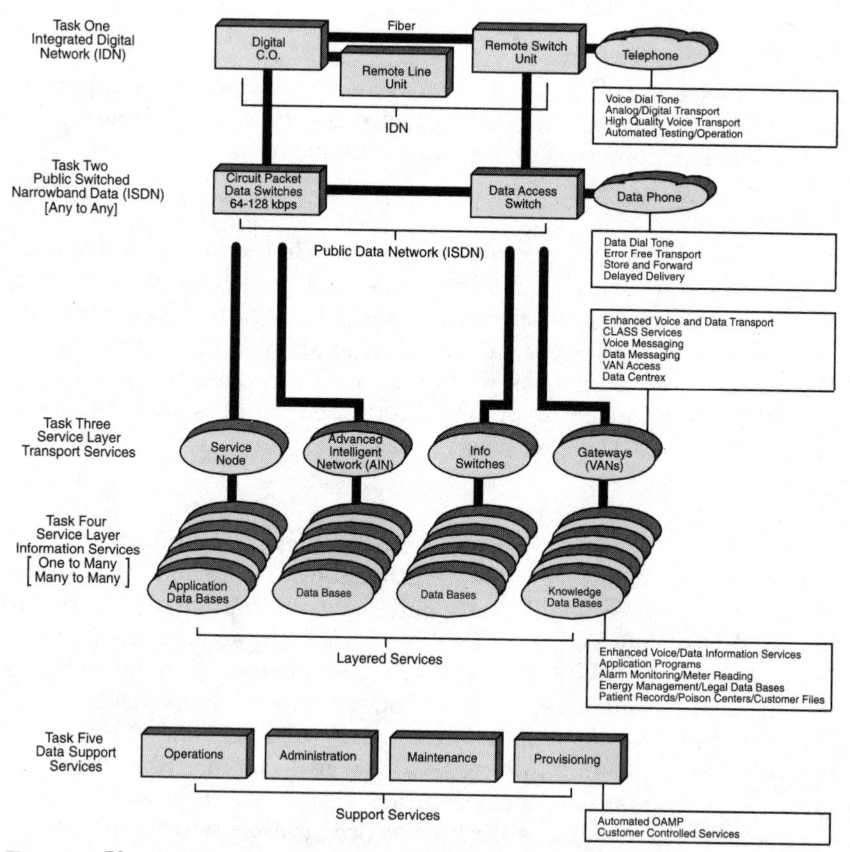

Figure 9-3. Plan for the future narrowband. This diagram depicts the layering of numerous data service platforms once the integrated digital network is overlaid with public switched data-handling capabilities.

NARROWBAND

User Needs

Audio Communications - • *Call Control Features* • *Specialized Call Control (Based on Calling Party)* • *Customized Call Control (CLASS)* • *Extended Call Control (AIN)* • *Stereo Quality Audio*

Data Communications - • *Standard Interfaces* • *Global Addressing* • *Error-Free Transport* • *Fast Call Handling* • *Shared Transport* • *Usage Pricing* • *Security and Privacy* • *Virtual Private Networking* • *Economical Fast Transport* • *CPE Signaling, Sensing, Polling* • *Multi-Party Messaging*

Video Communications - • *Small Screen (Eye to Eye)* • *Economical Quality Imaging*

Network Options

1) • 64K bps Digitized Voice Network Interface
 • 64K bps Circuit Switched Data Network Interface
 • 16K bps Signaling/Packet Channel

2) • 64K bps Circuit Switched Data Network Interface
 • 64K bps Circuit Switched Data Network Interface
 • 16K bps Signaling/Packet Channel

3) • 128K bps Circuit Switched Data Network Interface
 • 16K bps Signaling/Packet Channel

4) • 64K bps Digitized Voice Network Interface
 • 64K bps Packet Switched Data Network Interface
 • 16K bps Signaling/Packet Channel

5) • 64K bps Packet Switched Data Network Interface
 • 64K bps Packet Switched Data Network Interface
 • 16K bps Signaling/Packet Channel

6) • 128K bps Packet Switched Data Network Interface
 • 16K bps Signaling/Packet Channel

7) • 64K bps Circuit Switched Data Network Interface
 • 64K bps Packet Switched Data Network Interface
 • 16K bps Signaling/Packet Channel

8) • Two 64K bps Digitized Voice Network Interfaces
 • 16K bps Signaling/Packet Channel

Network Services

• *Second Audio Line* • *CLASS Services* • *Caller ID Services* • *High-Quality Stereo* • *Integrated Audio and Data ISDN Standard Interfaces* • *Digitized CPE to CPE Transport - Security - Privacy* • *Packetized Data* • *Universal Numbering/Addressing* • *Closed User Groups* • *Broadcasting* • *Priority Override* • *Delayed Delivery* • *Data Caller Identification* • *Polling/Sensing* • *Error Detection/Correction* • *Encryption* • *Dynamic Routing* • *Connection Oriented/Connectionless Transport* • *Data Networking - Protocol Conversion - Code Conversion* • *Image Transfer* • *Robust Transport - Survivable - Alternate Routing* • *D Channel Signaling* • *Data Packet Networking* • *Application Service Center Accessibility*

Service Applications

Voice - • *Multiple Lines - Small Business, Residence, etc.*

Data - • *Point of Sale* • *Alarm Monitoring* • *Environmental Control* • *Inventory Control* • *Credit Checking* • *Weather Sensing* • *Home Incarceration* • *Traffic Monitoring* • *Automatic Teller Control* • *Text Messaging* • *Group 4 Fax* • *X-ray Transfer* • *E-mail* • *Global Data Directory* • *Data Storage* • *Delayed Delivery* • *Private to Public Networking*

Video - • *Videophone (128K bps)* • *Desktop Videoconference* • *Multimedia Workstations*

Figure 9-4. The narrowband public information network.

INFORMATION TELECOMMUNICATIONS PLAN FOR THE FUTURE WIDEBAND/BROADBAND

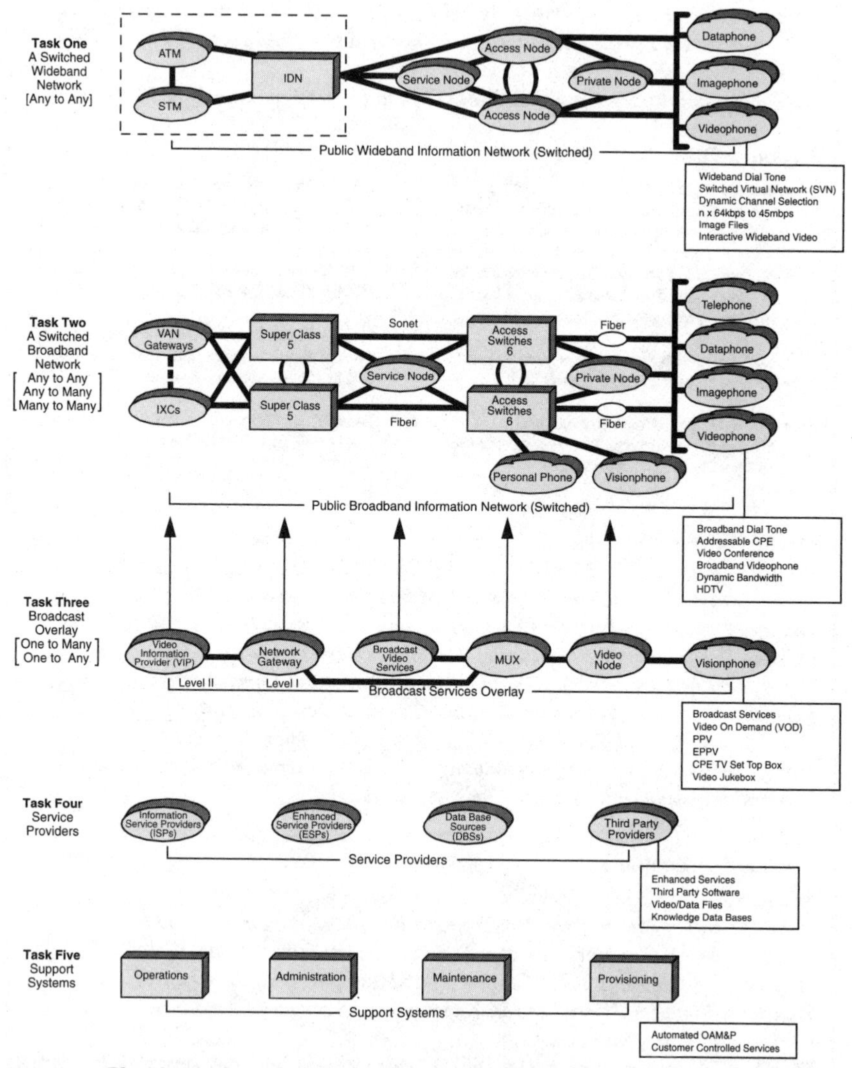

Figure 9-5. Plan for the future wideband/broadband. This diagram denotes the wideband network as well as its transition to broadband, augmented with the broadcast offerings supporting various broadband service platforms from a multitude of providers.

WIDEBAND

The wideband public information network addresses the following user needs, providing network options as well as network services for various types of service applications . . .

User Needs

• *Variable Bandwidth On Demand* • *Variable Destination Addressing* • *Switched Virtual Private Networking* • *Usage Pricing* • *Error-Free Transport* • *Non Blockage* • *Non Congestion* • *Fast Set-Up/Take-Down* • *Dynamic Reconfiguration* • *Full Range of Variable High-Speed Wideband Transport* • *Security/Privacy* • *Survivability* • *Global Accessibility* • *Universal Availability*

Network Options

• *P x 64,000 bits per second* • *Range - 128K to 45M bps* • *Availability - Standardize Up To 1.54M bps (Copper) - Customized Up To 6.02M bps (Copper) - Selective Up To 45M bps (Fiber)* • *Primary Rate ISDN (Switched) 23 B (64K bps) + D (64K bps) [Where each B channel service is identified via the D channel for integrated voice & data transport]*

Network Services

• *Global Addressing* • *Virtual Networking* • *Bandwidth On Demand* • *Precall Routing & Bandwidth Selection* • *Usage Sharing* • *Variable/Dynamic Bandwidth* • *Error-Free Transport* • *Security/Privacy Mechanisms* • *Survivability/Alternate Routing* • *Delayed Delivery* • *Broadcasting* • *Sensing/Polling* • *T1/T3 Networking* • *Fractional T1/T3 Networking* • *Primary Rate ISDN Networking* • *Frame Relay/SMDS Networking* • *FDDI Interfacing* • *Global Service Center Access*

Service Applications

• *Public WAN/MAN Networking* • *LAN-to-LAN Transport* • *Virtual Private Networking* • *Supercomputer-to-Supercomputer Transport* • *Workstation to Supercomputer Transport* • *Videophone - 1.5M bps* • *Multimedia Workstations Networking* • *Private-to-Public-to-Private Networking* • *Access to Local, Regional, National, & Global Databases* • *Customized Application Service Centers* • *CPE-LEC Access to IXCs, CAPs, ATPs, & VANs*

Figure 9-6. The wideband public information network.

BROADBAND

The broadband public information network addresses the following user needs, providing network options as well as network services for various types of service applications . . .

User Needs

• *Full range of voice, data, text, image, video, & vision information monitoring, sensing, polling, auditing, exchanging, interexchanging, storing, accessing, listing, searching, browsing, retrieving, processing, manipulating, & presenting - any to any - anyone, anywhere, anytime, any multimedia combination . . .*

Network Options

• *Full range of switched SONET transport as multiples of Optical Carrier-1 (OC-1) rate (50+M bps) with international interfacing at OC-3 (155M bps) and OC-12 (620M bps) from customer premise - overlaid with HDTV broadcast capabilities . . .*

Network Services

• *Broadband Addressing* • *Global Networking* • *N•W•B Networking* • *Interactive Switched Broadband Videophone (50/155M bps)* • *N•W•B Videophone Integration* • *Multi Channel Broadcast Television* • *Full Range of Wideband Services Extended To Broadband Rates*

Service Applications

• *Videophone (50/155M bps)* • *High-Definition TV/Conventional TV* • *Videoconferencing (50/155/620M bps)* • *Video Picture Windows - Wall Size Travel Logs* • *Video Simulation Systems - Virtual Reality* • *Video Storage Retrieval Systems - Video Files - Movies On Demand - Home Shopping/Catalogs* • *Remote Education - Broadcast - Interactive - Delayed Delivery* • *Specialized/Customized Educational Video Files* • *Workstations Access to Remote Databases* • *Supercomputer to Supercomputer Transport* • *Supercomputer to Distributed Intelligent Workstations Transport* • *Imaging Systems - Medical Tests/ Records* • *Global Broadband Networking - Distributed Computing - Holographic Meetings* • *Robotic Control Systems* • *Energy/Transport/ Defense On-line Feedback - Control Systems* • *Access to Universal Broadband Application Service Centers*

Figure 9-7. The broadband public information network.

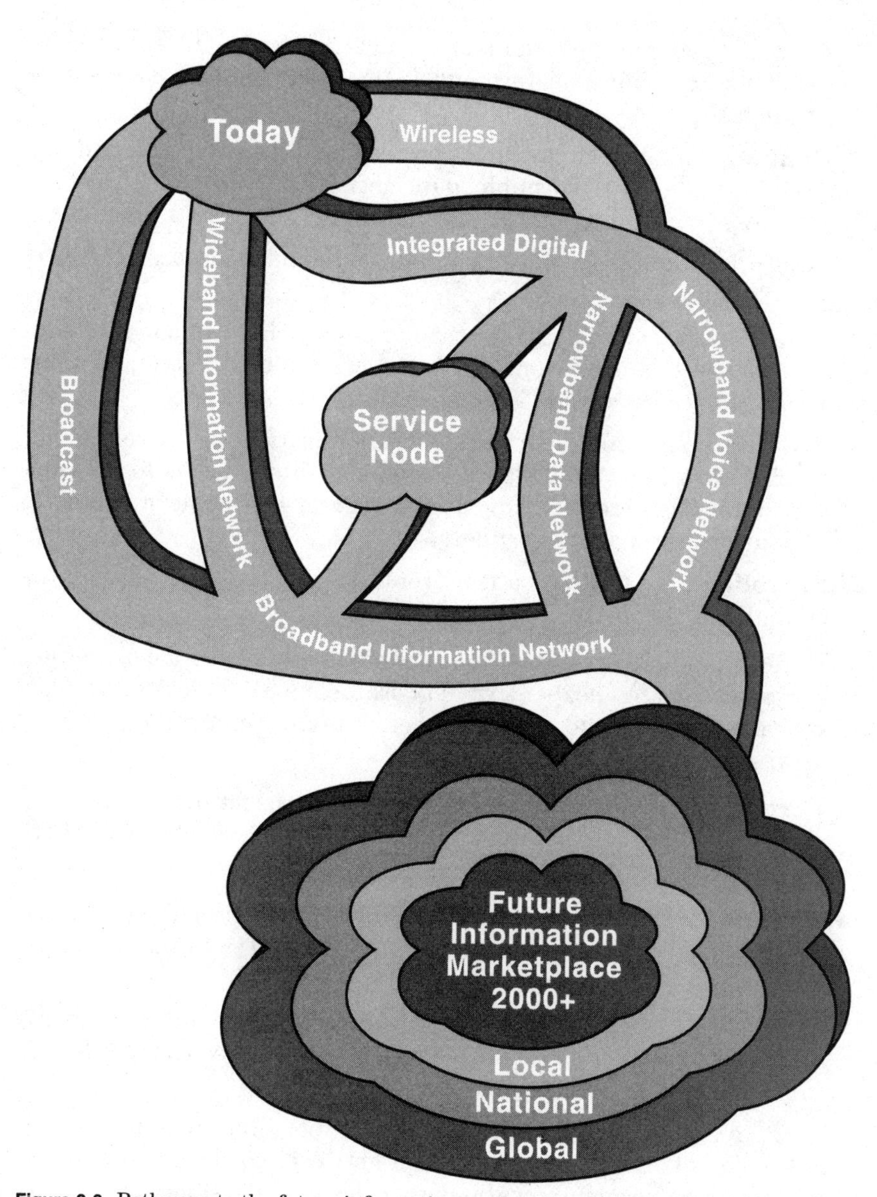

Figure 9-8. Pathways to the future information marketplace.

Plan of Action

To successfully achieve the plan for the future (Fig. 9-9), the following steps must be taken with appropriate checkpoints to verify progress:

- **Assess . . .** current physical plant to determine deployment capabilities for establishing narrowband ISDN-based public data network offerings.
- **Define . . .** digital circuit and packet transport switching and signaling requirements for public data networks.
- **Identify . . .** an appropriate array of narrowband data-handling services priced and packaged for promoting expanded data transport growth and usage.
- **Conceive . . .** a highly sophisticated advertising campaign to educate prospective users concerning the capabilities and applications of public data network offerings.
- **Provide . . .** enhanced data transport application service centers for helping data users find address locations, obtain access to information databases, identify advanced services, and achieve access to various data transport providers.
- **Establish . . .** OAM&P support centers specifically for serving the public data network.
- **Enable . . .** network access to numerous types of customer-premise equipment, using inexpensive network terminating devices to allow various versions and variations of stimulus/functional signaling arrangements.
- **Formulate . . .** appropriate interface standards for internetworking narrowband, wideband, broadband, and wireless offerings, both locally and globally.
- **Deploy . . .** a universal, fully available narrowband public data network throughout the operating region with access to the full range of IXCs, VANs, ISPs, and ESPs.
- **Educate . . .** both providers' and suppliers' personnel to correctly develop and deliver successful features and services to the information user in a timely manner.
- **Inform . . .** regulatory bodies of the exciting potential of future narrowband, wideband, and broadband networks to achieve the proper competitive telecommunications information infrastructure.
- **Initiate . . .** new topologies for fully switched, high-speed, fully survivable wideband/broadband services, using switching access points (nodes) closer to the information user.

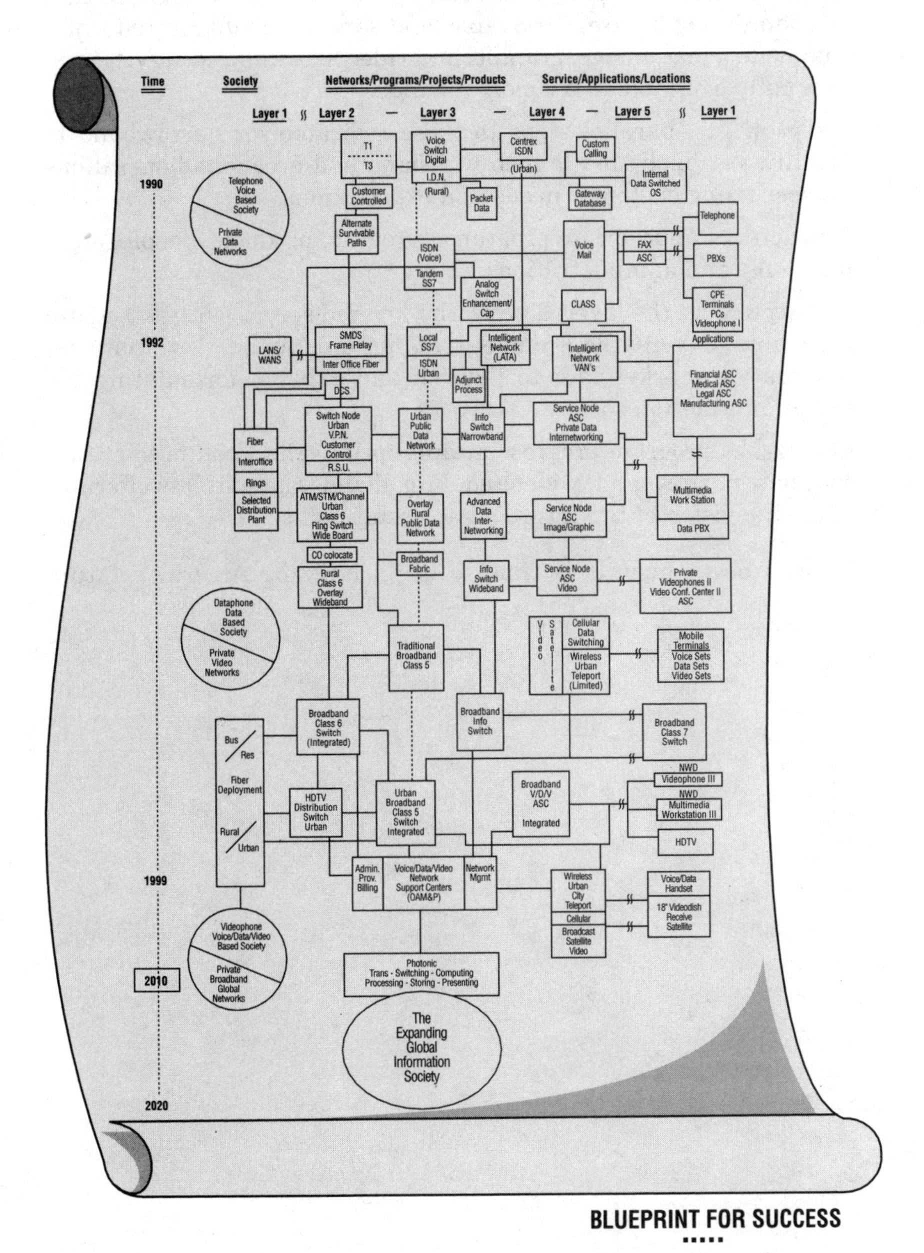

Figure 9-9. This flowchart demonstrates the orderly progression of network products and services required to achieve and sustain the future global information society.

- **Design . . .** the next generation of transport and switching systems based on this new topology and class-switching hierarchy, such that new families of secure, survivable, and serviceable integrated voice, data, video, text, image, graphic, and video features and services can be readily available in a timely manner.
- **Develop . . .** parallel steps to those indicated for narrowband to identify, establish, and launch wideband and broadband operations to meet rising customer needs and expectations.
- **Launch . . .** appropriate programs, projects, products, people, organizations, and administrations.
- **Construct . . .** the layered networks' layered services infrastructure to enable a versatile, competitive, private/public, local/national information marketplace to flourish and expand, formulating the global information society.
- **Check . . .** program progress at appropriate checkpoints to ensure that new narrowband, wideband, broadband, and wireless offerings make the vision of the future a successful reality.

So, let's begin now. It is time to begin building America's future today.

Chapter

10

A User's Perspective on the Information Age

Bishop Robert Carlson

Having recently moved from a relatively large urban archdiocese to a rural diocese with 35,000 square miles and 120,000 people spread out in small towns and villages, farms and ranches, and large reservations, I soon found myself driving more than 50,000 miles a year to preach and try to build community.

As I questioned how we might communicate more effectively, I began to investigate the field of computers and communications. I was surprised to discover that as the church has experienced major changes the past twenty-five years, so has the world of high technology. There have been striking advances in both computers and communications, and together they are now positioned to leave the laboratory research environment and have a striking impact on today's society. When future historians look back over the late 1990s and the early years of the next millennium, they will see technical breakthroughs in these endeavors having similar power to dramatically affect and change society as those leading to the nuclear age. Some will call these years the Information Age. Unfortunately, there are currently two major forces at work—each creating a series of thrusts that lead us in quite diverse directions with quite different conclusions.

Today we are at the crossroads of these diverse paths, where one approach is to use a few of the emerging technologies to build an entertainment network offering 500 or so channels to the home with the additional ability to call in a request for a particular movie, sporting event, home shopping catalog, or game. This form of broadcast network with its capabilities to access a "video jukebox" has substantial media support. It continues to hype this broadcast network's possibilities as the new information highway for the twenty-first century. What the media does not talk about is its greed. Having this new vehicle to offer "movies on demand" is considered by market researchers as a $13-billion revenue stream, of which 80 percent of the offerings are NC-17 or X-rated movies. This form of network is based on using the existing coaxial plant currently in place (modified with some fiber-optic transports), with limited risk—a quick return on investment.

The alternative path leads to a network enabling fully interactive voice, data, text, image, graphic, and video offerings, which provide not only voice capabilities (as we know them today) but stereo-level audio, data messaging, access to global databases (information files), and high-definition, high-resolution videophone and video conferencing, as well as broadcast television. (Here, entertainment services are part of the full offering, but the network is not being established simply to deliver entertainment.)

We are fully aware of the impact that existing voice networks have had on society since their origination 100 or so years ago. In the last decade, with the advent of digital technology, we have seen long distance rates diminish as both competition and better, more efficient technology automated the telephone companies' operations to provide greater transport and switching capability to handle more and more voice conversations. (They call these the new digital networks, where voice is changed into digital binary streams of 1s and 0s—thus preparing for the integrating and interlacing of voice with data and video.) Yes, these changes have indeed prepared the telecommunications world for deploying an exciting array of new services for helping to improve our mode of operation in implementing our daily tasks. We have seen the beginning of other additional advances, such as the cellular phone, etc.; however, we have not yet experienced the advantages of fully integrated multimedia offerings.

As the computer enters every aspect of society, there arises the need to interconnect disparate terminals together. This interconnection first took place in the office, where three or more devices were interconnected, creating a local area network (LAN) to interexchange information and access shared database files. As time progressed, needs expanded, requiring the ability to send and receive messages from further and further remote locations. These capabilities were provided by dial-up data calls, using low-speed modems over voice-grade telephone lines at transport rates up to 1000 characters per second or 10,000 bits per second.

You have heard of the Internet, which is also being promoted by the media as the answer to business's future information-handling needs. This highly successful academic system basically provides the data terminal with a universal address, using dial- up voice-grade facilities or an internal local area network to send and receive messages. It was originally funded by the Department of Defense with the state universities picking up the local access service center cost to allow relatively free individual student and researcher usage. Unfortunately, this network is not secure, not private, congested, and, if used for more and more applications, will be unable to handle its own success. Hence, neither offering, the cable movie network nor the low-speed, dial-up Internet, really meet the needs of an expanding global information society. They are simply an expanded entertainment network that facilitates the distribution of movies to the residential home and a relatively low-speed, obsolete data-handling network for the business community.

However, if we step back to assess the alternative opportunity, we find a very feature-rich network enabling new service offerings that can greatly enhance our work and leisure time. This is an opportunity now to build a very robust, very secure, survivable public data network to enable the universal interexchange of messages, either personal letters or announcements in text or graphic form at quite reasonable rates, that are 10 to 20 times faster and are substantially clearer than existing offerings. This can be accomplished by deploying over existing telephone facilities ISDN voice and data-handling capabilities. These narrowband offerings need to be provided across the country by the local exchange carriers (the telephone companies) at reasonable rates with access to inexpensive national and international data transport (AT&T, Sprint, French and German PTTs, MCI-BT, Wiltel, etc.).

An entire diocese can be interconnected to exchange information as well as enable visual conversations with the proper quality and resolution to achieve meaningful dialogue. The power of seeing each other, exchanging documents, and working together remotely on material that is simultaneously visually available to each other has tremendous impact on our mode of operation, thereby greatly enhancing our communication effectiveness. This can all be accomplished with small modifications to existing networks and services. We can then subsequently use more expanded technology to obtain fully switched wideband services. Here, information is transported around the world at substantially higher rates—thereby enabling a jump in capabilities from 10,000 bits per second over current voice-grade networks to the narrowband ISDN 128,000 bps and then to wideband's 1 to 50 Mbps. Here, the narrowband videophone's talking heads have substantially higher quality and resolution, enabling video conference capabilities for Diocesan Pastoral Councils, as well as the ability to see large events around the world, such as a Papal Mass at St. Peter's, or to be part of the World Youth Day in Manila.

Finally, the last phase in new technology, called broadband, will encompass both narrowband and wideband capabilities and extend the ability to see anyone, anywhere, anytime, into the realm of 150 to 600 Mbps with information transported over shared fiber-optic facilities at rates of 50 billion bits per second with ultimate speeds at terabit rates of 1000 billion bits per second. With these capabilities, I do not need to emphasize the forthcoming changes in the world as communication replaces transportation, as images become so clear that one's presence can be readily felt at distant locations, indeed, several locations at one time.

We would have a system that is not simply based on reviewing a multitude of movies in one's home, but a system in which data and video messages are easily and economically exchanged anytime, anywhere—locally, nationally, and globally. Just as interactive conversations over our voice telephone network today have greatly enhanced our quality of life, we will be able to have full multimedia conversations in which we share a letter, access a data file, see each other eye-to-eye, or have meetings across an electronic video conference table. In this new society, people will have the ability to live in more rural communities and still have access to all the sources of information usually only available to those in the highly congested urban cities. Similarly, many families will be able to use work-at-home telecommunications capabilities, enabling parents to be closer to their children, reducing the need for child-care facilities, and enhancing family life. New cities can be formed where people have more space to enjoy life and where homes can be built at cheaper land prices, thereby reducing economic pressures. The list of exciting positive possibilities goes on and on. However, due to the push for more and more entertainment networks, the communications industry has unfortunately almost abandoned its pursuit of this form of technical achievements.

In my diocese, as a first step, I am endeavoring to test a narrowband ISDN network eventually interconnecting all the parishes to exchange messages securely and privately as well as have reasonable quality (128,000 bps) videophone conversations. This communication capability, integrated with multimedia computer systems, will enable us to exchange local and distant information files, broadcast letters to selected individuals and parishes, and save and store information on CD-ROMs. In time, when other dioceses have similar capabilities, we will be able to continue this form of multimedia communication around the globe. Once the future fully switched narrowband, wideband, and broadband networks are universally available, this information exchange will become more and more powerful and effective.

In summary, it is essential to focus on building the right infrastructure: pursuing an upgrade of the existing copper telephone network and a phased deployment of fiber rings for public voice, data, and videoconferencing applications. Monitoring the flurry of telecommunications events in the headlines that concentrate on showy applications, I am concerned that there is too much attention on the fireworks with no one building the infrastructure. Without the proper underlying network, these initiatives will be short-lived and not meaningful to America's future prosperity.

While I am impatient to see a fully switched high-speed broadband fiber network deployed, I believe that upgrading the existing copper telephone network to achieve the narrowband and wideband ISDN networks, which would support many new data and videoconferencing applications, is a necessary step. These more readily achievable networks can immediately begin America's use of this new communications technology in anticipation of the emerging broadband fiber network infrastructure.

Implementation of the fiber deployment plan should begin in parallel to allow large users to realize its benefits as soon as possible to better use the increasing computer capabilities of the various service providers. However, developing this fiber infrastructure will take considerable time and resources. A plan that focuses on a phased approach seems most reasonable to get started, allowing the user market to develop, establishing a momentum in which success can lead to success, such that the narrowband/wideband copper phases can feed the broadband fiber phases. Such plans—when fully implemented—will meet the needs of my diocese and indeed the needs of America's businesses, schools, and homes for a secure and survivable, feature-rich, ubiquitous, switched, public telecommunications infrastructure providing ever-increasing switched transport at increasing rates.

This type of program requires not only a decisive multiyear vision (ten to twenty years) and detailed plan but the leadership to achieve it. America, especially rural America, must speak with one voice to the telecommunications community. I believe that we must take an active role in guiding and establishing the right narrowband, wideband, and broadband infrastructure, ensuring each is deployed everywhere and not just in heavily populated corners.

Chapter

11

The Society

For I dipt into the future,
Far as human eye can see,
Saw the vision of the world,
And all the wonder that would be.

TENNYSON

The plane leveled off at 39,000 feet and adjusted its westward course. I looked at my watch and noted it was now 7 p.m. We were scheduled to arrive in L.A. at 9 p.m. local time. When we left, the sun was just beginning to set in Chicago, and the plane seemed to be attempting to catch the sun as we sped westward at speeds nearing 600 miles per hour. It was a very pretty sight to see the sun continuously at sunset. After several hours the sun did make it down, and we entered the twilight hours of dusk. It was a pity; we almost made it to L.A. with a continuous sunset!

I always played this mental chase game on my evening flights west. It was similar to my younger days, when after a Sunday afternoon storm my father would load the family car with neighborhood kids and chase madly toward the end of a rainbow. We never did get the pot of gold. However, when we returned home, we all helped make ice cream and enjoyed it as much as if we'd received a gold coin each.

So much for daydreams, but the reminder of the elusive chase did turn my thoughts to the newspaper and magazine articles that I had read earlier on the New York to Chicago hop. The articles were concerned with the problems and direction of today's society. As I looked down through the clouds at the quilt-patch ground below, and later at the lights of large cities, I could not help but contemplate where all the people in that miniature-like land were going in terms of their goals, actions, and accomplishments.

"When farmers began harvesting the first domesticated plants about 8000 B.C., the earth's population was around four million. Today that many people are born every ten days. If the trend continues beyond the year 2000, we will have to grow as much food in the first two decades of the new century as was produced over the past 10,000 years and we will need dynamically growing infrastructures to support this growing global population." (*National Geographic*, March, 1991.)

As a planner of future telecommunications, I have spent considerable time projecting what life would be like in the 1970s, 1980s, 1990s, and after the turn of the century. It has been educational to watch the future develop, as or not as projected. Since more and more money was at risk, as the future became more complex and uncertain, I made many trips across the country to listen to politicians and fellow planners express their views of the good and bad possibilities for the future. I watched top management's chess game with strategic and not so strategic plans. I saw organizations and reorganizations come and go, as companies restructured for new markets or more powerful positions in older markets.

During this period, I have seen many sound ideas, programs, machines, and products changed, delayed, or not implemented due to various individuals' views and influence. Their actions greatly affected all our futures. Thus, I noted that each of us does have a definite effect on our own and others' lives. Many people have asked, "If we can go to the moon, why can't we solve other problems?" My conclusion has been that problems today (as always) are not really technical problems; they are people problems.

As a result, for the last several years I've attempted to look at society in terms of what its people want and what they have been doing about it or were willing to do. This showed what the future could be and why it was or wasn't being achieved. The result was some interesting observations in terms of the "big picture" outlook on America. As I looked a second time at the magazines and newspapers beside me, I concluded that such an analysis, which not only addressed where we want to be but also how we got to our current situation, would be both interesting and helpful, especially today. When people see everything that is going on in its proper perspective and understand why it is happening, then they have arrived at the point of beginning to solve the problem, or at least beginning to stop creating future problems. Hopefully, views expressed will be considered realistic, not pessimistic nor too optimistic, but applicable and acceptable for future considerations. We cannot, especially in today's informed society, expect everyone to agree or disagree with our conclusions. In fact, the best that we can hope for is to provide some thoughts that are seriously considered and challenged by others. That was the purpose of this analysis: to provide a platform for further thinking and discussion; but now it is to be used at a critical time for a critical decision.

Today, several years before the turn of the century, we are becoming more and more aware that we are all very much involved in a highly complex and interconnected world. Of course, there are degrees of involvement. The analysis doesn't apply to every country the same way. It has attempted to help each of us stop and take a hard look at ourselves, our current society, and its role in the global society.

Real-world forces and factors were identified and reviewed. It analyzed in detail the effect of technology on a changing society, in order to fully appreciate why things have happened and to understand new technology's role in society and its effect on future societies.

As the plane's engines droned on, I rested my head back against the pillow. A very pleasant stewardess had provided me with the pillow earlier. She greatly resembled my wife, who had also been a stewardess in Africa. She had initially worked there as a nurse, coming directly from many years of training in Ireland and Scotland. I recalled some of the experiences and observations she had recounted to me. I noted that people's wants and needs are basically the same everywhere. We all have different opportunities and outlooks, but a global perspective of our needs and desires would apply to many throughout the world. The basic needs are similar. Just the time frame for achieving them, may be different... Suddenly I felt very tired...

Once again my mind returned back to the scene below me. America is certainly a large country. It has been the cradle of many scientific discoveries that have helped create our current world. The other countries of the world have had their share of discoveries also, but the strength of America is that it has been able to take advantage of the diverse nationalities that make up its population. It has been called the melting pot for every known race and nationality in the world. The early immigrants, explorers, and inventors had, through their hard work and patriotic loyalty, left us with the wealthiest and strongest country in the world. But this is history. What is happening now? How did we get to our current situation? Who or what is controlling it?

If we were to study the various past civilizations of the world and attempt to define one major force that is directing our present society differently from past eras, we might conclude that it's "information communications." Today, more than at any other period in time, the direction or lack of direction of society is being determined by our use of technology. No longer are the world's major countries controlled by royalty or allegiance to government or party. Revolutions, wars, political corruption, and misuse of power have usurped government's power and given it back to the people.

Looking at history, the world has experienced sudden periods of change due to wars, establishment of new empires, introduction of new cultures, new advances in science, etc. At times, these changes have been violent or quiet and orderly, but none have been so global, radical, quick, and numerous as those of the past hundred years. During this period, mankind has experienced massive changes in environment, culture, work, and religion, due to numerous factors. In America, the engine of change is science, and the major catalyst is "individual freedom" under the American Constitution. These factors provided new inventions, which brought about the industrial revolution. This subsequently caused a large concentration of people at nodal points (major cities) across the country. Subsequently, these people struck for more benefits from the factory owners. These benefits allowed men and women to upgrade their living conditions and send their children to schools, including college. These children grew up and entered the work force. Due to their higher educational background, they were inquisitive and began to ask the basic question "why" concerning many issues. As a result of television, the older, now better informed adults asked similar questions. This past decade has been one of questioning by both young and old. They both now have more and more time to ask questions and think, due to these further advances in science, labor benefits, financial security, human comforts, etc. This has set in motion new controlling factors that influence our current society. Today, after the turbulent 1960s, inflationary 1970s, high-rolling 1980s, somewhat stagnant 1990s, followed by these last chaotic years, we are now somewhere in the middle of extremes.

Much of the world is still in desperate poverty and ignorance. It's extremely difficult to discuss the beauty of a coin collection with a man who is starving. However, this society, with its interexchange of information, has more people than in previous societies who have reached a point where they can listen, discuss, and greatly affect their own destiny and that of the rest of the world. . . .

The plane banked suddenly, drawing me abruptly back to reality as the pilot brought the plane through the clouds directly over the city of Los Angeles. Lights twinkled from each home, illuminating the city and making it appear to be one huge Christmas tree. Then, remembering my previous thoughts, I knew that each twinkling light held one or more persons who were somewhere pursuing their goals, fulfilling their particular wants and needs. As the plane landed, I realized that my current want was now quite basic—a good night's sleep.

The next morning, I awoke with renewed determination to ensure that my presentation would be as effective as possible in light of my thinking on the plane; it was indeed that very day that we would be discussing a critical opportunity at a very critical conference meeting concerning the future direction of society. I went down to breakfast and selected a comfortable-looking table in the corner. There I could observe the room and collect my thoughts while I waited for breakfast. I knew from past history that service was notoriously slow. Food in 2010 was quite expensive. However, the coffee and orange juice were available for me to help myself. Normally, this is all I need to get my mind functioning.

Since it was two hours before the conference, which was going to be in the same hotel, I had enough time to rearrange a list of facts and facets that describe life in society as it could have been today. Over the past thirty years, I had compiled the list from discussions, observations, articles in magazines and newspapers, and commentators on television. As I turned over the notebook cover, a piece of paper fell out; it had been inserted there for lack of a better place to put it. I remembered tucking it in the notebook on a recent European flight. I opened the folds and laid the paper out in full view while I sipped my coffee.

The first figure attempted to show how society at any given point in time is the result of technology's impact on our individual roles. These roles that people play are the result of factors and forces that influence, expand, or limit us. Of course, we always have free choice, but to a great extent the scope of operation is controlled by these operational factors. On that long plane flight, I had sketched two figures. One showed the cause-effect relationship and the other showed the cause-effect sequence on ever-changing societies. They showed that as we act out our roles, we change, modify, or produce factors and new forces that affect our future roles. This is the life cycle of our changing world.

I again noted to myself the importance of understanding this relationship because it does indicate that we do, even though sometimes slowly, change the major factors that are driving us. In so doing, we help determine society's present status and future direction. It's equally important to recognize that at certain points in time we have more or less control of our role due to an imbalance effect of a particular major factor. For example, a man's wants may be recognition and power, but, if he is a slave of a conquered race in the year 20 B.C., he may have difficulty fulfilling his want.

Thus, at a given point in time, we could have a relatively accurate big-picture scenario of society by noting the "top ten" facets that best describe the mode of operation of the society. To project the future direction of a society at a particular point in time, it's important to have a series of these looks at preceding points, for example 1927, 1937, 1947, 1957, 1967, 1977, 1987, 1997, and 2007. In so doing, particular scientific factors, or particular modes of operation (the results of collective individual roles) that cause desirable or undesirable impact on future societies can be noted. For example, it's interesting to note what the automobile and television factors initially contributed to society, and what they are currently contributing—good and bad.

I also noted that if we were to plot the course of events over the last 100 years, it may show how some supposedly uncontrolled factors were not necessarily totally uncontrolled and unpredictable; many events that led to this or that factor would not have occurred if we had taken a different path at some crossroad in time. Hence, when applying this theory, it's important to carefully assess the impact that various technological possibilities have had when correctly applied to market opportunities. It's also important to note the missing impact that could have happened, would have happened, but didn't, when technology is not used or is incorrectly applied.

It took only a half hour to arrange my presentation to indicate what could have happened, didn't happen, and why.

The Information Society: 2010

What
could have been,
should have been,
might have been,
but wasn't.

A look at what could have been:

1. One hundred miles from each major city, multigigabit information superhighway rings are deployed in each heavily populated state. Each state has at least one new city established by the year 2012 A.D. on these rings. The new planned cities are designed to "cap" the growth of older cities. Using telecommunication-information transport rings to communicate with separated communities, the entire ring complex becomes the new city of the future. By being distributed on the ring around the older city, a combined old/new interconnected relationship is established, forming a new infolopolis. Each new infolopolis is linked across the states and the world via super-high-speed national information transport networks.
2. Each urban city has information transport superhighways throughout, ringing its business, university campuses, industrial parks, shopping malls, financial centers, hospitals, large business complexes, educational centers, state agencies, federal bureaus, police networks, manufacturing plants, airports, suburban centers, business hubs, and residences together.
3. Unlimited information access is provided to both public and private databases and numerous advanced services.

4. Information movement is priced economically for both voice and data, narrowband, wideband, and broadband transport on a usage basis. Similarly, wideband and high-bandwidth services are priced on the service, not the bandwidth, basis.
5. Fiber is deployed to 80 percent of the older urban business community, 50 percent of the urban residential community, 40 percent of the rural business community, and 10 percent of the rural residential community. The next twenty-year plan will complete the upgrade.
6. New public data networks and public wideband networks are interconnected across America to move data, image, text, and video along with the traditional voice network. Broadband networks begin to integrate narrowband and wideband services in new fully integrated system complexes.
7. Private/public internetworking, interprocessing, and interservices needs have been resolved via the global interconnectability standards, thereby enabling orderly growth.
8. Each major city in the industrial world is linked together using global carriers and global VANs. In this manner, multinational private local networks are able to internetwork throughout the world.
9. Third-world nations have turned to telecommuting information services as a major source of revenue. In this manner, 25 percent of the entire world is networked together over two international grids, with 75 percent of these users on a universal services voice, data, video network and the remainder on a voice and data network.
10. Each major industry is fully networked together, with access available from each offnet user. In this manner, the U.S. business communities are a competitive global player in the Information Millennium.

As a result of these massive information networks and the resulting new cities, congestions and environmental problems were being brought under control. Each new city provided something of use, such as the new bullet trains to interconnect the cities.

However, what actually happened was—I then reviewed the long list of disasters. It would be interesting to see what was noted in the conference by the other speakers.

With a sigh of frustration, I closed the notebook. It's a pity that a society so big and powerful, with so many technical tools, is sometimes so helplessly wrapped up in itself. It's a shame that it took the problems of today to awaken us again to a broader need, the need to survive in the global information marketplace. I guess this is the reason for the conferences.

My thoughts were interrupted by the aroma of the now cooling breakfast. I remembered that my current "need" was food. I put the notebook away for future reference.

Chasing the Rainbow

I hurried into the conference room and took my place at the round 100-foot conference table, which was in the form of an oval, open on the inside to enable the speaker to enter the inner area. In this area, he or she could make their presentation using the latest technology that simultaneously showed images on three walls behind the table. Also, the new, but still very expensive, holography-technology devices were available in the center. This allowed all participants to easily see and hear the presentation. Each participant had a "mike" that was always on. Conversations were electronically changed to the listener's own language and provided through ear plugs.

There would be no separate discussions with aides or interested observers who sat in the four rows behind the table or in remote video centers around the world. This was to be a frank, open, working meeting with all discussions available for all to hear. Hence, we were provided with all the modern tools to communicate and resolve issues. Expense was not spared to ensure that at least this meeting used all of what could be. In the past, these meetings had just been used as a forum to express frustration and disagreement. However, today was to be the last day.

The conference had three separate topics to cover in one day. Now was the time for a decision. The country was waiting for a conclusion that was already two years overdue and time was running out. Hence, the agenda was straightforward—it was to review and summarize the following:

1. Review the situation (how we got here).
2. Analyze the options (what can we do).
3. Recommend a solution (where we go from here).

The first speaker rose and presented a current assessment of America's role in the global community. I made more notes in my notebook for later reference. We were all pretty well versed in all areas by this time, but it was necessary on this last day to summarize all aspects of the situation and recommendations. The speaker showed how America did or did not participate in many of the international markets.

She then presented a detailed look at the last five years. She showed the terrible shifts that had taken place. They caused the subsequent formulation of the new party and the election of its candidate to the presidency in 2008, with the promise of having a solution within this term of office. Two years had passed and time for discussion was over. The president planned to act and act soon, on the recommendation of this committee.

My mind returned to the charts being shown and the comments of the speaker. I jotted notes that summarized her key points.

The Communication Society of 2010

1. The traditional voice network structure is being overloaded by data, freeze-frame images, and highly compressed video transport. Traffic-line load control mechanisms are continually going into play during 10 a.m. and 2 p.m. busy hours.
2. The private community has built up separate overlay networks for each major business sector.
3. Global VANs have tied islands of U.S. industry to world markets, leaving information deserts in many areas. Several foreign countries such as Japan have recently completed their entire upgrade to broadband to provide the infrastructure base for unsurpassed economic growth. Japan has surpassed the entire Pan-European community as well as the Germans in this endeavor. Their only competition was Hong Kong (backed by China) and several other equally self-contained locations, such as Singapore and the new Hong Kong being constructed in northwestern Australia.
4. U.S. businesses have turned to more cost-reducing tactics in an effort to reduce American overhead as a result of the sellout in the 1990s of numerous U.S. firms to foreign owners. Hence, U.S. divisions became simple assembly plants for U.S. applications of foreign products. Little, if any, new product research is now performed in America, except for specialized software developments.
5. Four of the traditional "baby bells" have been purchased by financial houses as cash cows to generate revenue for foreign investment. Five to ten national consortiums have established parallel networks to provide specific services to various industrial and educational businesses, but the bulk of society is not interconnected. Large-city traffic congestion problems, environmental problems, reduction in manufacturing, limited research, limited advanced engineering, and lack of planning new endeavors have continued. This, in many cases, has expedited the trend of shrinking high-paid manufacturing jobs that require sophisticated technology.

Continued loss of world technological leadership to Japan and some European multinationals has kept the United States a consumer nation, forcing intense pressure on the dollar and causing inflationary changes while the economy is stagnant—stagflation. Unrest and moral decline have led to violence, requiring more and more police and military protection of the citizens. With intense debt, crumbling infrastructures, and zero business growth, the U.S. society, previously based on the most modern communication infrastructures, roads, education, etc., is now faced with its first real threat of internal collapse since the Civil War.

Of course, there are still twenty to thirty years before the patient dies, but it's quite clear to all that this patient is seriously, seriously ill. The only consolation is the ability to look at several European cousins, who are sicker. The Russian union disintegration is nearing completion, as global economic war takes its toll. Unfortunately, the opportunity to remain the leader of the free world shifted to the Japanese ten years ago, as they purchased more and more firms located throughout the world, thereby becoming a global empire.

As a result of a lost world market, there was, in America, considerable confusion, disillusionment, and anger, with a mood for change. The country was ready, really ready, for a change to something else, but what else?

I looked over the grim notes and remembered the last chaotic years. The next speaker rose and presented his summary of why we got into this situation and what we could have done to avert it. Instead of taking any more notes, I decided to summarize this area in terms of my previous thoughts.

How Did It Happen?

We don't plan to fail.
We fail to plan.

Loss of communications infrastructure

1. RBOCs' refusal to shift to long-term information infrastructures, due to pressure from Wall Street.
2. Refusal of Congress to provide incentives to upgrade public industries.
3. Continual unrealistic legal limitations for public information services, thereby encouraging RBOCs to spend money abroad.
4. Heavy losses encountered by RBOCs in foreign investments, requiring capital drain to salvage international investments.
5. Loss of ability to compete, as players lose desire to play the information game due to all the resistance, penalty, and play recalls.
6. Loss of . . .

The conference members broke for lunch. I found a few close friends and discussed our options. They also noted that the choices had slowly eroded to the point that we were really no longer free to continue the current pursuit any longer. This was to be the conclusion of the forthcoming afternoon meeting.

The meeting began with the chairman discussing the desperate state of the United States, especially since the turn of the century. He finally noted the mood of the people for reform. They were ready for a new form of society that returned some of the competition that had been lost over the years. He noted that the choices were bleak. He put forth the following possible options, noting that we had overrun the opportunity for the better alternatives:

The Options

If we don't plan,
we may overrun
our options.

MACGYVER

Communication networks

Option one. To rebuild the communication infrastructure, all U.S. firms must agree to return to using the public information network and pay a 2 percent profit tax to finance its construction, while residences pay a 5 percent basic communication-usage fee increase.

Option two. To obtain the information infrastructure, foreign investors must be given tax incentives to construct two parallel, massive local networks in each city to interconnect to global VANs.

Option three. A massive federal bailout program for the now failing communications industry to finance a total changeover to new systems. (However, the switching products can only be purchased from European and Japanese suppliers, since there are no longer any competitive, high-technology, "made in America," equivalent systems.)

Option four. Five new four-year programs for U.S. cities in which federal money will totally refinance change-out and conversion; federal ownership or an overseer of public communication networks returns for an interim period of twenty years, until the twenty-year upgrade program is complete.

Transportation highways

Option one. Establish . . .

The conference members argued for another hour over the pros and cons, but finally the hour grew late. They elected to recommend for the communications networks the fourth strategy: federal ownership or an overseer of the public communication networks for at least a twenty-year period. Later, after the options for the various other infrastructures were selected, with a touch of sadness and heavy hearts, we left the meeting. We had introduced major new factors that would require restricted individual roles. This would build a completely different society, perhaps one with considerably less freedom, which will inhibit the American creativeness we once had. It appears that people could not govern themselves. This had been the view of some of the early Greek philosophers. People could not govern themselves not because we could not, but because we would not restrict personal "I wants" in favor of others, unless forced to do so. It will be interesting to see what this new society will look like at some future point in time, such as 2029. I wonder how different it will be from what could have been. As I left the meeting, I wanted to have another chance; I wanted to turn time back, to try again, but in a different way, to reach the end of the rainbow.

The Choice

I awoke with a start—someone was trying to get my attention. Who was it? It was the stewardess. She awakened me to ask if I wanted anything to eat before they closed the galley. She had let me sleep through the meal. I looked out the window and reflected. It had all been a dream. It was not 2010; I was back in the 1990s. I was heading to another conference on data communication's effect on the business world. It was not a conference on determining the fate of a chaotic society. I smiled in relief, but then frowned as I remembered the society portrayed at the conference. It could happen. All the factors and forces were leading us in that direction. Maybe not quite that drastic, but some of the events were definitely possible.

As I looked down at the lights below scattered over the countryside, I noted that we are a big nation. There are over 280 million people down there. If only one-half of 1 percent or 1 million people or perhaps 100,000 or 10,000 or even only 1,000 or 100 or 10 began to really tackle the issues, they could set in motion new factors or change old factors to change our present course. There was time. We still have a choice, and the alternatives are not yet as drastic as only those that were available to the conference members in 2010. As I sipped some coffee, I pondered the questions. What can we do? Where do we start? How can we achieve the right infrastructures so as not to create a society in which no one wants to live?

Hence, now is a second chance, a new opportunity. At this "crossroad in time," we have a chance to choose the right path, to seriously address the issue of what price must we pay for inaction. As noted earlier, the final bill will come, even though we pretend it will not. Or as one leader in the field of economics said "there is no such thing as a free lunch."

New jobs can be created once America becomes again a strong competitive manufacturing-based (not service-based) society that creates and uses information successfully in the global economy. Inflation and stagflation can then be controlled by correct infrastructure spending and subsequent reduction of the national debt. We can't have everything. We think we can, but we can't. Bills do come and must be paid. Similarly, in rebuilding the country, we will not be able to do everything immediately, but we should be able to implement an effective five-part twenty-year program. (See Fig. 11-1.)

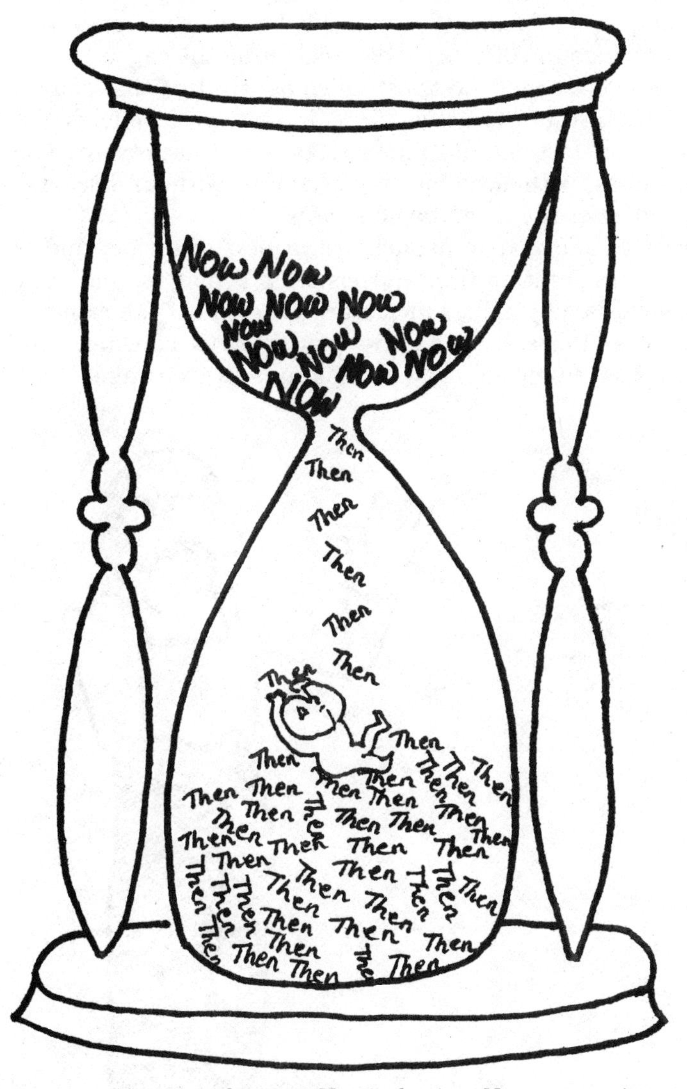

Figure 11-1. A crossroad in time. Now is the time. Now never waits.

The plane approached the Rocky Mountains. I could see the snow-capped peaks. I remembered the drought problems and energy problems and noted that nature still provides the answers, if we are willing to take a bigger look at our problems and resources. Problems of this magnitude require an effective government to resolve them. Hence, some issues may have to be addressed by a strategic plan for America. It may require changing some of the nonresponsible players and letting more competent leaders resolve the issues. People do affect other people. If those in power are seriously inhibiting others, then it's important to either change them or their influence. Therefore, it's important for us to carefully look at the cause and effect relationships (noted earlier) in addressing our problems. We can easily become frustrated if we only attempt to change the effects without changing the cause.

It may be much more beneficial to establish new factors that change or modify the cause rather than to continue to react to the effects. For example, reestablish lost morals and values, for a society without morals has no values; without values no discipline; without discipline no order; without order there can be no society.

As we attempt to add a new perspective, a new direction, and to change our mode of operation from indecision to action, we may find that this path surprisingly brings us closer to the end of the rainbow. My thoughts were interrupted as the plane's tires squealed and touched down in L.A. I pinched myself—this time I was awake.

A Final Note

Young men dream dreams
Of what could be.
Old men see visions
Of what could have been.

JOB

Adieu

Figure 11-2. Adieu.

Epilogue: Old Men Have Visions, Young Men Dream Dreams

The old man called to the boy to slacken the main sail sheet as he eased the boat off the wind to catch the slight wind shift and then rounded the buoy off the shores of Block Island. It was a clear day. The sleek 38-foot J boat lay out in front of him, forming a surging and crushing motion as it took the rolling ocean, occasionally washing a spray over the bow that blew gently into his face.

The boy in the yellow slicker was only ten years old, but he was already a seasoned sailor, having sailed with his grandfather and father for many years. Since the early years when he could first walk, he had sailed off Cedar Point in Long Island Sound. He wore a protective harness for safety in times of dangerous seas. As the old man looked at the boy, he noted that it was the same harness that he had used with his son so many years ago in sailing off Whidbey Island in Puget Sound.

His mind drifted to that day when his boy, not quite 12, had saved their lives. They had gone out on a tough day to see what the little 24-foot sailboat could do in rough weather. His mother had allowed the boy to go only if he had a harness on with a line secured back to his father. As the weather worsened, the boat heeled close to the water, letting the sea wash over the rails. Then, in a sudden wind shift, the whole boat rose out of the water onto its lee side, burying the opposite rail and snapping the mast. The lines tangled so that they were unable to release the dragging spinnaker, sometimes called the widow-maker sail. In a heartbeat, the boy yelled for slack on his harness and crawled out onto the foredeck with his knife poised. With the boat on its side, he cut the halyard to release the spinnaker, easing the strain and instantly righting the boat. This quick thinking and action saved both of their lives, as few survive more than forty-five minutes in the icy cold water of the Sound.

So here he was today with his grandson, in the Block Island race 3,000 miles from the Northwest, some thirty years later. He had had a tough time over those years, attempting to move American and world telecommunication leaders to build the right infrastructures to support and sustain a new information society. Several times he had been almost successful, but something always happened to block or divert his efforts, letting success slip from his fingertips, always just out of reach of his outstretched hand.

With a call to the boy, they completed a successful jibe and began the last leg home. The old man was thinking of what the telecommunications infrastructure for a future information society could have been, might have been, but wasn't.

* * *

"Jibe! Jibe! Jibe!" the man with the sandy hair and windblown complexion called to his college-bound son. It was not ten years since his father had died soon after sailing the last time with his curly red-haired grandson in the Block Island race. They had had a great outing. After that he himself had deliberately made considerably more time to sail with his son, further cementing the tight bonds of father-son friendship. They were not only father and son but also best buddies, as he had been with his father, and his father had been with his grandson.

Today, they were racing just off the port bow of an adjacent sister ship, sailed by a long-time challenger from New York's Larchmont Yacht Club. As boys, they had crewed together when not sailing with their fathers, but now both boats were neck and neck in force-10 winds, nearing the Fastnet Rock and lighthouse. The winds began to increase as they turned and headed back to England. The two boats had come over from America to race in the Cowes race week, followed by the Fastnet race.

The skippers had also competed against each other for years in the telecommunications industry as they pursued the game from different directions, with different perspectives, and different solutions. He came at it from the vantage point of building a new telephony infrastructure for interactive voice, data, and videoconferencing calls; his colleague from that of the cable entertainment industry. Unfortunately, he had had the worst of it. There simply was no interest in the States to build anything; everything was in the mode of use and abuse. Several of the LECs for which his father had worked were long gone these past four years as takeovers and mergers had taken their toll, especially since none of the LECs had elected to build on their core telephony base but chose the route of pursuing multiple diverse opportunities here, there, and everywhere. Hence, though he had done well financially, his life had been a constant period of attempting to survive as jobs continued to vanish from a shrinking market. His father's dreams had never been fulfilled. Global competitors, who did embrace his vision, became extremely tough players, causing a shift and redistribution of power from American businesses to the far regions of the globe. As his father's work remained in the world of what is and what isn't, his own life efforts remained in the world of what is and what isn't.

* * *

The red hair swirled in front of his eyes as he crossed the starting line in the Newport-to-Bermuda race. He finally had his own boat and crew. It had been a long pull since his father died in the Fastnet race. He would have died as well if it had not been for his father's long-time comrade and competitor, who had seen their boat go down in the height of the storm. All but his father had been saved. His father had been struck by the flailing rigging while trying to free a member of the crew. Now here he was, twenty years later. He yelled to his son to trim the mainsail as he rounded up into the wind. It was exciting to match his wits and strengths with the always formidable foe—the sea. At least he knew where he stood in this arena. Both his father and grandfather had attempted to move America forward in the telecommunications arena, attempting to provide building blocks for the future. Unfortunately, many did not share their visions and views, electing to pursue other arenas and other endeavors. But now, some thirty years after the turn of the century, well into the new millennium, new interest was being aroused to pursue those past visions. So as the boat lurched forward to meet the challenge, the young man daydreamed of what could be, what should be, what might be, what would be.

Appendix

Key Provisions of the Telecommunications Act of 1996

Telecommunication Services: Development of a Competitive Marketplace

Interconnection

Of primary importance in the Telecommunications Act is the establishment of an open competitive marketplace. To achieve this, the act establishes the following obligations and duties for telecommunications carriers and local exchange carriers (LECs) to open access to their existing public infrastructure:

1. Interconnection
2. Resale
3. Number portability
4. Dialing parity
5. Access to rights of way
6. Reciprocal compensation

Additional obligations of incumbent LECs include:

1. Duty to negotiate
2. Interconnection
3. Unbundled access

4. Resale
5. Notice of changes
6. Collocation

Negotiation, arbitration, and approval of agreements

As numerous entities approach incumbent LECs and Regional Bell Operating Companies (RBOCs) for interconnection, unbundling, resale, and collocation, the act has outlined an escalating series of steps these groups will take to resolve their inevitable conflicts and to gain agreements:

1. Voluntary negotiation
2. Mediation by PUC
3. Arbitration by PUC
4. Litigation by Federal District Court

Universal service

The act calls for the establishment of a joint federal and state board to define the services supported by the universal service support mechanism and to review details of how that support mechanism should operate. The act also made special provisions requiring basic and enhanced service availability for health care providers, schools, and libraries while ensuring that such services are provided at discounted rates. It establishes the following six key *principles of universal service* upon which the FCC's joint federal-state board will base its policies for the preservation and advancement of universal service:

1. Quality and rates. Quality service should be available at just and reasonable rates to all.
2. Access to advanced services should be available in all regions of the nation.
3. Access in rural and high-cost areas to basic and advanced telecommunications services.
4. Equitable and nondiscriminatory contributions to the advancement of universal service.
5. Specific and predictable support mechanisms.
6. Access to advanced telecommunications services for schools, health care, and libraries.

Infrastructure sharing

The act permits joint ownership of public switched network infrastructures and services.

Special provisions concerning Bell Operating Companies

In addition to the above provisions, a Bell Operating Company (BOC) must meet the provisions on the following interconnection checklist to begin providing in-region interLATA services (out-of-region interLATA services are allowed immediately by the act):

1. Interconnection
2. Access to network elements
3. Access to poles, ducts, conduits, and rights of way
4. Unbundled local loop transmission from the central office to the customer premise
5. Unbundled local transport from the trunk side of the switch to anywhere
6. Unbundled local switching services, including "transport, local loop transmission, or other services"
7. Access to 911/E911, directory assistance services, and operator call-completion services
8. White page directory listing for competitor's customers
9. Access to phone numbers for assignment to other carriers' exchanges
10. Access to databases and associated signaling for call routing and completion
11. Interim number portability via remote call forwarding, etc. (until number portability is established)
12. Access to services and information to allow a competitor to achieve local dialing parity (ability to route calls automatically, without the use of access codes, to their service provider of choice for two or more service providers)
13. Reciprocal compensation
14. Availability of telecommunication services for resale

In addition to complying with this special checklist, the BOCs must satisfy the in-region test by virtue of the presence of a facilities-based competitor or by the failure of a facilities-based competitor to request access or interconnection within 10 months. The BOC must also provide timely information on planned deployment of telecommunications equipment to interconnecting carriers.

The Communications Act also removes the restrictions on manufacturing with certain conditions that had been imposed by the MFJ on the BOCs. A BOC can engage in manufacturing through a separate subsidiary once it has been certified to provide in-region interLATA long distance.

Electronic publishing by Bell Operating Companies

BOCs can only engage in electronic publishing through a separate affiliate or an electronic publishing joint venture.

Broadcast Services

Broadcast spectrum flexibility

The act limits initial eligibility for advanced television services (ATV) licenses to those who currently own licenses or permits. It allows the holders of such licenses to offer ancillary or supplemental services that are in the public interest.

Recovery of licenses

If the FCC grants a license for advanced television services, the licensee will surrender the original license for the FCC to reallocate or reassign.

Public interest requirement

Public broadcasters maintain their obligation to serve the public interest.

Fees

The act allows a licensee to use some of the spectrum allocated for ancillary or supplementary services for a fee. The FCC will develop a program to assess and collect a fee from such licensees who use the designated spectrum in this manner.

Evaluation

The FCC will reevaluate the use of the spectrum within 10 years based on:

- The willingness of consumers to purchase the TV receivers for advanced television services.

- The assessment of alternative uses of the spectrum, including public safety.
- The extent to which the amount of spectrum assigned can be reduced.

Broadcast ownership

The act reduces or eliminates restrictions on radio and television station ownership. Specifically, it:

- Eliminates any provisions limiting the number of AM or FM radio broadcast stations which may be owned or controlled by one entity nationally.
- Eliminates the restrictions on the number of television stations that a person or entity may own nationwide and increases the national audience reach limits for television stations to 35 percent.
- Relaxes the *one-to-a-market* ownership rules by extending its waiver policy to any of the top-50 markets in the country.
- Eliminates many other restrictions, permitting a person or entity to own or control a network of broadcast stations and a cable company.

Direct broadcast satellite service

The act extends the current legal protection against signal piracy to direct-broadcast service. It also clarifies that the FCC has exclusive jurisdiction over the regulation of direct broadcast satellite (DBS) service.

Cable Services

The act made the following changes or additions to the 1992 Cable Act:

- LECs may obtain a controlling interest in, or form a joint venture with, a cable system. There are however, several restrictions.
- LECs are prohibited from purchasing more than 10 percent of a cable operation within its territory.
- Cable operators are prohibited from purchasing more than 10 percent of a LEC within its territory.
- LECs and cable operators whose telephone and cable serving areas are in the same market are prohibited from joint ventures that offer service within that market, except in designated rural markets.
- LECs may provide cable service to cable service subscribers in its telephone service area through an open video system. As an operator of an open video system, an LEC would qualify for reduced regulatory burdens under the act.

- The Communications Act ensures that a cable operator that offers telecommunications services is not required to obtain a franchise for the provision of such telecommunications services. This limits the authority of the local franchising authority to regulate the service the cable operator offers.

Obscenity and Violence

Updating the Telecommunications Act of 1934, the Telecommunications Act of 1996 establishes criminal penalties, including fines and imprisonment, for those who make, create, or solicit and initiate the transmission (by means of a telecommunications device) of obscene, lewd, lascivious, filthy, or indecent material with an intention to annoy, abuse, threaten, or harass another person, especially when that person is a child.

The act establishes fines for transmitting obscene programming on cable television. It allows cable operators to refuse to carry certain programs and paid programs or advertisements that contain obscenity, indecency, or nudity. It goes on to protect any provider or user of an interactive computer service from any liability on account of voluntarily restricting access or availability of material that it considers to be obscene, lewd, etc.

The act creates provisions for the FCC to establish a television rating code and requires the manufacturing of televisions that allow the blockage of programs.

Effect on Other Laws

The Telecommunications Act effectively supersedes the MFJ/AT&T consent decree and GTE consent decree. The act also preempts local taxation on direct-to-the-home services, e.g., direct broadcast satellite.

FCC and State PUC Obligations

Throughout the act, much of the detailed implementation burden is left to the FCC and ultimately to the states, including the FCC's obligation to

- Establish all regulations to implement the act within six months.
- Develop access standards, determining what network elements should be made available for open access.
- Manage numbering administration by forming a committee to administer telecommunications numbering and the North American numbering plan and to make them available.
- Establish a federal-state joint board on universal service to determine what services are covered under the federal universal service support mechanisms and a timetable for implementation. The FCC

will implement the recommendations of this board within 15 months of the enactment of the Telecommunications Act.

- Establish rules for interexchange and interstate services within six months, requiring the rate charged by a provider for interexchange services in rural and high-cost areas be no higher than it charges in urban areas.
- Establish regulations eliminating market-entry barriers.
- Establish regulations for infrastructure sharing.
- Review BOC applications to provide in-region interLATA service and form a determination within 90 days.

and the state PUC's obligation to:

- Act as mediator.
- Act as arbitrator.
- Hear disputes and petitions and resolve issues within nine months from the date on which the LEC received the request.
- Determine pricing standards, establishing just and reasonable rates for:
  Interconnection and network element charges
  Charges for transport and termination of traffic
  Wholesale rates

based on recovery of cost and a reasonable profit.

- Review all interconnection agreements, approving or rejecting them within 90 days of submission.
- Review statements by BOCs and LECs of compliance with the Telecommunications Act's interconnection provisions, responding within 60 days. (They can get extensions.)

State's regulatory authority

The act does make provisions for the states themselves to impose the requirements necessary to preserve and advance universal service, protect the public safety and welfare, ensure the continued quality of telecommunications services, and safeguard the rights of consumers. A state may adopt regulations, not inconsistent with the FCC's rules, to preserve and advance universal service.

Exemptions

The act's sweeping measures affect all but a few service providers; the two main exceptions noted in the act are designated rural telephone companies and wireless/cellular companies.

FCC and Courts Rules and Rulings on the 1996 Telecommunications Act

Subsequent to the signing of the act, the FCC began to pass rulings detailing the implementations of the act's three main goals, which are to:

1. Open the local exchange and exchange access markets to competition.
2. Promote increased competition in the telecommunications markets that are already open to competition (including long distance).
3. Reform the Universal Service System to preserve and advance universal service as the local market moves from a monopoly to a competitive arena.

The August 8, 1996, rulings by the FCC addressed the first goal—that of opening the local exchange market to competition by setting forth arbitrated rules and guidelines for: interconnection, unbundling network elements, access and collocations, transport and termination, pricing methodology to be used by the state, and default price ceilings and percent discount ranges.

Specifically the rulings declared the following:

Interconnection

Incumbent LECs must provide interconnection at any "technically feasible point" in their network, including:

- Line and trunk side of the local switch
- Trunk interconnection points for a tandem switch
- Central office cross-connect points
- Out-of-band signaling transfer points
- Point of access to unbundled elements

Unbundling network elements

The FCC defines a "minimum set" of seven network elements to which incumbent LECs must provide access:

- Network interface devices
- Local loops
- Local and tandem switches (including all software features provided by such switches)
- Interoffice transmission facilities
- Signaling and call-related database facilities

- Operations support systems
- Operator, information, and directory assistance facilities

The FCC notes further that *the incumbent LEC cannot limit nor restrict access of use of these elements!*

Access methods

The FCC requires incumbent LECs to provide *physical collocation* of equipment "necessary for interconnection or access to unbundled network elements" [p. 45476 Federal Register/Vol. 61, No. 169, 8/29/96]. LECs may also provide *virtual collocation* if they demonstrate the need for and adequacy of this alternative, to the state.

Pricing methodology and pricing ceilings

The FCC "strongly encourages" states to establish arbitrated rates for interconnection and access to unbundled network elements, using forward-looking cost methods for terminations, transport, unbundling, and interconnection.

While the rates for interconnection and unbundled network elements, are to be based on costs, wholesale rates for resold services will be based on retail rates less "avoided costs." The default discount range for resold services is set at 17 percent to 25 percent of the flat-rate basis.

LECs cannot "de-average the cost of providing interconnection nor unbundled elements" based on classes of service (e.g., business versus residential); it can, however, de-average these costs based on geography.

The FCC states that these aggressive rules have been established for resellers who otherwise have no bargaining power with incumbent LECs. By establishing an aggressive pricing methodology based on the TELRIC (total element long-run incremental cost) of providing a particular network element, plus a reasonable share of forward-looking joint and common costs (including risk-adjusted cost of capital and depreciation rates), the FCC sought to reduce the ability of incumbent LECs to "engage in anti-competitive behavior."

This thus allows new entrants to receive services based on their cost, as provided in a fully competitive (i.e., most efficient) marketplace, versus their cost in a monopolistic (i.e., less efficient) environment as well as reflecting a reasonable *utilization* of the network elements. Here, per-unit costs are derived from total costs, using reasonable "fill factors" based on a "reasonable projection of actual total usage of the element." The incumbent LEC bears the burden of demonstrating the "business risk" that it faces in complying with these prices.

The FCC further specifies that rate-of-return and other rate-based costing and pricing tools are not valid for the states to use when they set rates for interconnection and access to unbundled network elements.

The incumbent LEC's, new entrants', and courts' response

Following the issuance of these rules, the incumbent LECs, especially the RBOCs, responded in earnest with a legal appeal to a federal appeals court. Focusing primarily on the pricing issue, they argued that the FCC's ruling was unfair, that it did not accurately reflect their actual cost of providing the various services and network element nor of supporting interconnection. They further argued that the FCC's rules would offer an unfair advantage to their reseller and new market entrant, and was effectively a losing business proposition for the incumbent LECs. One RBOC maintained that such unfair compensation would result in its inability to continue to invest in its rural markets as well as its urban ones. The appeals court agreed to stay the implementation of the FCC ruling until sufficient time had been allowed for the LECs to more fully argue their case. Then, the FCC, joined by AT&T, MCI, and other major players seeking to enter the local markets, appealed the appeals court decision and raised the question ultimately to the Supreme Court, where Judge Thomas upheld the lower court's decision to delay implementation of the FCC rules until further testimony could be heard.

Observations

So it goes! When considering these activities in light of the greater goals of their act, one recognizes that we now indeed have *a new game—a fully competitive game!* In this new game, what is missing from the "discussion" among the LECs, courts, FCC, states, etc., is:

- How to *protect the old network* while establishing this new competitive game
- How to *encourage and build the exciting new networks* that the vast technical advances currently allow, thus establishing a new network infrastructure architecture capable of handling a fully competitive game as well as *new public data, video, and multimedia markets* at increasing narrowband, wideband, and broadband rates.

One must pause and consider what network infrastructure will really result from the current "negotiated" solutions, where the existing copper plant is totally opened up. While the LECs, courts, FCC, and states toil away—trying to open up the existing architecture, containing a network which was not designed for this new open and competitive game—it has become exceedingly evident that they are, in fact, attempting to solve the wrong problem. Their success in establishing new services and exciting new network capabilities for a new competitive information marketplace is ultimately achieved by changing the inherent limits of the current underlying infrastructure, in its present form. The act, if improperly implemented, can result in chaos; if properly implemented, it can provide *incentives* to anyone who wishes to build a new network infrastructure (or enhance the existing copper plant) with new competitive interfaces. Further, as the actions of the LECs, FCC, and others verify, they have been focusing their attention on exploiting the existing limited infrastructure in ways it was never intended to be used (or indeed abused). Hence, this is not the way to proceed with such a tremendous opportunity for *openness*. A proper interpretation of the act, which encourages the use of new technologies to achieve a fully distributed and integrated array of new transport and switching (narrowband, wideband, and broadband) offerings, is presented in "Part 3: The Competitive Information Infrastructure" of this book, which outlines the structure and networks required to address these concerns.

The competitive marketplace cannot exist without a new, underlying, robust public information infrastructure. Simply put, using and abusing the current copper plant will not achieve meaningful advances for America and will restrict its citizens from competing effectively in the New Information Millennium.

Appendix

B

Information Technology

Changing technology is challenging us to achieve significant advances in society. It was once said that if we simply understood a new technology's terminology, we would be three-fourths of the way to an appropriate market application solution. This still applies today and is even more appropriate as technology makes faster and greater advances. Today's technical possibilities have become the portal to tomorrow's marketplace opportunities. In fact, technology could be viewed as the bricks and mortar used to construct the highways and byways of the new network information infrastructure, upon which future information service vehicles will move to and fro to satisfy applications. As seen in the complexity of Figs. B-1 and B-2, our information highways' support structures will be constantly changing to formulate a more expanded, versatile broadband structural base.

In the widely acclaimed television series "Star Trek," it is readily apparent that their exploration of the universe would have been quite limited and humdrum without the "transporter," which enabled Scottie to "beam" everyone up and down from here to there. Similarly,

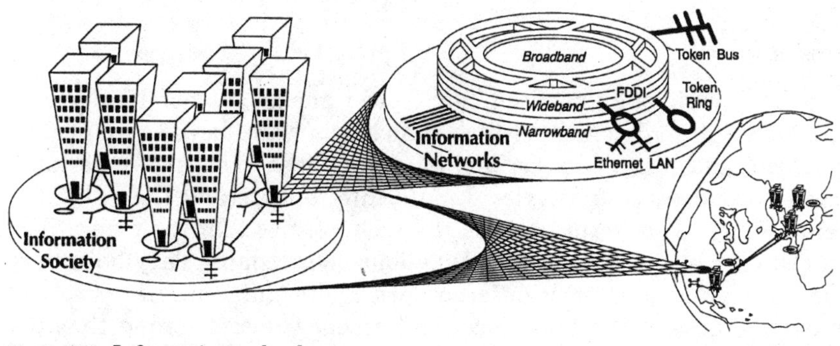

Figure B-1. Information technology.

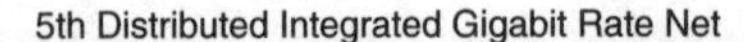
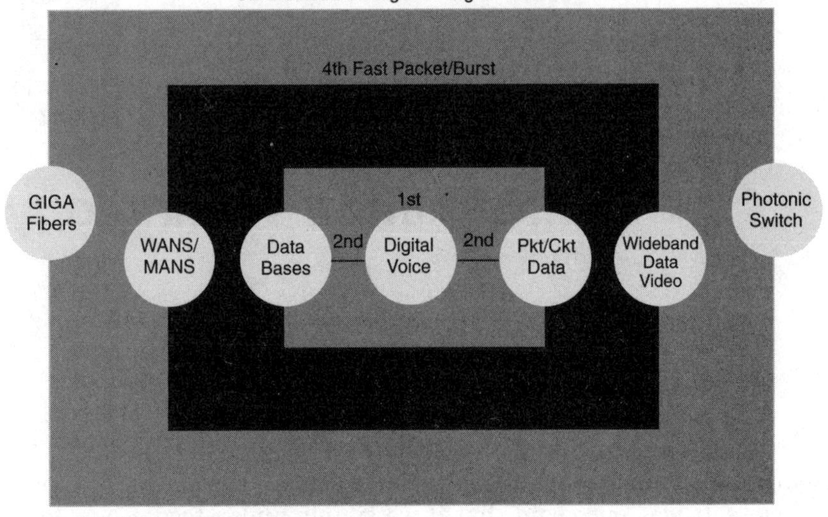

Figure B-2. ISDN technology changes.

the capability of communicating directly to anyone from anyplace was as essential as the ability to talk to the computers and instantly receive information from a different database. This made their existence in the new universe quite plausible and survivable. But what was also essential to their existence was the relationship of its participants to their technology. It was the key to their success. Technology was not feared, but embraced, as each person attempted to use technology to his or her maximum capabilities. As we "timorously" step into the Information Age, we need to pursue technology and increase our understanding of its possibilities and opportunities, as well as its limitations and restrictions. With this in mind, let us embark upon this pursuit by first considering an interesting analogy that was used by a visionary executive as he noted the need to pursue the changing, challenging communication and computer technologies.

Analogy

In today's world of high-priced, high-performance automobiles, it's a lot more difficult to determine which vehicle you want, beyond simply color, style, and price. Any serious buyer of these expensive cars must move beyond color and style to more fully understand the value the vehicles deliver in terms of both the technologies that the cars utilize and the use of the automobile. Take four-door sedans. They have very similar styling, but very different prices. To differentiate cars, you must understand the differences in horsepower and torque that the various one-liter, three-liter, and even five-liter engines deliver. This is often expressed in terms of number of cylinders, valves, turbo chargers, multi-point fuel injection, super chargers, etc. From these terms, the customer needs to translate these technical possibilities into results such as a more quiet and comfortable ride, less work on the engine, longer-lasting engines, reliability, faster acceleration, and better gas mileage.

What's Under the Hood

No longer does a customer simply buy a car for its looks; the customer must also consider what's under the hood. To determine if a car is worth the asking price, it's important to lift up the hood, know what technologies are involved and then translate these technologies into the vehicle that meets your own particular application needs.

Only the Shadow knows

For example, an advertisement of a Dodge Shadow said, "The new Shadow is available with a 2.2-liter, 4-cylinder engine that uses a variable-nozzle turbo charger (VNT), which reduces turbo lag for quicker throttle response. This means that since turbos are quicker but more raucous, the VNT Shadow is a sneaky, fast, budget hot rod."

The Yellow Brick Road

Similarly, as we look at the information vehicles that will be moving along the "information highway," as we follow "the yellow brick road" to the information marketplace, we need to more closely understand the technical possibilities and the differences that the narrowband, wideband, and broadband networks will have in terms of the features and services that will meet the requirements of different applications. For example, today when we talk about the public voice network, we are now including the capabilities that are provided by the SS7 signaling CLASS family of services, so that we can now offer selected call-transfer capabilities in order to block incoming calls and transfer selected calls to, say, a car phone. Or, if you want to leave a message for some particular individual, then that selected call can be routed to that waiting recording. Alternatively, incoming voice messages can be translated into data messages on a selected basis and routed over the public data network to some other point in the world.

Data Networking

We need to understand what these networks can deliver, technically, in order to appropriately provide the right technology to the right application. For example, data networks. When we speak of data networks today, we need to understand that there are 30 or so data-type features that are attributed to a data network. At one time the RBOC thought that, to meet the needs of data users, all they had to do was put a packet switch into the network and indicate to their potential data customers that the RBOC now had a data network offering. However, they missed the opportunity of adding broadcasting, polling, delayed delivery, guaranteed error rates, alternate routing, three-attempt limit, selected data interexchange carrier, protocol conversions, code conversions, bit and byte interleaving, encryption, audit trails, etc. Those added features would have ensured the correct delivery of the message by the network. Perhaps this oversight was due to a lack of understanding of the new data-industry terminology.

ISDN Networks

Similarly, when we discuss ISDN (Integrated Services Digital Network), we must learn to realize that there are three different types of networks; narrowband, wideband, and broadband. For the n-ISDN (narrowband) Basic Rate Interface (BRI ISDN), we're also discussing access to a public data network at 64-kilobit rates in digital form from user to user. This then provides an opportunity to no longer have the error rates that are involved in analog-to-digital conversions. n-ISDN enables the network to move information in either a circuit-switched data mode or a packet-switched data mode. This then requires us to understand the various technical differences and opportunities of circuit and packet switching in terms of what the standards are that interface terminals to the network or between networks, such as X.25 and X.75. We also need to understand what a packet is, how long it is, etc., and what the differences are between connection-oriented and connectionless movement of packet information. This need for technical knowledge is further indicated as we begin to interconnect LANs over wide area networks (WANs), using T1 switched services and T3 switched services. This then leads us to Primary Rate Interface ISDN (PRI-ISDN), which can also be called Wideband ISDN (W-ISDN), followed by Fractional ISDN (f-ISDN). Here we will interconnect wide area networks using *N* number of 64-kilobit channels, up to T1 and perhaps T3 rates. When we move further into the arena of internetworking private networks with the public network, we must learn to understand what FDDI is, what it delivers, and what a cell relay ATM (asynchronous transfer mode) switch is and how different it is from an STM (synchronous transfer mode) switch. What are the various protocols that are available within a FDDI network versus a SMDS network? Are these networks connection oriented or connectionless?

Finally, as we get into the world of broadband movement of multiples of 50+ megabits, we must learn to differentiate what services are truly at the 155-megabit rate versus the 600-megabit rate versus the 2.4-gigabits rate. For example, we have learned that we can digitally move x-rays at 128K (which is two 64-kilobit channels) successfully at low bit error rates (BERs) of 1 in 10^{11} instead of 9,600 bits per second, at 1 in 10^4. This allows us to move a large volume of information so that doctors are able to give a diagnosis versus simply having a consultant look at the x-ray. On the other hand, when we have thousands and thousands of doctors looking at and moving x-rays with patient records, we need a much faster transport vehicle, one that will move the lower speed packets of this medical information more quickly so there is a minimal delay in the actual transport mechanism. This will save time at the originating source. For example, the lab technician would have a great deal of problems waiting for x-rays to move around the system at 128 Kbps, if he or she is attempting to meet the needs of many doctors versus simply sending a few x-rays here and there to a couple of remote doctors. So, the more successful we become, the more information we will move. This leads to transport-volume problems, which provide a completely different set of transport issues that must be handled by higher speed and more powerful networks, which can also provide better resolution of imaging and video services.

As we start looking at secure and survivable (S&S) transport, as the world becomes more and more dependent upon the automatic movement, processing, and analysis of information, we then move into the world of audit trails, passwords, access keys, and encryption mechanisms, as well as self-diagnostic, self-healing automated recovery-type systems that ensure that the networks are up at all places, to all people, all the time. (See Fig. B-3.)

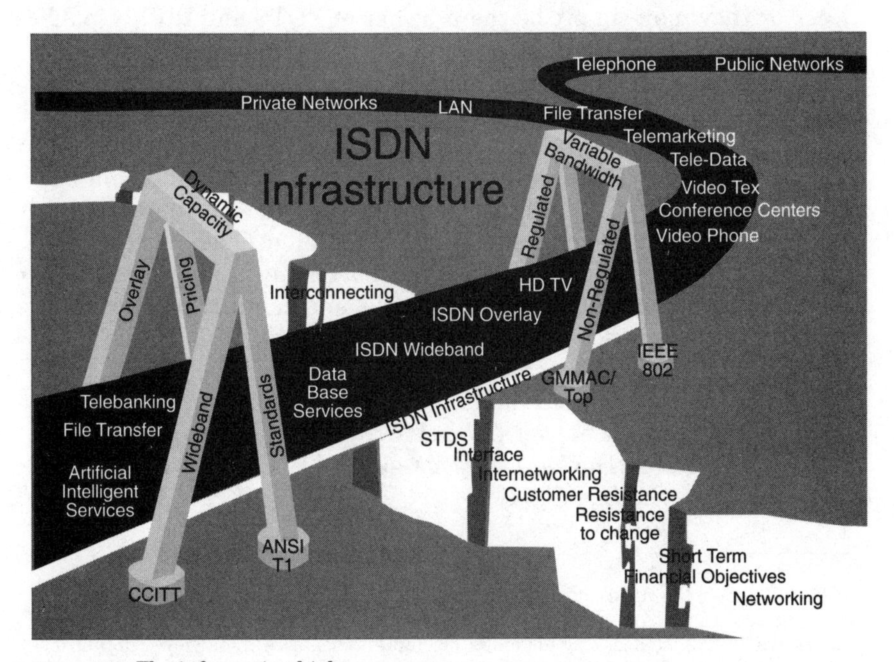

Figure B-3. The information highway.

The Information Highway

As we travel down the "yellow brick road" searching for the "information marketplace," we must understand the structural support technologies that future information services vehicles will require for the different information highways. In truth we will find that these interconnected information pathways, together with the private CPE applications that are interconnected through these information pathways, formulate the "information marketplace." Therefore, we need to understand the differences and technical possibilities of the various "pathways" of the "yellow brick road" as it leads us into the forthcoming information society.

With this in mind, let's find our way, at a leisurely pace, through many of the voice, data, and video technologies in order to formulate a reference from which to pursue future networks' service strategies. Since both technical and nontechnical readers will be with us, I will use footnotes, inserts, and references to Appendix C to further pursue the more complex aspects of the various issues. In this manner, those of us who have further interest in specific topics can pause along the way for a better, more involved look at more complicated facets.

POTS and PANS versus POTS and PIPES

It has been noted that providers of the public network need to shift from plain old telephone services (POTS) to PANS (pretty awesome new services), or they may simply be the provider of POTS and PIPES (public information pipeline equipment services). Let's understand what the base network infrastructure is today and what we mean by simply becoming a PIPE network versus a PANS.

It's an interesting exercise to examine the telephone company's product line and how it makes money. When we look at an equipment supplier, we see its products. Its designers tell us what new products are under development; its manufacturing plants tell us what current products are made; its marketing folks tell us how many products were sold, and its finance people add, multiply, divide, and subtract product-based numbers to tell us how much they made yesterday, last week, last month, and last year. They also attempt to predict the future, based upon numeric extrapolations and market projections.

So what is the public network? What are the private networks? Where are they both going? How are they merging?[1] Let's begin with a look at the public network. (See Fig. B-4.)

---

[1]We will assess the future of the integration of P&P in terms of twelve networks, discussed in depth in Chapter 6. But first, we need to understand what today's visionary leaders are saying. Hence, this appendix attempts to coalesce their views, definitions, opinions, and concerns as expressed throughout the 1990s. (See the Acknowledgements at the back of the appendix.)

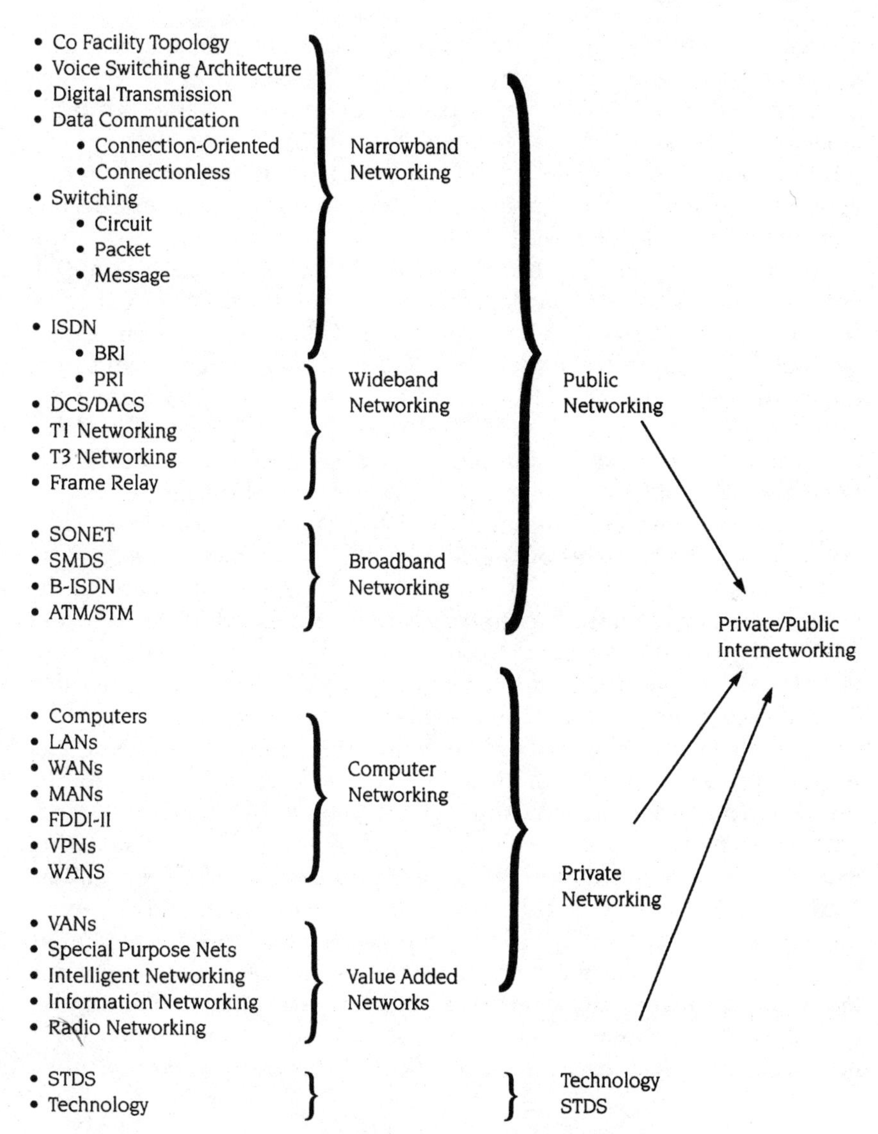

Figure B-4. Technology networking model—a technical review.

Public Voice Networks

Traditionally, telephones were connected over voice-grade twisted-pair lines to a wire center. The wire center may be a 50,000-line central office (CO) or a community dial office (CDO) serving several hundred subscribers. The CO or CDO provides services to its subscribers (customers) via a copper-pair (24-26 AWG thickness) cable within a 2.5-mile (12,000-foot) radius. Primary feeders between CO and CDO may consist of as many as 3,600 copper-wire pairs. Secondary feeders from CDO to remote terminals (RTs) such as a distant office park consist of as many as 1,800 copper pairs. From the secondary feeder, RT pairs trunk off to distant individual subscriber locations over a 9,000-foot geographical radius. Territories served by remote carrier terminal sites (RTs) are called carrier serving areas (CSAs).

It may be helpful to update this traditional picture by noting that, in actuality, the feeder plant extends much greater distances than 12,000 feet; in practice most subscribers exist within 10,000 feet, many within 2,000 feet, but the length can run 18,000 feet or more. Distribution area (DA) nodes home in on CSA points. They are usually located 2,000 to 3,000 feet from the CSA, and contain SAC (serving area connector) equipment to interface directly with 200 to 600 homes, which are approximately 1,000 or so feet from the DAs. Smaller 100-pair distribution points, formally called "B" boxes, have, for the most part, been replaced by DAs. Major businesses in the area normally "home" directly on the CSA's RT and are within 2,000 feet of its location.

Subscribers connected through the CDO or remote terminals have access to all the features of the central office, to which they are ultimately linked. As noted, the present wire-pair media are limited to the voice frequency bandwidth of approximately 4 kHz, which is sufficient for POTS (plain old telephone services) but not for high-speed data services. Loading coils are used to improve baseband (4 kHz) transmission for distances longer than 18,000 feet. Wire pairs without loading coils are capable of supplying wider bandwidth than 1.2-, 2.4-, 4.8- and 9.6-Kbps data rates. Special conditioning (C1 and C2) and transmission line balancing enabled 14.2-Kbps, 19.1-Kbps, 56-Kbps, and 144-Kbps data services, as well as supplementary short-range video. The digital loop carrier (DLC) or digital subscriber loop carrier (SLC) consolidate traffic into a prearranged T1 transmission path upon conversion of analog voice to digital (A/D), functioning as a time slot interchange (TSI) system. In this manner, pair gain is accomplished by putting many voice conversations on a single line, returning it greater distances back to the main office, where digital switching takes place or it is converted back to analog for older switches or transmission media.

Digital

But what do we mean by digital and T1? Analog voice signal frequencies and levels are sampled 8,000 times a second. Their status is coded into an 8-bit word called a byte, where one bit indicates the status of the conversation (on hook or off hook). In this manner, $8 \times 8{,}000 = 64{,}000$, or 64 kilobits per second represent the voice conversation. These can be multiplexed into a higher speed transport system in which the conversation is provided one of several transmission slots (like seats on a train car). In fact, using the T1 transmission system, there are twenty-four slots (words), or 24×64 Kbps = 1.536 Mbps.

In reality, another way of looking at it is in terms of frames, where a single T1 frame consists of twenty-four samples (8 bits) from each of twenty-four voice conversations. These are packaged together with a framing bit used for synchronization to complete the frame of 24 × 8 = 192 + 1 = 193 bits. Since each conversation is sampled 8,000 times per second, there are 8,000 frames sent each cycle. This means 8,000 frames × 193 bits per frame = 1.544 million bits per second that are sent to actually represent the twenty-four conversations. The difference of 1.544 – 1.536 Mbps is the single framing bit sent 8,000 times a second as 8 Kbps. Frames are clustered in groups of twelve to become super frames (SF) and then in groups of twenty-four to become extended super frames (ESF). AT&T uses channel banks to achieve this analog/digital conversion and multiplexing, called D4 and D5. Signal and control codes have been developed in the last framing bit so that the sixth and twelfth frame form special call-status messages. In fact the last bit, sometimes called the F bit of the frame, can send repeated numbering patterns for synchronization of timing and provide several 4K signaling paths to provide an array of signaling information, out of band from the voice information. Thus a hierarchy of transmission systems can transport a high number of multiplexed voice calls, when changed to digital 64-Kbps channels. The analog to 64-Kbps digital conversion is called pulse code modulation (PCM). Using different coding techniques called adaptive PCM (ADPCM), the voice could be represented at a subchannel rate of 32 Kbps. Other compression techniques, existing today, can rob bits to further reduce the voice representation to 24 Kbps, some to even 9.6 Kbps. However, there is (some may disagree), less quality in these representations, as more and more of the bits are removed. Hence, the industry has settled on 64 Kbps. In the future, two ISDN "B" channels over a twisted pair can represent audio stereo quality, achieving an effective 7-kHz bandwidth capability for high-fidelity music.

A single 64 Kbps is called a DSO, while the T1 rate of 1.544 Mbps for twenty-four voice conversations is also referred to as a DS1; similarly, T2 (ninety-six voice conversations), which is 4 T1s, 6.312 Mbps, or DS2. Though used throughout Japan, in actuality, the U.S. industry's T2 multiplexors M12 skipped to M13 for T3 rates of 28 T1s (for 672 voice conversations) at 44.736 Mbps, commonly referred to as 45 Mbps, DS3, or FDS3 (fiber). Hence, T1 and T3 transmission facilities have become the backbone of the digital transmission plant of the telephone companies over the 1970s and 1980s. Europe's Common European Postal and Telephone STD group (CEPT) has worked on a similar approach and elected to use the El rate of thirty VF channels and one channel for supervision/control signaling, as well as one channel for framing synchronization, as a ((30 + 1) + 1) × 8 = 256-bit frame × 8,000 = 2.048 Mbps, known as "30 B + D."

The growing need to change existing analog plants to support digital T1 and data has always been confronted with hidden *bridge taps.* These taps are hard to detect when changing analog voice-grade lines to digital, where they degrade T1 and data. Bridge taps are parallel connecting nodes somewhere between the normal end-points of analog lines. They were attached anywhere along an available analog line to provide new service. The balance of the circuit (line) is not cut off, but remains in place as a dangling open pair of lines. These taps do not usually affect voice conversations, but their crosstalk frequencies cause unacceptable data bit error rates (BERs). Hence, bridge taps, echo suppression, load coils, noise crosstalk, and water in the cables cause many designers to refer to the real-world plant as "hostile" to data. Traditional analog-switching system-line circuits dealt with the BORSCHT functions. (B means battery supply to subscriber line; O means over-voltage protection; R means ringing current supply; S means supervision of terminal; C means coder/decoder; H means hybrid (2-wire to 4-wire conversion); T means test.) These BORSCHT functions must now also be accomplished, but somewhat differently, in the digital environment.

To accomplish digital transmission, the echo, intersymbol interface, impulse noise, and crosstalk problems must be overcome. Echo cancellation is achieved by an adaptive device that monitors the echo impulse response effects of the line and generates an accurate replica of this echo. It then mixes the replica with the incoming signal to cancel the echo interference. A modem uses adaptive equalizations to continually analyze and compensate for variations of transmission quality in the subscriber loop.

Data

As data was integrated with "digitized voice" channels to share digital transmission facilities, the major concern was loss of framing synchronization due to a long pattern of zero bits, because the T1 density rule requires at least 1 bit per 15 data bits. Since the eighth bit of each 8-bit byte channel had been used for voice status signaling, it was also available for data. Therefore, for every 7 bits of data that was sent per channel, the office channel units (OCUs) added 1 bit for data calls in the eighth bit. When the channel was idle, it inserted a zero in the eighth position and all ones in positions two through seven, thereby ensuring no consecutive eight zeros. Therefore 7 bits of data inserted 8,000 turns per second becoming an 8×7 or 56-Kbps data link in an $8 \times 8 = 64$-Kbps channel. Thus, T1 could send 24 channels of 56-Kbps data, or 1.344 Mbps. It is interesting to note that the 56-Kbps data channel can also accommodate lower rates up to 48 Kbps, such as five 9.6 Kbps, ten 4.8 Kbps, and twenty 2.4 Kbps, as dataphone digital services (DDS) were transported on T1 facilities.

Next, there was a need to obtain what is called clear-channel 64-Kbps data-handling capabilities; so T1 can transport twenty-four channels of 64-Kbps data, especially as ISDN services became available. To achieve this, two techniques were introduced: one was called B1 Polar with Eight Zero Substitution (B8ZS) or Binary 8 Zero Substation. It provided bit stuffing and substitution changes, while a less expensive ZBTSI (Zero Byte Time Slot Interface) method required overhead bits in the Extended Super Frame format about the location of all zero bytes.

New techniques and very large scale integration (VLSI) circuits now enable adaptive equalizers, enhancements to echo cancellation techniques, as well as advanced filtering and coding to enable high bit rates digital subscriber line (HDSL) rates to achieve 800-Kbps and 1.6-Mbps (using two pair) transmission over nonrepeated, unconditional copper loops that conform to Carrier Serving Area (CSA) rules for maximum cable lengths of 12,000 feet of 22-24 AWG cable and 9,000 feet of 25 AWG cable. This expanded capability enables the LECs' 70 billion dollar plant investment in metal loop to be used for more and more DS1 wideband services, as well as ISDN primary-rate services, which are also at the 1.544-Mbps rate.

Data Communications

Charles A. Peker noted the following points in his excellent assessment of types of data movement, "To Connect or Not to Connect" (*Telephony*, June 11, 1990).

In considering the movement of data, there are two categories of service to consider: connection-oriented network services (CONS) and connectionless network services (CLNS). The current voice telephone network is an excellent example of a connection-oriented network. As in establishing a voice call, the functions for a CONS data call perform connection establishment, data transfer and connection release. Connectionless data communication eliminates the first and last steps and simply transfers the data. There is no connection relationship established across the network. Using a connection-oriented network protocol such as X.25, first the originator sends a packet to the network, requesting a connection to another network user. The packet contains the connection request, destination request, and other information such as billing. The network determines a path to the destination terminal, as well as its status, resolves billing considerations such as reverse charges and notifies the originator that the connection is established. Data exchange begins using a logical count number that eliminates further connection-overhead analyses, thereby increasing the payload. The network detects errors, sequences the packets, and resolves congestion using sequence number, acknowledgments, and retransmissions, as well as alternative routes and delay delivery mechanisms. In this manner, the network takes over and adds value to the data transport.

In a connectionless network, the originator simply forms a packet that contains the information to be transferred, putting the destination address in the packet and sending it to the network. The network treats the packet as a one-of-a-kind message. It forwards it to the destination, not caring if it is lost or in error. It lets the communication between the two end terminals resolve transport difficulties, and the network simply provides the transport function. Depending upon the amount of burst and the length of the message, there is a crossover in advantages from the extremes of CLNS to CONS. For a continuous flow of isochronous data from which the receiver derives time of arrival, the network paths need to be reserved in advance. In this manner, virtual or physical connections can be established to avoid delays and missequencing.

Data transmission in a connectionless (CLNS) environment requires each packet to carry the entire overhead, while in a CONS environment subsequent packets carry less overhead. Ethernet, FDDI, and Switched Multimegabit Data Services (SMDS) are examples of connectionless protocols versus connection-oriented protocols, such as SNA (IBM's System Network Architecture) and token ring. Data internetworking becomes an issue as these network links have limited functionality. Therefore the integrity of the data is the responsibility of the end-to-end generators and receivers of the protocol, where they use such protocols as the current standard inter-LAN protocol TCP/IP (Transmission Control Protocol/Internet Program), as an equivalent level-four protocol of the Open System Interconnection (OSI) architecture. As we will discuss later, customers may in practice interconnect several connectionless LANs over a virtual private network in which the infrastructure is connection oriented. B-ISDN work is largely connection-oriented, while the first broadband service that will eventually migrate to B-ISDN will be the connectionless SMDS. But we are getting ahead of ourselves. Let's first take a look at how we got to where we are today in switching.

Switching

In 1878, the American District Telephone Company of Chicago obtained a patent for the multiple switchboard, and Western Electric obtained its manufacturing license. The switchboard used two banks of jacks: one for incoming, the other for outgoing. It was first in use in 1885. The two firms then grew together to become known as AT&T's "Ma Bell." In 1889, your friendly undertaker, known to GTE's employees as "Father (A. B.) Strowger" patented the step-by-step switch installed in 1892 in La Porte, Indiana.

From this beginning came subsequent systems such as: StrombergCarlson's Rochester rotary system and AT&T's automatic panel exchanges for big systems, both in 1910; crossbar (including the Marker) in 1924; L. M. Ericsson's LME 5000 for small capacity rural areas, in 1930; No. 4 crossbar, in 1941; No. 5 in 1948; No. 4A in 1950; EMC rotary switch system in 1955; the experimental electronic switching system using stored program control, Morris, Illinois, in 1956; the GTE Autovon core-reed wired logic military switch in 1968; and then the AT&T No. 1ESS and 2ESS stored program systems in the 1970s. The shift to digital began with the 1975 AT&T #4ESS digital toll and the 1978-82 Northern Telecom and GTE local integrated digital systems; next, in the 1980s, AT&T's digital ISDN switches and DCSs, and then in the 1990s, Fujitsu's broadband switch and AT&T's photonic switch, and so it goes.

Hierarchical Routing Structure (Class)

From 1910 to the 1990s, a switching structure has been developed in the local communities to remove the hundreds of wires going from each private location to its subscribers. In the 1920s, Wall Street was a maze of telephone lines. There were up to 29 separate noninterconnected companies in New York City. This led to stock brokers having as many as five to ten phones on their desk in order to reach customers served by the many different telephone companies. As the United States grew, so the telephone systems grew to be more sophisticated, as they handled more than fifteen million long distance messages a day from a comprehensive network of more than 300,000 long-haul trunks, which interconnect 1,600 long-distance switching offices and 30,000 local offices served by 1,900 telephone companies. Under the switching plan, switching and trunking arrangements were based on a hierarchical routing discipline noted as: end offices, "Class 5;" toll centers, "Class 4C;" toll points, "Class 4P;" primary centers, "Class 3;" sectional centers, "Class 2;" and a regional center, "Class 1."

Collectively, the Class 1, 2, and 3 offices constitute the central switching points of the distance-dialing network. A central switch point (CSP) is a switching center at which inter-toll trunks are connected to other inter-toll trunks. Each separate switching unit must be assigned its own classification within the hierarchical routing plan. The switchboard and trunks must meet VNL transmission-loss requirements. It is not necessary that Class 5, 4, or 3 offices must always "home" on the next higher ranking (class number offices). For example, a Class 5 may have direct trunks to another Class 5, or be served by a "classless" local tandem in an urban area of numerous Class 5s. Traditionally, a Class 5 will home on a Class 4, 3, 2, or 1; a Class 4 on a 3, 2, or, 1; Class 3 on a 2, or 1; and a Class 2 on a Class 1. High-usage trunks carry most but not all of the offered traffic in the busy hour. Call processing is performed at the local Class 5s to determine where to route the call, based upon the called party address, by using either a ten-digit or seven-digit translation. As time progressed, international prefixes have been added to note country and city codes, as well as distinguish where direct dialing or operator services are needed. ISDN now provides for fifteen-digit address codes.

Digital Switching

Digital was first introduced in transmission. It provided the ability to remotely collect and concentrate traffic by employing analog-to-digital conversion, multiplexing, concentration, and contention. Digital Subscriber Loop (DSL) systems, like the Subscriber Loop Carrier (SLC 96), enabled ninety-six voice lines to be multiplexed onto four T1 lines (with spare span control to move to alternate T1 lines if a transmission outage occurs). The T1 facility connected the remote line units to a host/base central office. However, many state regulatory groups required the ability to locally switch calls for police, medical, and fire needs in the event the central host machine or the transport facilities were disabled (down). This gave birth to GTE's Remote Switch Unit (RSU) as a key integral part of the Integrated Digital Network (IDN) plan for digital conversion. At its conception, there was much discussion as to whether to call the RSU a "Class 6" network switch point because there was indeed local switching, not just concentration. But, since it did not "home" on multiple, different, distant Class 5 central offices and was merely an extension of the line stage of the host digital switching matrixes, with nonstandard interfaces, this standardization was not pursued in order to give suppliers proprietary interfaces between host and remotes (base and satellites). One advantage to the provider was to obtain "clustered" system prices in order to encourage early change out of the entire area served by the base.

Unfortunately, these anticipated pricing advantages did not come to pass. Providers were prematurely "caping" base systems. This enabled more economically competitive systems to collocate and home their proprietary remotes in the same areas as the other proprietary remotes were connecting to the initial digital base, thereby losing much of the advantages of clustering. Due to the intense competitive local environment, providers also wished to place Remote Switch Units on customer premises; but due to NTCE regulations, the RSUs could then no longer be owned by the BOC operating company. Once owned by the customer, software updates could not be controlled and maintained if the customer did not elect to purchase them. Since this would prematurely freeze the feature set of these systems, few if any RSUs were customer owned.

During the 1990s, private network designers saw an increasing need for expanding internal switching on customer premises from PBXs and LANs to new internal broadband switches. These new switches will need to access multiple providers, as IXCs and VANs establish their local point of presence within the local monopoly. Initially, there is a need to interconnect distributed office complexes around the city as directly as possible from LAN to LAN or from internal switch to internal switch, using dynamic switched bandwidth at wideband and broadband rates. Hence, pressure has mounted for more and more public special services lines and trunks to be switched, especially as their provider's (considerable) maintenance costs continue to rise toward the anticipated year 2000 target of $100,000 per person per year, and as the time delay for the customer for the set up of special circuits increased to 30 days. Hence, point-to-point special services traffic is going through a rapid transition to become switched services, as new fiber is being deployed with its new high capacity capabilities.

This then leads to new pressures on the traditional hierarchy based upon voice, data, and video traffic congestion, due to the use of early voice-switching and transport-routing tables capabilities. Many of the knowledgeable players recognize that there will be less traditional class levels in the upper layers of the network as direct fiber trunking will provide much higher capacity. Also, some see the need for new class levels for satellite-to-satellite switched services and national switched radio networks for personal communications (PCNs). Conversely, on the local side, more class levels are needed as fiber ring switches are established closer to the user and traditional urban central offices are replaced by superswitches. Some believe as high as a four-to-one reduction will occur so that a typical urban area having 40-50 offices could only need 8 to 12 superswitches. These super Class 5s will have a remote front-end Class 6 switch that can "home" on other Class 6s or any Class 5 superswitch, as well as directly interface to a POP. Following this thought process, the customer premise systems became the new Class 7s and the ring switches for these super highways became the Class 6s. They are located at carrier serving area's remote terminals (RTs), or they replace small Class 5s. Tandem-type systems will in time no longer be needed, as superswitches home directly on each other or the POP or use the ring switch to reach the IXC. So it goes in a world of continuing changes; technology is challenging us to proceed in new directions.

Switching Fabrics

Switching systems today come in many types, with many types of fabrics, such as: circuit, packet, and channel!

Circuit switching

Traditional switches are circuit switches that effectively hold the path from an input port to an output port during the entire conversation or data transmission. The switching matrix for these machines was designed in increasing-size stages, so calls were given more and more multiple paths through the switch to the desired port. Three-stage, five-stage, and folded-five-stage were the terms to denote this build up, as each additional stage was designed to ensure minimum blockage was achieved. When we talk about long-holding-time data calls affecting a switching system, we mean this blockage that inhibits voice calls from passing through the matrix (especially its first and second stages) to the designated delivery trunk or line. In many instances, deloading of the switch could reduce the chances of blockage or redistributing (engineering) long-holding-time data callers to different first-stage groups, so they had less impact on any single first-stage matrix.

Digital time space time (TST) channel switching effectively selects and dedicates a path for an incoming channel that has been converted into an incoming PCM digital stream (in time) to be continually moved from this incoming channel position, held in a transmittal buffer (in space) and then inserted in an outgoing digital stream (in time) to its destination. There are, of course, many variations of this, such as TTSTT, TTSSTT, etc. Digital technology has enabled switches to become more and more nonblocking as greater numbers of paths were made available on microchips and wired backplanes.

The second design limitation for switches is their processor limitations due to extremely short-holding-time calls. They can bury either the front-end line stage (first stage) processors or the call-processing processors that perform functions such as number translations, MF multifrequency (office to office) signaling, call supervision and control, etc. A high-volume credit check or bank-transaction user could generate an abnormal amount of call-handling attempts, causing improper loading on these voice circuit switching systems that were designed for three-minute to five-minute holding-time calls at five residential calls per day and twenty business calls per day. These volumes translated into Erlang or CCS traffic that the switch is engineered to handle. (A. K. Erlang, the father of queueing theory, in 1918 assumed a Poisson distribution of calls arriving in a given time and an exponential distribution of holding times. This means that longer calls occur less frequently than shorter calls. One Erlang is equal to one full hour of phone conversation or $60 \times 60 = 3{,}600$ seconds. CCS, 100 call seconds, or 100 seconds of telephone conversation, is another method of describing traffic, where 1 hour of telephone traffic is equal to 36CCS ($60 \times 60 = 3{,}600$ divided by $100 = 36$). CCS (hundred call seconds) is converted into Erlangs by multiplying by 100 and then dividing by 3,600.)

Message switching

Message switching was the term used for store-type and forward-type switching systems that usually took a block of data information using some form of analog modem to modulate (amplitude, frequency, phase) transport frequencies to carry the data to a message center, where the data was received, demodulated, stored and then forwarded to another intermediate location or its destination. Delay was introduced as nodes talked to each other, asking for acknowledgment (ACK) of correct reception of information or request repeat (RR) of information until it was properly received. Various types of codes were used to represent the data such that the alphanumeric codes were represented in 1s or 0s in patterns of 5, 7, or 8 bits, depending on how many symbols were sent. ASCII was a popular 7-bit code that could represent 128 symbols indicating the alphabet, numeric symbols, and control commands for personal computers. The other main method of coding was IBM's EBCDIC (Extended Binary Coded Decimal Interexchange Code). It was their large mainframe 8-bit encoding scheme, allowing 256 symbols to be represented. Information was usually sent asynchronously, meaning independent of network clock timing, to input buffers (storage devices), which, when full, were then synchronously emptied into the network under the network clock. Alternatively, isochronous data was transmitted where the timing or synchronizing signal was devised from the signal carrying the data.

Packet switches

Packet switches evolved from early 1960 military message switches, where frequency shift key (FSK) systems delivered segments (pieces) of the message using a header, payload, and tail-type format for self-contained, closed networks. The header was designed to not only carry destination address information, but also to enable the receiver to sync onto the message isochronously. The payload usually contained a vertical parity bit for each character sent so that the number of 1 bits, plus the parity bit, ensured that an odd number of 1s were present in each 8-bit (byte) position. At the end of the payload, tail bytes were added so that all the horizontal 1s for each of the eight vertical positions would add up to an odd count. There were two tail bytes so that every other byte of the payload was checked in this manner. These vertical and horizontal parity bits and bytes were a method of ensuring that the information was transmitted successfully in the hostile telco plant environment of noise and crosstalk. Hamming fire codes and cyclic recovery codes were used to perform error correction and detection without resending the information. This then leads to various schemes of how long the payload should be to ensure that undetectable multiple errors did not occur. The short-packets versus long-packets issue challenged the amount of extra overhead time required to establish where each packet needed to go.

From these early systems, numerous private systems were developed using their own internal protocol of ACKs and RRs as they moved information along leased lines. Next, internal buses were deployed so that packets of information were simply sent to numerous terminals located a short distance from the mainframe. Here, each terminal looked at the header to see if the message coming down the pipe was for that terminal. Using two separate buses, information could be sent on one and received on the other as a four-wire, full-duplex system. Otherwise, a two-wire, single-bus, send-receive half-duplex was the cheaper, lower speed, but more complex alternative.

Therefore, from these basic concepts came many private networking implementations where these computer-to-terminal networks were called local area networks, such as DEC's Ethernet or IBM's token passing-type network called Token Bus or Token Ring. (The token enables the terminal to send its information on the bus/ring; The token is then passed to the next terminal to enable it to send its information.)

On the switching side, packet switches became more and more sophisticated as multiple-node networks were configured across the country so that each location directly accessed other locations over leased lines, microwave links, or satellite circuits. Thereby, N nodes had up to $N - 1$ direct paths to each other; the networks became more store-oriented and forward-oriented as packets were moved from one node to the other over less direct links. This gave way to multiple-nodal-point networks, where packet switches were deployed to network the data to the desired receiving location, usually at rates of 4.8 or 9.6 Kbps over dial-up circuit-switched paths, or perhaps at 14.1 or 19.2 Kbps over conditional leased lines. The protocols were designed to continually ask for retransmissions, assuming a hostile environment of bit error rates (BER) of 1 in 10^4. However, even as the facilities became better at error rates of 1 in 10^7 or even 1 in 10^{11} as T1 type facilities were utilized, the information was really sent at 9.6 Kbps, but packaged in 64-Kbps channels. Considerable overhead was added as end-to-end terminals or intervening nodes checked and rechecked messages for errors, talking back and forth, taking up time, and, in reality, dropping the throughput rate of 64 Kbps to 9.6 Kbps.

Narrowband Networking (n-ISDN)

The purpose of the Integrated Digital Services Network (ISDN) was to effectively build on the Integrated Digital Networks (IDN) of the late 1970s. The "S" was added to indicate both voice and data services. This really required a merging of voice circuit switching and data packet switching into the same switch. Hence, the multifabric machine blossomed, which could effectively take information in one side, split the information, depending on the type of content, and bring it to one type of matrix or the other, effectively switching it to a separate or integrated outgoing service. The key to this strategy was the out-of-band signaling, which did not take up the capacity of the content channels to perform the supervision and control of the call from source to destination. The interoffice CCITT SS7 signaling-system service could move called-party information to and from offices, while the customers were provided several types of interfaces to send and receive information.

Basic Rate ISDN

Basic rate ISDN provides the customer with a four-wire send-receive interface as a 2B + D transport bus, where B is 64 Kbps and D is an out-of-band 16-Kbps signaling and packet network. Hence, the 2B + D interface provides a 2 × 64 Kbps + 16 Kbps, or 144-Kbps information stream so that a customer can use one 64-Kbps B channel for a single voice conversation and simultaneously use the other 64-Kbps B channel as a data link, as well as have the D channel as a 9.6-Kbps data link. So a customer could talk and look at changing information at the same time. The 64-Kbps B channel could be either circuit switched like a voice call and be held up for the entire call duration, or it could be packet switched in a header-payload format at 64 Kbps. These "bearer" services could be also changed for each call so that the B paths could both be used for voice calls or both for data calls, thereby providing tremendous flexibility to the customer. The 16-Kbps D channel was initially launched with a D channel 9.6-Kbps packet network providing an internal interface protocol called X.25 to the user and X.75 to the network. In this manner, low-speed data networks could be added as separate fabrics to the current systems without blockage and without affecting call-processing capacity.

All in all, this presented several changes to the outside plant, as the lines from the central offices (COs) to the Network Terminating Device 1 (NT1) were required to move information at 160 Kbps for distances as long as 18,000 feet. Using special coding called 2B1Q, distances of 12,000 to 14,000 feet have been satisfactorily achieved. The CO interfaces to a NT1 network node for American interfaces, which in turn interfaces to a NT2 interface, which can reside on the customer's premises. The output of this NT2 node is a four-wire system-handling internal 192 Kbps to and from the customer. In fact, using the CSMA-CD type collision protection mechanism, eight terminals can be addressed on the same bus from the NT2, sharing the ISDN port. The "D" channel will be able to provide a whole host of supplementary services such as alarm messages, environmental control messages, meter reading, terminal network signaling, etc. Since "D" channel signaling can interface at the switch to SS7 type intelligent networks, conversations can exist between the customer and remote databases or information sources as value-added services are requested for terminals. The network switches may request routing instructions for particular destinations, for virtual network requests to establish (change) predefined (changeable) private network configurations.

Primary Rate ISDN

The primary rate interfaces (PRI) provide the customer a 23B + D channels interface consistent with the T-1 rate of 1.544 Mbps. Here the "D" channel is a 64-Kbps signal control vehicle for enabling the many users of a PBX to send their specialized control and ANI (automatic number identification) messages to the central offices, as PBXs begin utilizing ISDN primary rate capabilities. As primary rate services become more switched, we see the advantages of channel switching *N* number of channels of information here or there, as fractionalized primary rate capabilities are shared. PRI-ISDN will consolidate DID, DOD, Inwats, Outwats, and FX, and tie access trunk traffic into one, at different traffic capabilities such as at a P.01 (1 call blocked out of 100 call attempts) grade of service with various CCS capacities (centa call seconds). Note that the primary rate interface delivers twenty-three channels at 64-Kbps rates rather than the traditional 56 Kbps + 8 Kbps in band signaling. Depending on total CCS usage of the various services, we only need a limited number of PRI tie trunks to replace the more numerous separate trunk groups.

The 2B + D = payload of 144 Kbps + 16 Kbps = 160 Kbps works nicely on a two-wire pair between the CO and the NT1 using 2B1Q coding. However, going from the ISDN basic 160-Kbps rate to the primary rate, 1.544 Mbps causes more crosstalk and attenuation. This causes a smaller CSA range (18,000 feet), over which the DSL (digital subscriber loop) operates. It is reduced to the range of 1 mile or approximately 5,000 feet on a standard 26 AWG pair. This is considerably short of the 9,000-foot CSA guidelines. 2B1Q line coding greatly increases system gain by using (two-pair) dual full-duplex wires that operate at a 500-Kbps rate, requiring splitting and recombining to the DS1 rate. This brings the distance up to the 8,000 feet, near the 9,000 feet of the CSA.

Non-ISDN

Non-ISDN terminals can access the network through terminal adapters (TAs) that will provide asynchronous-to-X.25 conversions, as well as interface 2.4-, 4.8-, 9.6-, and 56-Kbps packages to the 64-Kbps synchronous interface. ISDN standards now exist for basic rate (BRI) and primary rate (PRI) for the various R, S, T, and U interfaces of the network (noted in Fig. C-1 in Appendix C as the inputs and outputs of the NT2 and NT1 points). Personal computers can now add one or two chips and be able to directly interface to these ISDN interfaces.

(Figure C-4 in Appendix C shows an example of AT&T's ISDN chips for each OSI layer of interface.)

Hence, the n-ISDN standards are being finally resolved as the IDN switching systems are equipped with ISDN interfaces. The users will now be able to utilize these multiple interfaces for both voice and data services. However, it should be noted that for packet switching capabilities, the IDN systems must add internal 64-Kbps to 128-Kbps packet-switching data-networking capabilities to their current systems in order to provide both a circuit-switched and packet-switched public data network, as a narrowband ISDN (n-ISDN) offering.

OSI Layers

The International Standards Organization (ISO) has adopted an open systems interconnection (OSI) model, which has seven layers, from physical to application. The first three layers (physical, data link, and network) are communications-oriented; the fifth through seventh layers (session, presentation, and application) are computer-oriented, with the fourth layer (transport) being the interface transition layer.

One view of the OSI model is noted in Fig. B-5. IEEE 802 LAN Networks have also been constructed on three layers of communications services as they proceed from physical to data link to network. The challenge has been to tie these TCP/IP, Ethernet, Token Ring, Token Bus, MAN, and FDDI layers to the OSI model. Internetwork protocol interfaces are needed to interconnect to X.25, SMDS, and Frame Relay, as well as FDDI II.

OSI, TCP/IP, SNA, XNS, and DNA define standards for protocol and message formats for compatible network-to-network connections. We have seen the ISO OSI standard seven-layer model. TCP/IP is a layered structure set of routing protocols developed for switched-packet networks used by the military and Internet. SNA is a mainframe-oriented IBM layered *defacto* standard similar to OSI; XNS provides interconnection between Xerox products, including Ethernet LAN and X.25 public networks; DNA is a proprietary DEC architecture for DECnet products similar to XNS. (See Fig. B-6.)

A detailed analysis of these subjects will be pursued in the reference section on ISDN. However, for the moment, let's pause to formulate a more global perspective of what is actually needed and why.

When we plug our telephone into the ISDN wall connector, we are using a physical interface to transport our voice information. The type of wires and the information on each connector pin is defined in the interface standard, similar to data modem interfaces defined by RS232C. When we dial our destination code (called party number), we begin establishing call set up information, containing destination identity and the communication requirements (priority, etc.). Call processing

LAYER	FUNCTIONS
7 Application:	User interface for applications to networks, such as: ISO 9595/9506 CMIP, ISO 9549/CCITT x.500 Directory Service.
6 Presentation:	Reformats, converts code connect message to common format, such as: CCITT X.400/X.410 file transfer, graphics, videotex.
5 Session:	Establishes coordinates end to end dialogue process; such as X.225 OSI connection oriented protocol.
4 Transport:	Insures end to end reliable data flow, quality of services. Performs PAD functions for network; such as: CCITT X.224 transport protocol.
3 Network:	Routes and switches message/information through the network; such as: X.25 packet level protocol and X.75 packet gateway.
2 Data Link:	Packages data and performs error handling control, maintaining internode integrity: has two parts; upper LLC (Logical Link Control) (802.2) organizes data flow and error control recovery; while the MAC (Media Access Control) performs addressing, enables shared transmission rules and error detection, using the following access methods - Circuit Switches: TDM, FDM; Packet Switched: CSMA/CD (IEEE 802.3), Token Bus (IEEE 802.4), Token Ring (IEEE 802.5), FDDI (ANSI X3T9.5), SMDS (TA 772/773 Bellcore), BISDN (ANSI T1S1); Hybrid (ATM/STM) (CKT/PKT), MAN (IEEE 802.6), FDDI-2 (ANSI X3T9.5), FDDI MAC/CCITT HDLC.
1 Physical:	Provides interface connection and network transmission interfaces; such as: physical media synchronization, clocking, RS232, RS499, CCITT V.35, transmits bit streams in different topologies, such as ring, bus, star networks.

Figure B-5. OSI model.

Example TCP/IP router interconnection LAN to LAN:

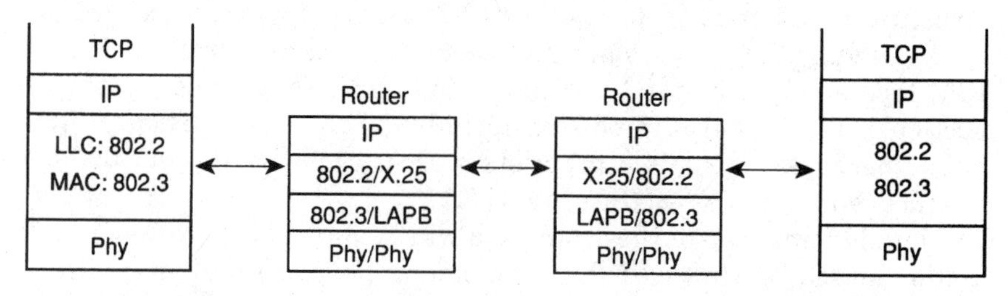

Figure B-6. TCP/IP model.

then defines what to do with the call and looks ahead to see if the calling party is busy or idle via the intelligent network. In the data packet world the disposition of call takes place by setting up the packets and headers to go from here to there. For connection oriented calls, interfaces and links are identified and established across the network and between the final network node and the receiving terminal. Next, the network layer protocols are used to communicate between ends as information is routed from A to B over the physical links as our phone connection paths are established and the voice messages are delivered. In the case of data gateways, protocol conversions and call connections take place to move the information from terminal with protocol X to network with protocol Y to terminal protocol Z. Once the information flow is achieved, then end-to-end terminal and computer systems effectively ensure the quality of the conversion as their information is packetized and depacketized, checked for error, and, if necessary, re-sent. This is equivalent to voice users adjusting their voice communications to be adequately heard at the other end.

For the voice conversation, it's left for the human to adjust to the various needs of understanding the conversation as our minds probe stored information to be accessed and utilized during the conversation. Parties may be speaking in different languages, requiring mental conversions from English to French, or whatever. The application becomes the subject of interest, ignoring the mode of communication. Two or more persons may be discussing a given topic as their minds probe for more and more information to reach understanding, agreements, and solutions.

Another analogous view of the functions performed at each layer of the OSI model is that of mailing a letter in the postal system. Here, the application is the writing of the letter at the application level, putting it into a format at the presentation level, mailing the letter in an addressed envelope at the session level, sorting at the post office, putting the letter in the mailman's bag, and then on to delivering it at the physical level. Whatever analogy you may use, the key is to separate the tasks at each level, with the detailed transport tasks at the first three layers and the creation, presentation, and packaging for distribution at the higher levels. For this reason, each lower layer hands off to a higher layer for a "broader" service; hence, the nuts and bolts of each protocol and transport scheme is usually expressed for the three lower layers, physical, data link, and network, as each performs the tasks just a little differently, requiring gateways, routers, and bridges in-between to provide appropriate interface conversions, translation maps, etc. (See Fig. B-7.)

Similarly, computer systems may need to perform code conversions, coordinate information as it arrives and ensure it is distributed correctly; a computer may need to store information; it may need to manipulate and process the information into a graph or image. These are the tasks of the upper layers of the model. In reality, it's in the final layer, the eighth layer, the undefined layer, where the money is made, as the users actually utilize the transmitted information to solve a given problem. Just as the meaning of what's said is contained in the details of the voice conversation, the information payload's payback is in the solutions it brings to society, and that makes it all worthwhile.

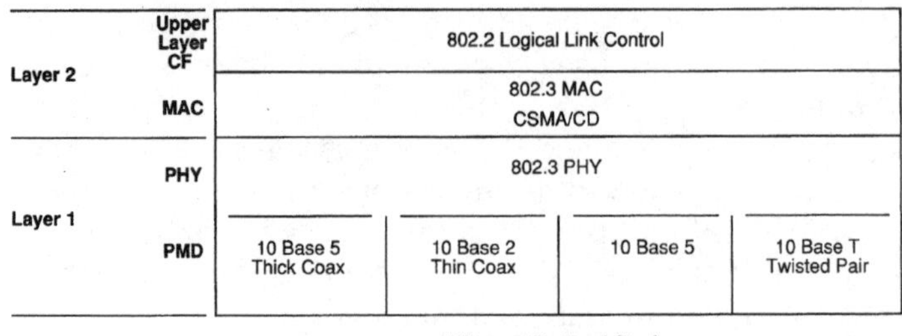

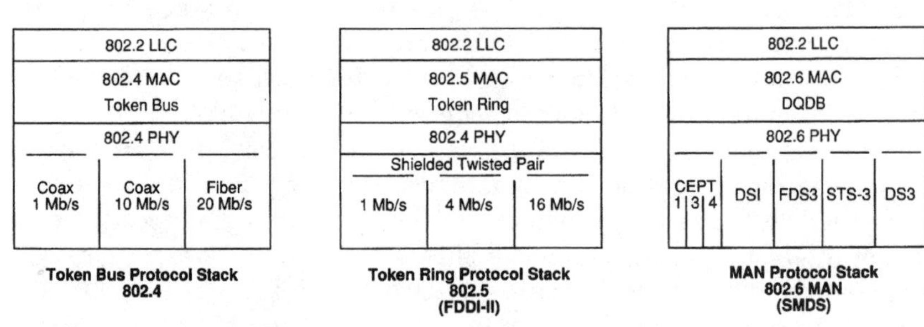

Figure B-7. IEEEOSI stacks.

Wideband Networking

Let's take a look at wideband transport and switching using Primary Rate Interface ISDN (PRI-ISDN), T1, F-T1, f-ISDN, DCS, DACS, and T3 technologies.

Digital cross connect systems (DCS)

A digital cross connect system simplifies the provisioning of circuits, the process of network administration, disaster recovery, and remote test access. As a computerized facility switch, it redirects circuits on a facility basis. This allows the positioning of facilities and swapping of circuits, as well as test and verification access for the network to break in and rearrange channels; this provides automated mainframe capabilities for enhancing operations, administration, maintenance, and provisioning (OAM&P) flexibility. Placed at key points in the network, DCSs can help cable cuts and critical central office outages by providing physical connections to always achieve the desired logical interconnection through the network. As near-term, real-time nonblocking switch matrixes, self-diagnosing and self-healing switch architectures, enhanced bit error rate (BER), and near real-time interaction with operating systems provide unparalleled reliability, there is a new role for digital cross connects to play in the restructuring of the network. Survivability, control, and restoration have become very important after the two to three weeks of outage caused by the October 1988 Hinsdale fire and then the 1990 Hinsdale cable cut, where 64,000 trunks and 27,000 special serving lines were affected for 12 hours. (In the latter case, even after a 12 percent to 15 percent restoration rate per hour, the telco was hit with a $150 million class action suit for the cable cut.)

T1 networking

Hence, the desire for faster and easier operations, the need to remove the costs of manually supplying special services, the desire for customers to have more control of their networks, and the dropping price of T1 facilities led to the deployment of digital cross connects such as AT&T's Direct Access and Cross Connect Systems (DACS). Some cross connects have the ability to perform DS1/DS0 (DCS 1/0) or DS3/DS1 (DCS 3/1) switching to add and drop Digital Signal level DSO (64 Kbps), DS1 (1.5 Mbps) and DS3 (45 Mbps) channels, as well as open up the access for testing. Subsequent systems allowed the customer to dial up and request a T1 facility in order to efficiently control (reconfigure) their own network. This led to the need for advanced T1 networking, to achieve such features as: compressed voice transfer, enhanced data services, compressed video, teleconferencing, automatic selected priority links, best-case routing in case of link failure, end-to-end diagnostics, circuit redundancy with automatic link restoral and impressive mean time between failures, aggregate trunk rates, drop and insert, passthrough, DS1 framing, D4 channelization, DCS compatibility, bit and byte interlocking, and dynamic bandwidth allocation.

Fractional T1 (F-T1)

As one technical leader noted, Fractional T1 is actually a misnomer. The cost of the services is fractionalized, not the service itself. Customers pay for only the bandwidth used, not bandwidth delivered. F-T1 is based on standard T1 digital service parameters. T1 is a digital multiplexed service; twenty-four channels are provided on one 1.544-Mbps line, and each channel occupies a 64-Kbps time slot, generally known as DSO. T1 multiplexors formulate various voice and data channels for transmission over a T1 line using D4 or D4-extended superframe formats. Digital cross connects systems (DCSs) and network management systems enable multiple partitions and even subpartitions to coexist on the same backbone network, as large and small users are combined in the pipe with everyone else.

Fractional ISDN (f-ISDN)

With ISDN's Primary Rate Interface (PRI), the current set of IXC services that can be accessed over the same F-ISDN interface can be expanded to include, for example, AT&T's Megacom, Megacom 800, Multiquest, SDN, Accunet switched digital service 56 Kbps, as well as 64 Kbps, 384 Kbps (H0) and 1.536 Mbps (H11). PRI will enable a uniform numbering plan for customer networks that can now be subdivided into smaller subnetworks or user groups. User groups may be created by restructuring connectability to selected users or data transmission equipment, where call screening can be performed by called-party number, calling-party number or through automatic number identification (ANI).

Therefore, DTE (data terminal equipment) can obtain MB + D on deducted 384-Kbps, 768 Kbps, and 1.536 Mbps, as f-ISDN is provided initially on full duplex, point-to-point, multiple 56-Kbps and 64-Kbps circuits. We can expect to see switched digital rates of clear 384 Kbps and 1.536 Mbps with diagnostics on the "D" channel to pinpoint the location and cause of disconnection. This will enable selective rerouting on the basis of whether the disconnect was a result of network facilities or simply the interface of the remote DTE.

T3 networking

As noted, users have grown from 9.6-Kbps to 56-Kbps to T1 rates. New applications are driving users to higher and higher bandwidth usage. Larger increments of bandwidth become practical when a user's cost is less than a corresponding number of smaller bandwidths. The crossover point is the number of lower bandwidth levels at which it becomes more economical to use the next increment. For example, depending on geographical location, this crossover point from 56 Kbps to T1 may be 4 to 6 DSOs, and from T1 to T3, it may be 8 to 14 DS1s. Transmissions have seen rapid advancements from T1 (1.544 Mbps) capacity to TIC (two T1s) and T2 (four T1s) to the 44.736 Mbps of T3, which is equivalent to 28 T1s or 672 voice conversations. For private network users, a switch from T1 to T3 would increase their capacity four or five times, and, by the mid 1990s, would expand telco direct transport revenue by several billion dollars.

T1 mesh networks

Most T1 multiplexed networks originally used multiple connected nodes in a mesh topology, with each node connected to three or four neighbors; in the 1990s, T3s are replacing T1s' mesh networks using new network topologies. As success brings higher volume, T3 ring topology becomes the network of choice for private networking, for applications, such as CAD/CAM/CAE and image-video processing. However, getting at subrate bit streams such as 9.6 Kbps is a problem for T3 networks, which usually only treat T3 bandwidth, as indicated earlier, as twenty-eight separate T1s. By using Syntran (a Synchronous

Transmission DS-3 protocol) it is possible for a MUX to add or drop DSO channels directly out of the T3 bit-rate structure.

DS3 switching

The DS3 format consists of frames of 4,760 bits; there are fifty-six groups of 84 bits each that convey the 4,704 bits of information signals. The remaining 56 bits are called overhead, required to successfully deliver the information. These bits are portioned into M bits, P bits, X bits, C bits and F bits. Each DS3 frame has two bits called X bits, which have been designated for alarm control. C bits are for parity and error control. F bits are framing bits, which have a regular pattern for fixed intervals. Hence, if a back hoe, ice storm, or hungry rodent causes a loss of a span of DS3 channels, after several seconds of outage there is significant risk of disabling a local office. But it's possible to recover within a few seconds by using DS3 cross connects to route interrupted traffic into spare capacity as alternative routes.

Private T3 network planners have used T3 multiplexors, referred to as add-drop multiplexors (ADM), which support several T3 interfaces that enable an alternate path to bypass a failed link or node. Unfortunately, many of these multiplexors do not have remote configuration or diagnostic capabilities. They cannot be efficiently networked to provide alternate paths, direct DSO access, or drop and insert DSO channels without digital crossconnect (DCS) 3/1 type systems at each node where such features are needed. Next generation T3 multiplexors will support a greater number of inputs to make full mesh (or ring) T3 networking complete with alternative routing available for applications such as: video teleconferencing, digital imagery, data vaulting, routing, diverse routing for disaster recovery, LAN-to-LAN interconnects, high-speed host-to-host transfers, as well as virtual private networks, load balancing, SLC 96 hardware consolidation, and enhanced T3 hubbing arrangements for use in metropolitan fiber networks and teleports.

Frame relay

Frame relay is a data-link layer protocol that defines how frames of data are assembled and routed through the packet network. Due to its higher performance than other wide area packet switching technologies, such as X.25, frame relay is a viable alternative for supporting bursty, high-volume data traffic such as LAN-to-LAN links. The frame relay has a total of 48 bits of overhead, which is four to five times less than the X.25 implementations. Frame relay is so named because it defines how frames are relayed across a series of predefined switches. The order in a single transmission is maintained as it traverses the network, while X.25 routes packets over a number of different lines, requiring reordering at the distant receiving device. After receiving a frame, the switch examines an 11-bit data link connection identifier (DLCI) to determine the frame's destination and then set up the connection between switches.

The DLCI creates a permanent virtual circuit that is similar to a leased line between devices. This virtual circuit defines a specific path for all frames to follow between two devices. For devices that communicate less frequently, a switched virtual circuit can be created, similar to X.25 use of ISDN Q.931, signaling protocol to create virtual circuits. This would send signaling from the frame relay interface over an ISDN "D" channel to instruct the ISDN switch to set up a connection between two nodes. The connection is torn down when transmission of all the frames is complete. Errors are checked at the network's last switch, not from switch to switch, relying on better grade digital interoffice trunks for less errors. Errored frames can be resent via transmission protocol directives. Therefore, the frame relay network has less delay than the public X.25 data network delivery; some tests indicate that the reduction is from the 200M seconds of X.25 to 20M seconds for frame relay. Also, congestion control features can tell input devices that the network is nearing capacity in order to slow down the rate at which frames are forwarded to it.

Some call frame relay the ISDN X.25 replacement; some question fastspeed cell relay versus frame relay. (See Appendix C.) Frame relay is widely accepted by private networks. Since their bridges and routers are based on similar packet structures, they can more easily adapt to it. T1 DCSs will have to add a packet engine to interface to frame relay, as it's an evolutionary step beyond X.25. Private-network users like frame relay and will use it, since it helps meet their bandwidth-on-demand objectives. (See Fig. B-8.)

Therefore, frame relay is an ISDN packet-mode-bearer service for data. Two versions exist as outlined in CCITT I.122. These are permanent virtual-circuit services and switched virtual-circuit services based on CCITT Q.921, HDLC variable-length packets and LAPD multiple virtual circuits. (See Appendix C.) Data is transmitted from 16-Kbps to 2-Mbps bandwidth, thereby enabling bandwidth on demand. It can be supported on ISDN 64-Kbps B channel and 16-Kbps D channel and on 56-Kbps (with subrate multiplexors for 4.8 Kbps

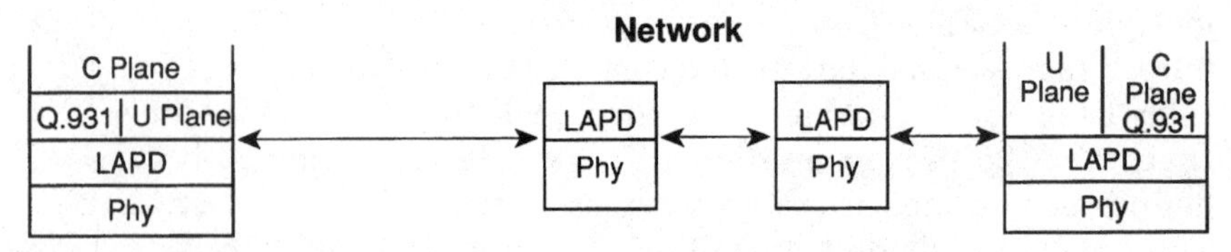

Figure B-8. Frame relay.

and 9.6 Kbps), 128-Kbps, 384-Kbps and 1.5-Mbps LAN bridges. Frame relay provides near instantaneous communication for bursty data traffic, enabling bandwidth on demand, higher speed, and higher throughput compared to X.25. Frame relay matches Ethernet, Token Ring, FDDI LAN attributes, and it ties nicely to T1 hubbing arrangements for private lines, thereby enabling simple networking of identical LANs via bridges and interoperability with different LANs via routers, as well as cost-effective access to LANs via gateways.

Frame relay provides users with capacity, flexibility, control, universal access, survivability, and convenience at a competitive price. It replaces full mesh router networks that use leased 56-Kbps circuits to directly interconnect routers to each other; instead, frame relay uses a virtual wideband network of single leased circuits (T1 typical) to a frame relay service. Note that the shorter distance to a frame relay hub and the reduction of indirect circuits enables purchase of single, higher capacity mediums. Frame relay is for the wide range of users who require low to moderate speed but more flexibility and reach. These users have exhausted 56 Kbps, are worried about the vulnerability of a private network, or are looking for more economical alternatives to 9.6-Kbps leased lines. In time, frame relay will be able to interface to ATM B-ISDN. SMDS is for a different market than frame relay; where frame relay would nicely satisfy E-mail and fax users, SMDS is for long-holding-time (several hours), high-volume, high-capacity users such as video-image, and CAD-CAM users, who are at the next frontier of the network, driving toward early use of ATM switching and B-ISDN.

Broadband Networking

Using SONET, SMDS, FDDI, ATM/STM, W-DCS, B-DCS, and frame-relay-type technologies, we set the stage for the challenge of the 1990s—the broadband information infrastructure.

Synchronous Optical Network (SONET) will take full advantage of the fiber's broadband capabilities; while DS1 (T1) and DS2 (6.3 Mbps) (T2) can be achieved on the copper plant, as speeds jump to DS3 (45 Mbps), we must go to coaxial or fiber. Since any copper line over 18,000 feet has been loaded with coils to add induction to the transmission, passing high-speed data requires deloading and conditioning. This tends to be expensive, which encourages the telco deployment of fiber. SONET will package the lower-order signals DSO, DS1, and DS1C (two DS1s), and DS3 asynchronous bundles called *synchronous payload envelopes* (SPEs). These are translated within a synchronous transport signal protocol referred to as an *electrical synchronous transport signal* (STS), having a 51.84-Mbps signaling rate for STS-level 1 (STS-1). Its optical equivalent is referred to as *optical carrier level 1* (OC-1). European and U.S. standards groups have chosen STS-3 (or OC-3) at 155.520 Mbps as their agreed-upon, common building block. See Fig. B-9 for the buildup from OC-1 to OC-48. The European synchronous digital hierarchy (SDH) has been standardized at 155 Mbps (STM-1), 622 Mbps (STM-4), and 2.5 Gbps (STM-16), with an upgrade to 10 Gbps forthcoming.

SMDS

Switched Multimegabit Data Service will enable wide area data connectivity over a public, high-speed data network. The SMDS service provides for multimegabit throughput over a 1.5-Mbps (T1) and 45-Mbps (T3) interface with access classes of 4, 10, 16, 25, and 34 on the 45-Mbps interface. The SMDS services use the 802.6 standard for metropolitan areas networks (MANS), and it provides a SMDS subscriber network interface (SNI) that will use the access protocol Distributed Queue Dual BUS (DQDB), which allows single or multiple customer premises equipment (CPE) access over separate paths for transmission while receiving fixed-length (53 byte) data packets (dual bus), where only one station transmits data at a time in specific intervals (slots) of transmission. In the future, the 155-Mbps SONET-based interface will also be used, as standards groups pursue a 149-Mbps SMDS protocol that will be compatible with FDDI-II as well. (See Fig. B-10.)

The network flow could typically be from a customer's system that interfaces to a LAN. The LAN interfaces to a gateway router that ties to the public switching network through a T1 or T3 interface directly through central offices to a hubbing location where a central office system has a SMDS subnet, which may connect directly to a host computer or workstation; or the SMDS switching system will enter the switched network to a distant SMDS and then on to the destination.

Optical Carrier	Line Rate M B/S	DS1 Equiv.	DS3 Equiv.
OC 1	51.84	28	1
OC 3	155.520	86	3
OC 9	466.560	252	9
OC 12	622.088	336	12
OC 18	933.120	504	18
OC 24	1244.160	512	24
OC 36	1866.240	1008	36
OC 48	2488.320	--	--
OC 96	4800.000	--	--

Figure B-9. SONET.

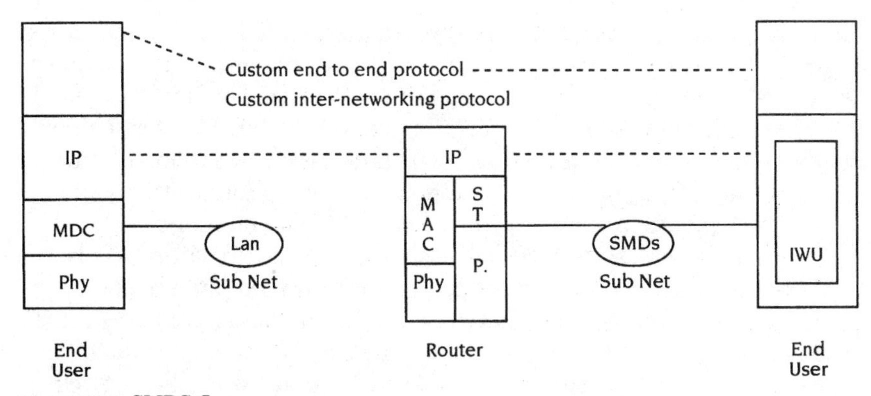

Figure B-10. SMDS flow.

Therefore, the purpose of the SNI interface is to communicate with the appropriate layers of the local area network router and customer equipment. There is also the need to interface with second-generation LANs such as a FDDI operating at 100 Mbps, as well as frame relay, which will initially operate at rates up to 1.5 Mbps and 2 Mbps. Therefore, initial SMDS interfaces will support the T1 and T3 rates. Then multimegabit rate SMDS will complement the existing FDDI LANs. In so doing, users of FDDI will be able to extend their networking capabilities (which are limited to 60 miles) across a metro area and eventually across the globe. Similarly, SMDS service and frame relay interfaces could be complementary to one another, if signaling and other issues are addressed.

SMDS allows the customer of private networks to have group addressing, similar to LAN multicustomer capabilities. For security, SMDS provides for address screening and validation as it checks to ensure the message is coming from viable locations. To make the network secure, SMDS also screens incoming calls to ensure that they are on the receiver's private network list. SMDS is a packet-switching network that operates in a connectionless mode, handling variable-length packets up to 9188 octets (bytes). Every packet has a header containing the address of its destination in the form of a ten-digit number (future 15) address, based upon the E. 164 ISDN numbering plan. Packets may be in different lengths because of the various protocols used by customer equipment. A trailer is added to mark the end of the packet. Since packets are sent in this header-payload-tail datagram mode, no connection is set up. They operate in a connectionless manner so that there is no prior history of previous packets. Network error control is left to the router or subscriber equipment. The packets are divided into uniform units called segments for transmission over the SMDS network. Each segment is bracketed by its own header and trailer. The transmitter looks for idle slots to send a packet segment, if it has gained permission according to distributed queue procedures. Here, they must first request bandwidth and honor previous requests on a first-come, first-serve basis, as they effectively wait in a queue. In receiving packets, data processing devices on the bus look for segments with headers that contain their own address in addition to start-of-message signaling information. When a match is found, this and subsequent segments are copied until the end of the message.

In this manner, the SMDS operates in a layered structure where the SMDS Interface Protocol (SIP) determines how the subscriber equipment communicates to the network. SIP is a connectionless protocol that consists of three protocol layers whose functions are such that: layer one provides physical layer bit-level transmission; layer two provides framing and error detection mechanisms, and SIP layer three Protocol Data Units (PDUs) provide connectionless destination and source addressing and overall header and trailer information. This leaves customers' internetworking protocols completely independent of these transport mechanisms. Similarly, higher layers can then communicate using customers' end-to-end protocols separate from SMDS, thereby making SMDS easy to integrate with TCP/IP protocols and support systems.

The market for SMDS is seen as tying together small and medium-sized businesses who cannot support separate private networks. SMDS also becomes a major vehicle for extending private enterprise networks to remote branches, as well as to these small-to-medium networks. In this manner, a medical network for an internal radiologist group within a cluster of hospitals can also send x-rays to remote specialists or even to the patient's family doctor. The medical network can also interface to a banking network for billing transactions. Many new uses for interconnecting LANs are enabling wide area networking to continue to blossom, as more and more LANs continue to demand interconnectivity at a rate that will double every two years over the 1990s.

B-ISDN

Initial broadband ISDN (B-ISDN) technologies will deploy synchronous transfer mode (STM) or circuit time division multiplexing that uses octet units to form the *N* bits per second channels, noting the position of arrival. The alternative technique for B-ISDN is asynchronous transfer mode (ATM). It has also been called fast packet and cell relay. Asynchronous transfer mode (ATM) has become an integral new switching technique for integrating conversational voice and bursty data traffic, using a fixed-length packet cell of 48 bytes + 5-byte header.

The actual input data stream can contain a variable number of cells. Hence variable-length messages can be easily handled by the same switching fabric. Early application of ATM will be to switch SMDS traffic based upon 802.6 MAN requirements. Note: SMDS also uses a 5-byte header and 48-byte payload with a few differences in field content. There will be a need for interoffice protocols for ATM switches called *Inter-Switching System Interface* (ISSI), as well as cross-LATA IXC switches, using a protocol called *Interexchange Carrier Interface Protocol* (ICIP) to complete the local SMDS offering.

ATM identifies each channel by the header (label), consisting of the virtual channel identifier (VCI) and a virtual path identifier (VPI) to speed up switching, using low-cost broadband network elements. Since bandwidth of an ATM virtual channel is not affected by the rate at which cells are inserted, the bandwidth is not linked to direct channels of 64K, 1.5M or 45M bits per second. Different virtual channels in the same transmission stream can all have different effective bandwidths (input rates) without affecting the ATM response to variable bit rates (VBR) for data, constant bit rates (CBR) for voice, or connection-oriented or connectionless data-transport traffic. Hence, the access interface may be required to handle 64 Kbps, 1.5 Mbps, 19.42 Mbps, 51.84 Mbps and 155.520 Mbps. Various SONET multiplexes and switches provide wideband multiplexing to change incoming DS1 streams of SONET Virtual Tributary 1.5 Mbps (VT1.5) into STS-1 outputs; alternatively DS1s, DS3s, and STS-1s are switched to various DS1s, DS3s, and STS-1s. Therefore, DS3s and STS-1s can be either broadband multiplexed to higher level STS-Ns or switched to DS3s and STS-Ns. SONET OC-N rings will deploy add-drop multiplexors (ADMs) in conjunction with wideband (W-DCS) or broadband digital cross connections (B-DCS) to add or drop off DS3s. As a front end to switching, ATM terminals can concatenate various DS-, DS1, DS3, and STS-3c (where c stands for concatenation) rates into input streams of ATM cells.

STM is based, on the other hand, on switching fixed-rate channels such as the traditional circuit-switched 64 Kbps. Different rates would require multiple switching fabrics, each designed to switch channels at one specific rate. As conversation video, high-definition TV, long graphics, and broadcast services expand, STM will have a major role to play. The age-old issue of bits and bytes challenges the type of architecture of Broadband ISDN switching of the future. Delays, overhead, and congestion in moving datagrams of connectionless transport are challenging X.25, and lower-speed bursty traffic is challenging high-speed, high-volume traffic. In order to form internetwork operability for multiple voice, data, and video applications, there is increasing pressure to resolve the following issues and controversies: ATM versus STM, frame relay versus SMDS, SMDS versus FDDI-II, and connection-oriented versus connectionless data handling.

Similarly, decisions are needed for merging continuous bit rates with variable bit rates, fixed-length packets versus variable-length packets, and packetizing or not packetizing long connections (requiring extensive packet assembly and disassembly at the application layer, which some refer to as the band aid layer). We have seen heated discussions on different approaches, even for the hybrid compromise. However, we need these agreements to ensure that the different systems can better communicate with each other. After many years of noninteroperability of dissimilar ISDN systems, the Corporation for Open Systems International (COS) was able to have a consortium of switch vendors and computer firms and users reach agreements so that any CPE terminal is portable to the different ISDN switches. ISDN services need to be consistent. In fact, leading technical planner Jim Conlisk, in his vision statement on ATM versus STM, may have had the last word:

A vision

A vision of the future
A noble thought indeed
Without ATM indentures
A hybrid is my plea

ATM's the solution
'Tis the cry we often hear
Without a transition
Responds a skeptic's ear

It's information transport
The game we need to play
And information transport
The "vision" we need to say

Not ATM nor STM nor hybrids
Hold the view
Though technology enablers
Do always pipe their tune

It's information transfer
That ought to be our due
And always market drivers
To set our vision true

Technology enablers
Though mighty they may be
Only provide the infrastructure
To support our market needs

The real challenge will be to see how well we work together, as noted by Aaron and Decina in their observation on IEEE committees:

Pure packet players,
parcelling puny packets,
pretending parsimonious packets payoff,
please preach partnership programs,
promising past present prospects,
permitting prodigious packets,
promoting people Progress,
producing peaceful prosperity!

This indeed is the challenge of the new millenium.

Private Networking

Every two years, computer processing power takes another quantum step forward, with larger memory sizes close behind. Over the last twenty years, we have seen rapid evolution from stand-alone mainframes with batch processing. They first changed from having fixed partition sizes, which inhibited working on multiple variable-size programs, to variable partition sizes, with foreground and background processing so the system could be spooling in and out multiuser's programs and sectionalizing the programs at the same time that it was processing several other programs in the main computers. Early programmers either worked in neighboring "open bullpens" or waited in their software design work areas for the results of their last "shot" to be returned to them by carts, carrying boxes of their IBM cards. Growing tired of three or four "shots a day," small local processors were used to perform preexecution computations looking for syntax and semantics errors of wrong context, improper keypunches, missing commas, etc. Test time was precious; workers hated to miss a shot due to a missing comma in the set up stack.

Next, programmers in distant buildings used modems to modulate/demodulate data for off-line remote keyboards or printers. Here, information was sent to storage tapes or drums in central files on either direct links, using balanced lines, or through dial-up (unbalanced) facilities at rates of 600, 1,200, or 2,400 bits per second, using amplitude or frequency modulation techniques on transmission carrier frequencies to superimpose the 1s and 0s representing the alphanumeric characters or numbers. "Big time" was the introduction of phase quadrative modulation techniques to move 9.6-Kbps information at error rates of 1 bit in 10,000. And so it was over the late 1960s and the early 1970s.

Then in the mid-1970s, computer designers designed a computer for themselves. It had a remote front end to the mainframe. At the same time, LSI (large scale integration) technology had begun to enter the commercial world due to the collapse of the industrial-military complex in the late 1960s, causing a form of technology transfer as military computer designers obtained commercial design jobs. Soon, LSI became VLSI (very large scale integration), as drawers of equipment circuits became cards, cards became chips, and chips became elements on other chips. Similarly, software changes came in leaps and bounds from bootstrap loaders written in machine language, to highspeed, primitive instructions, to macros, to Basic Assembly Language programs (having shift left, shift right instructions), to commercial RPG (report program generator) and COBOL, as well as the scientific Fortran, and on to higher-level Program Language One (PL 1) general programming languages, followed by list processing special purpose languages (LIST), and then on to the world of object-oriented programming, Objective C, C + + and distributed program and relational database management systems. Next came high-speed reduced instruction set computers (RISC) and specialized miniature chips for specific applications—ASICs (application specific integrated chips). These highly complex chips, with hundreds of thousands of logic elements, cost about $50 in the 1990s, whereas the very first integrated circuit (IC) chips of the 1960s, containing two logic elements per chip, cost $50 in lot purchases of 1 million plus.

So technology changes. As computer power changes, computer applications change. As computers move from mainframe to front-end minis to small micros having the power of previous mainframes, we have seen instruction lengths collapse from 32 bits per instruction to 4 bits per instruction, as the emphasis went to speed. Various storage device capabilities increased as we shifted from tapes and drums to disks and chips, becoming more and more economical, as noted by the 64K bit chip change to 256K bits to 1 million to 4 million and on to the 20-60 million bit chip storage objectives. This has enabled more powerful instruction sets to return to 16-bit and now 32-bit instruction lengths for micro-sized desktop personal computers with the power of a mainframe.

Hence, as personal computer power increases from 2 MIPS (million instructions per second) to 20 MIPS, and mainframes approach supercomputer 200+ MIPS capabilities, pressure mounts and mounts to interconnect all these computers. Since switching systems are made from computers, their capabilities are paralleling computer growth. Therefore, transport is only limited to the medium. Now fiber, with its unlimited capabilities of adding more and more spectrum, has shown the capacity to first handle 6,000 voice conversations, and then 18,000, using wave length division multiplexing technology of numerous frequencies. In the future, we will have the capacity to handle over a hundred frequencies. As we probe the color spectrum, bit rates jump from megabits to gigabits to terabits. We can then leave the binary number system of 1s and 0s and go to the decimal system, as 10 frequencies are simultaneously sent down the same fiber, thus providing a giant step in computer power.

Again, this brings more and more pressure to interconnect the world's computers. Unfortunately, the RBOCs resisted recognizing what was happening, with their classic 1970s and 1980s microscopic view of a voice-only world with 3 percent data transport. That limited view led to their demise in the data communications business. Their limited point-to-point dataphone data sets offered 2.4, 4.8, and 9.6 Kbps, using the EIA RS-232-E (formerly C&D) STDs developed from the 1960s mil-standard specs for the military modulators and demodulators. These offerings were not to be the data solution for the 1970s. These feature-poor (and expensive) offerings encouraged computer users to privately build their own internal networks. Later, users bypassed limited public network solutions.

DEC and other equipment manufacturers encouraged their terminals to interconnect on high-speed buses, similar to the high-speed buses internal to computers. This led to the Ethernet-type local area networks (LANs), which IEEE standard committees developed into the 802.X standards, thereby creating a new world for CPE transmission manufacturers. Many new networking transport protocols emerged for host-remote communications, such as IBMs BTRAM, VTRAM, SNA, and SAA. Soon, new forms of distributed computer technology emerged, such as: cluster controllers, remote processing, distributed databases, etc. However, with few exceptions, most computer technology manufacturers, such as IBM, only internetworked their own systems and terminals. Some smaller firms did see the opportunity and rose to the occasion, but then it became a goal for the larger firms to keep their interfaces proprietary to keep terminal "clones" or other systems from interfacing to "their system." Each company developed their own version of a local area network. However, in time, the realization set in that the more systems that interfaced to a company's system, the more computer power was needed, and therefore the more processor power the customer purchased. Hence, more and more computer manufacturers participated in the Institute of Electrical and Electronic Engineers (IEEE) standard groups, providing a series of standards for local communication, 802.3 Ethernet, 802.4 Token Bus, 802.5 Token Ring, and 802.6 MAN. This desire to interconnect led a subsequent move to WANs (wide area networks) and MANs (metropolitan area networks).

LANs

Local area networks (LANS) are short-distance private networks for business buildings or a university campus. Initially, the terminals could usually only communicate 100 to 1,000 feet from their computer without some form of electrical signal regeneration or use of some transport modem. More and more terminals throughout a floor or building began wanting to access a file system or a printer or share electronic mail; using an internal bus system, several types of LANs were developed to enable different users to interchange information. As noted, 802 IEEE committees on local area networks established the 802.1 standard for overall architecture of LANs and internetworking LANS. (See Fig. B-11.)

Several standards were developed to enable multiple terminals to use the same facility, recognize messages sent to their "address"; send messages to other terminals, or request information from attached processors. Token-based systems such as Token Bus or Token Ring enable terminals to send information only when they have been given the "ok" to send. This is passed to them as a special combination of bits, which gives them the right to transmit on the high-speed LAN channel. When finished, the token is passed on to the next terminal. IBM uses token ring LANs that may be physically wired in a circle or star, but passes tokens logically in a circle. DEC, Intel, and Xerox are

802	LANs
802.1	Architecture
802.1B	Network Management
802.1D	MAC Level InterLAN Bridges
802.2	LAN Data Link Layer
802.3	Physical Layer Ethernet Bus Topology (CSMA/CD) Access Method
802.3	1 Base 5 Baseband Ethernet 1M b/s 500M Twisted Pair
802.3	10 Base 2 Baseband Ethernet 10M b/s 185M Coaxial Cable
802.3	10 Base 5 Baseband Ethernet 10M b/s 500M Broadband Cable
802.3	10 Broad 36 Broadband Ethernet 10M b/s 3600M Broadband Cable
802.4	Physical Layer (10M b/s) Token Passing Access Method Bus Topology
802.5	Physical Layer (4 or 16M b/s) Token Passing Access Method Ring Topology
802.6	MANS or DQDB (Distributed Queue Double Bus) previously called Queued Packet and Synchronous Exchange (QPSX)

Figure B-11. IEEE 802.

promoters of Ethernet LANs using twisted wire or coaxial cable operating at speeds from 1 to 10 Mbps. The IEEE 802 physical and data link MAC protocols are adaptable to the lower layers of the International Standards Organization's (ISO's) Open Systems Interconnection (OSI) model. The OSI model was discussed earlier in relation to its physical, data link, and network layers that concern data transmission and routing. The OSI model will be used throughout the 1990s to help provide more standard interfaces in order for the many LANs to have more internetworking capabilities. TCP/IP has been the *de facto* internetworking LAN standard, especially for the military's 100,000+ computers. Interoperability among different computer systems is essential. In the past, DECnet and SNA were vendor-specific, allowing interoperability only among DEC or IBM computers for their specific host-terminal-oriented networks.

Interconnecting LANs have many problems, such as routing, protocols, congestion control, network design and management, load balancing, configuration control, accounting, and security. As the "LANification" of America progressed over the 1980s, terminals were connected to hosts via LANs rather than data switches. PCs were networked through PC LANs' minicomputers and workstations, while mainframes used LANs as the "background-host-to-host" and "host-to-front-end" processor-access transport. LAN interconnection devices consist of bridges, single or multiprotocol routers, and hybrid bridge-routers or brouters. The new "intelligent" multiprotocol bridge-router is the new nodal processor for packet-switched networks. In time, SNA's synchronous data link control (SDLC) and high-level data link control (HDLC) protocol connectivity will be bridged over multiprotocol backbones via these nodal processors.

LAN interconnectivity is a multidimensional, complex undertaking, interconnecting the multimedia, multiaccess method, multiprotocol, multisite LANs.

Bridges and routers. *Bridges* interconnect LANs that conform to 802.1 and 802.3 industry standards from various vendors. Bridges are data link relays that are independent of the higher layers and forward access control packets. They function nicely on LAN-to-LAN specific protocols, but have serious shortcomings in different types of LANS.

Routers are programmable for each network-software-specific protocol to be routed. Complex mapping and addressing schemes of two physically different networks make routers more acceptable for networking diverse LAN types through the use of the network layer functions (OSI model) of hierarchical addressing, packet fragmentation/reassembly and congestion control for different data-link characteristics. Routers store the full station addresses of local stations, together with information to direct traffic to all the remote locations. Routers also control the endless looping of packets.

Brouters is a term used to describe a bridge with some router capabilities. The new nodal processors will better achieve this combination, as they handle both the routable protocols via router formats and non-routable protocols via bridged functions to eliminate parallel logical and physical networks.

As LAN operations address peer-to-peer applications of file transfer, file access, electronic mail, remote terminal access, remote printer access, and distributed computing services on multivendor platforms, open systems networks based upon OSI and TCP/IP are being widely implemented. Open network protocols, together with open operating systems, will be the objective of the game. Similarly, network management will be key for remote control; it's needed to enable the networking of networks.

CO LAN is a virtual circuit network service typically for asynchronous data up to 19.2 Kbps using existing centrex loop plant for DVM (Data/Voice Multiplexing) and X.25 packet switching.

Many LANs support a multiplicity of protocols, including TCP/IP, DECnet, SNA, XNS, Netware, Netbois, and Apple's LANs' Local Talk and Apple Talk. To successfully manage LANs, there are a variety of management protocols, including SNMP (Simple Network Management Protocol) for sending and rearranging information. This is the *de facto* standard for bridges and routers versus the OSI general purpose connection-oriented protocol CMIP (Common Management Information Services) and the API CMIS that uses the full series of the seven-layer OSI protocols.

Other management protocols are CMOT and SNA NMET. Alternatively, PC LANs built around client-server models are specialized products for small groups of users. These originally stand-alone systems are also being connected to other LANs during the 1990s. As more and more software is written, standard APIs (application program interfaces) are needed to ensure that software can be easily ported to new hardware platforms. All in all interconnecting LANs is a complex game, especially as more and more LANs are connected to WANs and MANs.

WANs

Wide area networks (WANs) for interconnecting local services served by LANs, PBXs, cluster controls, etc., will be key to the internetworking of private and public networks. These wide area networks can be achieved by leasing direct links from the telephone company to remote locations or by using the new ISDN 64-Kbps clear channel "B" links to formulate nailed, dedicated, permanent, semipermanent, or switched paths through the circuit-switched network to distant locations. Earlier, WANs were constructed over leased circuits or dialed-up paths sending 9.6-Kbps data via analog modems. Dial-up paths could be reconfigured several times a day or as needed, if error rates became intolerable.

Wide area networks can be implemented in packet-switching networks through virtual data networks (VDNs), as connection-oriented data switches establish virtual networks that can change to accommodate whatever block of capacity is needed, either as a function of time of day or dynamically. In this manner, T1 networking and T3 networking, using DCSs 3/1/0, will enable wide area networks to be established to provide the customer with easy, fast, changeable access to higher-speed facilities. The DCS (digital cross connect systems) reduce the long lead time for expensive, dedicated special-service communications such as FX lines that "home" on a neighboring exchange (foreign). This is seen in the trend for dropping DDS (dataphone data services) and leased voice grade facilities, while T1, T3, and fractional T1 is increasing over the same period. Therefore, wide area networks will be a major vehicle for virtual private networking (VPNs). This has been further seen as numerous data networking applications blossomed in the 1990s, applications such as CAD/CAM, CIM/CAE, simulation, consolidation of databases, distributed databases, distributed applications, distributed client-server capacity over wide areas, access to supercomputers, electronic reprogramming, and computerized document file transfer.

FDDI

The other half of the equation will be the LAN-MAN interfaces, where VPNs are achieved using FDDI, SMDS, and frame relay capabilities to obtain high speed data transport. With 100-Mbps data throughput off a 125-Mbps clock and built-in management capabilities, Fiber Distribution Data Interface (FDDI) adequately met the users' growing data transmission requirements through the 1990s. It was originally viewed as a "back-end" network to link mainframes using multinode fiber (2KM between nodes). New FDDI product standards as an evolution of the Token Ring LAN (IEEE 802.5) will enable linking lower-speed Ethernet and Token Ring LANs with FDDI to bring the power of FDDI directly to the user's desk. FDDI is the high-performance local area network (LAN) technology developed by the American National Standard Institute's (ANSI) X3T9.5 committee. The FDDI network consists of two counter-rotating token passing rings. Each ring runs at 100 Mbps and can span 100K meters to include hundreds to thousands of terminals. FDDI's second ring can move data, as well as provide backup in case of primary ring failure. The total network can support speeds up to 200 Mbps for standard layer one: physical layer medium, multinode and single-node SONET interfaces (PMD), and Physical Layer FDDI Protocol (PHY). FDDI can also support layer-two FDDI Media Access Control Token Ring (MAC) and 802.2 Logical Link Control (LLC), as well as SMT (System Management).

FDDI-II

ANSI's X.3 committee will specify how FDDI will run over single-mode synchronous optical networks (SONET) to extend FDDI networks over very great distances; the TIXI committee of the Exchange Carriers Standards Association is considering an FDDI-II compatible 149-Mbps hybrid slotted ring.

New transport mechanisms enhance the original FDDI capabilities to now include a hybrid ring control (HRC) technique for supporting both asynchronous packet data and isochronous circuit-switched data on FDDI LANS. Used in FDDI-II, an extended version of FDDI for data, HRC will allow voice and video traffic to be integrated with packetized data on the same FDDI-LAN. It would work by dividing the ring's bandwidth into 16 blocks of 6.144 Mbps each. Each block can carry either packetized or nonpacketized data. A header in each block informs the switch which is which. However, blocks do not have addresses. To send nonpacket data between two stations on the ring, the originating ring contacts the intended recipient through a channel reserved for packets. Once a block has been allocated to transmit nonpacket data between two stations, the same block continues carrying data from station A to station B and back to A, until the connection is broken.

VANs

Value added networks (VANS) will multiply as more and more data handling services, intelligent networking services, information network architecture-based services, application service centers, information switches (Info Switch) services, service nodes, metropolitan nets, teleports, PCNs, DBSs, cellular systems, and satellite VSATS are dispersed throughout the network. Virtual private networks will therefore not be limited to the public networks' switched and nonswitched facilities. Private networking will use more and more VANs for special handling and special voice, data, and video services. Value added networks such as Tymnet and Telenet nationally overlay switched packet data transport capabilities parallel to the switched voice network. We need only to look at what has happened in Japan to see a blossoming future for VANs.

Therefore, let's first review Japan's VAN services. Japan's VANs are of two types. Type 1 carriers own their own circuits, while Type 2 carriers lease circuits from Type 1 carriers, and then resell them. Type 2 carriers are further classified as special or general. The Special Type 2 carrier category covers international VAN (I-VANs) operators and companies that offer nationwide services. They must obtain approval from the Ministry of Posts and Telecommunications (MPT) for their services. General Type 2 carriers need only inform the MPT of their existence. As of 1990, Japan had 66 Type 1, 864 General Type 2, and 29 Special Type 2, for a total of 952 VANs. The market was reported to be as high as several trillion yen ($7.8 billion) a year and growing. As banks have excess capacity to offer time-shared services, 329 firms offer online accounting and 60 provide warehouse inventory monitoring. I-VANs provided by Special Type 2 VANs will provide more than simple packet switching, electronic mail, and facsimile services. In the 1990s, they have provided all types of enhanced services, including electronic banking and voice mail. However, NTT's ISDN pricing for INS-NET 64, INS-NET 1500, INS-P (packet switching), DDX-C (circuit switching), and DDX-P (digital leased lines) has become a key issue as VANs attempt to competitively link producers, wholesalers, and retail outlets so users can enjoy greater efficiency in manufacturing, ordering, delivery, and stock management.

In America, "local VANS" are being established in the RBOC's local monopoly to provide parallel high-speed data transport (movement) to bypass traditional transport or services offerings. RHCs estimated one and one-half billion dollars in losses in 1990 as a result of facility bypass, in which a customer obtains service from an alternate carrier and bypasses the local BOC (one billion in losses from switched facility bypass, and one-half billion from private line facility bypass). They also lose approximately two billion or so annually from service bypass when a BOC customer leases a private line from the carrier and bypasses switched access services. Competition will become increasingly intense by communications companies, alternative access vendors (AAVs), cable TV providers, power companies, computer companies, other LECs, plus governmental and educational private networks.

Examples of services provided by local VANs or IXCs are noted by AT&T's Software Defined Network (SDN). Essentially for voice services, SDN was initially a replacement for WATS or tie lines for large business customers (Advanced 800 Service). It let customers set up virtual circuits on their national network. Circuits are set up only for the duration of the call, so the charge is based on usage. SDN is not the same as a DACS, but they both will evolve into ISDN. Advanced 800 Service uses direct services dialing capabilities to route a call, for example, to west coast or east coast telenetworking centers, depending on time of day. Thus, calls can be rerouted based upon area code, time of day, day of week, and customer preplanned alternate service situations. Another AT&T local (VAN-type) network is SDDS (Software Defined Data Services), where ISDN data handling capabilities can be provided by access other than from the LEC. Similarly, Telenet and Tymnet have established their local POPs, which provide local VAN-type data-handling services with various pricing structures for different numbers of packets of data handled.

The RBOCs were hampered in providing a complete data network, due to their regulatory limitation of not crossing LATA boundaries. Hence, a data network from the Mayo Clinic to the Cancer Research Center in Minneapolis would require both the RBOC and an IXC, or VAN participation. Even though the LATA-to-LATA link may figuratively only be less than one "inch or centimeter." This gave local VANs a distinct advantage for "Intrastate Inter-LATA" traffic. Some VANs simply leased conduit space, others lease "dark fiber" (fiber without electronics); others use direct microwave. Some are promoting wireless radio data networks within a metro area as an alternative. Similarly, direct broadcast satellite (DBS) networks use higher-spectrum radio frequencies to send movies or news programs directly to the home to 24-in. receivers. In fact, as noted in Chapter 6, each carrier can select their medium of choice to offer their particular multimedia voice, data, and/or video services, via an overlay on the local network that ties directly to regional, national, or global networks for more universal networking.

Virtual data networks (VDNs) were achieved using data packet networks (DPNs) such as Telenet and Tymnet, where dynamically changeable routes were established and capacity expanded, as needed, while pricing (costs) was based upon usage. Hence, logical networks could be established between distant sources and links. Here the actual physical network configurations utilized survivable paths with expanded capacity that was only allocated when needed. To be truly virtual, initial configurations were "nailed" or "dedicated" as "permanent" on circuit-switched networks or reserved on priority connection-oriented switched-packet networks. Connectionless data may, in high-congestion traffic situations, be penalized to ensure availability. As permanent virtual networks (where the service is established via an operator) compete with switched virtual networks (where the service is established dynamically by the customer per call), the differences begin to blur as the systems for establishing permanents become more automatic and immediate.

Other VANs such as the National Research and Education Network (NREN) are being established to provide high-speed connections between the nation's major research and academic institutes. It even extends to local communities and individual customers. Here the expanded NREN will be "universalizing" the enormous knowledge base available throughout the academic world. NREN is being developed as a major enhancement to the existing Internet, the huge "network of networks" that currently enlists U.S. universities, government facilities, and research institutes, as well as a large number of international facilities. Internet also operates in conjunction with the National Science Foundation's NSFNET. More services are being added by IBM and MCI in a nonprofit entity called ANS. There are extended proposals to deliver many, many services into the local exchange, including CATV. This suggests the specter of a dedicated national network, conceivably placing it in a competitive stance with certain types of RBOC services. Some view this as a grand collaboration of independent networks, others as building the so-called data "super highways" of the future. There are many perspectives about how these complex "wheels with wheels" networking structures will overlay on existing public networking facilities as ANS's expansive plans call for switching platforms to be set up as part of a completely separate network infrastructure, independent from existing facilities.

Intelligent Networking

An overlay hierarchy of signal switching nodes and service centers will enable Bellcore's intelligent networking to provide real-time online customer access to extended services such as 800 and 900 database lookups, as well as access to customer line information databases (LIDBs) that define the customer's line information service needs for alternative transport and service providers. In addition, virtual private network databases can be dynamically changed to reconfigure private networks.

Intelligent networking uses the Common Channel Signaling System known as CCITT SS7 (Signaling System Seven) to transport data messages regarding the call. It's based upon the architecture developed for AT&T long lines that enables the signaling system to look ahead before establishing a call across the network to see if the called party is not busy and to identify the paths to take to reach the called party; as the call paths are developed, there is less concern about being blocked somewhere between point A and point B during the call set up. Tests indicate that SS7 can reduce average call completion times from 10 seconds to 3 seconds for the IXCs. The network is such that signaling paths from Class 5 or Tandem switches, called *service switch points* (SSPs), generate query messages at 56-Kbps rates to a signal transfer point (STP). (See Fig. 6-8 in Chapter 6.) There at the center of the network is this SS7 signaling network routing switch (STP), which translates the destination code contained in the message and switches the signal to the appropriate SS7 link. STPs came in pairs with parallel paths from each central office or Tandem. *Service control points* (SCPs) interface directly to the STPs over SS7 signaling channels. SCPs are databases that centralize the information for call disposition relative to enhanced services. As noted, these databases will provide credit card services, as well as VPN and 800/900 services.

The signaling system can facilitate credit card validation and control. ISUP (Integrated Services Digital Network User Part) is defined by CCITT recommendations Q.761 and Q.764 to determine set up and take down of trunk calls on the SS7 network. There are seven ISUP messages (IAM, COT, ACM, ANM, REL, RLC, and EXM) that establish address complete, answer, release, release complete, and exit messages to the CCS7 network through the ISDN primary rate (PRA) standard utilizing Q.931 protocol. The CCS7 ISDN application protocol TCAP enables transaction baud services to access SCP remote databases. Q.931 is the message signaling protocol of the PRI D channel. Similarly, CLASS services such as repeat call, return call, priority call, call forwarding, select forwarding, call block, call trace, and identicall will be readily available using the SS7 signaling channel. A *service management system* (SMS) is the operating system through which network operations and service personnel manage SCPs and their related service applications' programs and databases, using the X.25 network protocol in their internal networks.

As noted in Siemens' excellent book *Intelligent Networks (IN)*, the long-term goal of intelligent networking is to introduce new services and change existing services without having to modify internal SSP software. Only parameters or databases would need updates. The transition will take place in stages, where IN/1 does require updates to SSP switching-system software and the SCP in order to support the new services. Migration from IN/1 to IN/2 implies significant changes to current SSPs to accommodate new services, with interim stages such as IN/1 + where the SSP provides increasing flexibility in accommodating rapid service creation. Rapid service creation is such that a network operator can identify a need, create or acquire a corresponding service and deploy it within a market opportunity window, while maintaining the integrity of the network. This concept has been the goal of the BOCs, PTTs, and independents since they purchased their first switch and asked for their first new feature. They quickly learned it was a long and expensive wait to obtain new features and services for their customers. Hence, one of the very first requests to Bellcore from their new owners was to determine how to make switches dumb in order to interface to smart (intelligent) RBOC service platforms (SCPs). Not only should the RBOC's network operations people be able to write new features, but the SCPs should also be a platform from which any vendor can write and market new services that can interface and run on different hardware systems and not affect the network integrity. (Nice goal if we can achieve it.)

Therefore, a SCP (service control point) facilitates new IN services as they are introduced into the network. It is both a computer and a modified switch that can access databases efficiently and reliably. It provides a software platform for rapid service creation through user programmability and program portability. Its services and programs are updated by the Service Management System (SMS) and by the network providers and their customers. The SMS allows service subscribers to control their own service parameters, such as the time of day when an 800 number should be routed to a specific office. The STP is part of the Common Channeling Signaling Number Seven (CCS7) network. It switches CCS7 messages to different CCS7 nodes. The STP is usually provided by traditional switches, while the SMS is normally a commercial computer such as an IBM/370. The SSP serves as the network access point for service users. Intelligent peripherals (IP) provide enhanced services controlled by the SSP or SCP—for example, announcements, speech synthesizing, voice messaging, speech recognition, and database information made accessible to the end user. The IP is usually accessed from the SSP on a circuit or packet basis via ISDN. Other vendor feature nodes (VFNs) or services providers (SPs) are outside the network using ONA open network architecture interfaces to provide similar SCP and IP services such as 800 alternate billing services, emergency response service 911, private virtual networks, area wide centrex and pay per view (PPV).

Application service centers (ASCs) acting as autonomous service nodes are constructed above the network, where services are provided on an everyday basis. Calls are switched to these centers, work is performed, and then the calls are reinserted back into the network. At no time is the network switch waiting for the ASC. There is no interruption of call processing or sending signaling information to the IN nodes and waiting for a response before the calls can be rededicated. Alternative adjunct (adjacent) processors provide call processing software on an extended basis to the switch as multiprocessing moves to distributed processing but still functions as part of the internal system, not as an external system. This requires considerable testing and careful software control to ensure that the switch software integrity is not violated. (Many designers thought the original IN objective was to simply enable customer or network operators to change data parameters within the switch-line side of a customer database or change a separate nonprotected auxiliary database using recent change mechanisms of the supplier. Changing data is one set of problems; writing software in the midst of millions of complex supplier instructions is quite another type of problem.)

As noted earlier, AT&T's SDS and SDDS are forms of Application Service Centers. Providers such as AT&T's Software Definition Network (SDN) will also use SS7 signaling. As indicated, SDN lets customers set up virtual circuits on the public network. Circuits among scattered locations are set up only for the duration of the incoming calls; so the change is based upon usage. Software Defined Data Networks service centers will provide a whole host of enhanced data handling services above the local network, quite independent of LEC data handling.

In an effort to address not only the voice world via IN/1 and IN/2, but also the data and video world, Bellcore is defining a new information networking architecture. As switching changes from narrowband ISDN to broadband ISDN and as advanced, enhanced service centers such as the ASCs develop, the challenge to INA is how simple and how dumb to keep the network switches, and how to provide intelligent but simple service nodes that can deliver any service anytime within the window of opportunity. Broadband trials will be the methodology to help establish and resolve these objectives. In the 1990s, Bellcore has monitored trials to determine: How open is open? Without introducing danger, how much should switch manufacturers open up their software for anyone to insert instructions ? How much sophisticated protection software must be written by the supplier to protect outside users from bringing down the network? How much does protection software inhibit the outside would-be programmer from achieving the level of penetration needed to achieve the desired feature? Or have we generated a paradox where goals cannot be achieved? (We would all like to die and see what's on the other side and then come back and work especially hard at doing the things necessary to achieve heaven; but, unfortunately, it doesn't work that way.)

The challenge of the 1990s is to resolve this dilemma, perhaps with a background switch-processor that can interface to offnet systems that wish to change database information, perhaps by switching special calls to separate switches, such as information switches, where enhanced services are added independently of network switching and calls are reestablished after feature implementation. (Yes, there is some delay in double switching, just like LANs interfacing to routers, to gateways, etc., but at least it doesn't bring the internal LAN down if the gateway fails.) So it goes. That is the challenge.

Radio Spectrum Technologies

With technology achieved in just decades, systems now multiplex thousands of channels instead of hundreds, occupying tens of megahertz channel space instead of tens of kilohertz and operating at frequencies of 10 GHz, instead of 100 MHz.

Wire to wireless

The desire to transport more and more voice channels began with the frequency division multiplexing grouping, where many analog channels shared the same transmission-medium circuit. Early multiplexing systems used an open wire-pair carrier frequency of 200 kHz to carry 4 one-way voice channels; twisted pairs enabled the carrier to rise to 300 kHz for 12 channels, but twisted pairs' real spectra come with T2 with a bit rate of 6.3 Mbps providing 96 voice-grade channels. Coaxial cable could conduct carrier frequencies in excess of 100 MHz with frequent repeaters. Later, coaxial cable supported a digital carrier of 274.2 Mbps or 4,032 circuits. FDM grouped voice channels at different frequency carrier bands so that 60-108 kHz (48-kHz band) could handle 12 voice channels, the 312-552 kilohertz (240-kHz super group band) handled 60; the master group 564-3,084 kHz (2,520-kHz band) passed 600 voice channels; the L4 multiplexer jumbo group at .5-17.5 MHz (17-MHz band) handled 3,600 and the L5 for three jumbo groups at 3-60 MHz (57-MHz band) passed 10,800 voice channels.

During the mid-1990s, numerous spectrum band communications technologies such as cellular telephone, personal communications networks (PCNs), CT-2 and CT-3 (advanced cordless voice and data phones), mobile satellites, air-to-ground telephones, sound broadcasting from satellites, direct HDTV broadcasting and mobile data satellite networks clamored for frequencies in the United States and throughout the world.

Hence, the 1- to 3-GHz range allocations are desired by terrestrial mobile telephones, PCNs, mobile satellite services, satellite sound broadcasting, and space exploration. Allocations below 1 GHz are wanted by high-frequency broadcasting (international shortwave) and low earth orbit (LEO) satellite systems for radio location services such as Motorola's "Iridium 77" satellite LEO global satellite mobile telephone services.

Above 1 GHz, allocations are desired for high-definition TV from satellite and personal access satellite services in the 20- to 30-GHz range, where 60 to 200 MHz would be dedicated for PCNs. The United States has terrestrial fixed and mobile allocations in the 1,429- to 1,525-MHz band (where there could perhaps be a reallocation of spectrum for nonshared spectrum services). In addition, the United States has the 1,710- to 2,690-MHz band, where there could be an opportunity for sharing up to 200 MHz.

Radio applications are beginning to blossom. These include applications such as Mobitex, Mobile Data Network's private packet-switched 8,000 bits per second service that could be tied to ISDN X.25 public data networks. Similarly, when the CT-2 telepoint service is combined with paging systems, they make a nice two-way communication system, once hand held units reach the $99-$199 level and the Common

Air Interface (CAI) standards are resolved. CT-3 will add data-handling capability. Similarly, wireless PCs within a work area are a new possible application for local radio networks.

The microwave band contains frequencies too high to be carried by coaxial cable and too low to be considered infrared. The microwave band is a portion of the spectrum with almost unlimited room for communications, but, initially, only up to 30 GHz has been used. Spectrum-based communications use microwaves almost exclusively for microwave relay and repeater satellites. Communication carriers occupy two distant bands, the C band (4 and 6 GHz) and the KU band, 11 and 12 GHz, 20 and 30 GHz. Cellular radio operates in the UHF band at around 800 Hz and uses line-of-sight propagation similar to microwave relay. Fiber optics took a quantum leap from Ka band (tens of gigahertz) to infrared rays (hundreds of terahertz), where optical communications systems operate near the low edge of the visible spectrum. Here narrow beams of coherent rays from laser diodes (1,300 nm and 1,500 nm) are carried within a glass fiber, where they are propagated, refracted, and reflected exactly as though they were a viable beam of light.

The use of spectrum is critical. Although its amount of use has increased ten thousand times, so has the amount of information carried by it. Systems such as Direct Broadcast Satellites (DBS) compete for frequencies to carry entertainment or news, where one color TV (DBS) signal takes the same channel space that supercomputers could use to transmit millions of bits.

Hence, spectrum and its best use for society will always be a continuing challenge into the 2000s, especially as we attempt to construct new HDTV direct broadcast systems as well as universal PCN systems.

Advances in radio technologies have allowed for more capacity over the 1990s; for example, where cellular networks migrate to totally digital time division multiple access (TDMA) technology with its 6:1 capacity gain, it will effectively increase its allocation from 50 MHz to 300 MHz. Similarly, spread spectrum technology by overlay techniques could provide capacity improvements 20 times that of the current capacity.

PCN may be a heavy user of spread spectrum technology due to the inherent interference protection that allows closer spacing of base stations, which results in more frequency reuse. (Meanwhile, mobile radio is spectrum-starved as obsolescent analog radio techniques require 30 kHz of spectrum for each analog cellular radio cell.)

One of the most significant advantages offered by digital technology is its capabilities for increased calling capacity through various multiple-access technologies. One of these is the Code Division Multiple Access (CDMA) spread spectrum system, which uses a more wideband channel as compared to TDMA or frequency division multiple access

(FDMA). The key advantage of CDMA is that the additional interference can be tolerated over narrowband systems. The technology allows for closer spacing of adjacent calls.

Under Part 15 of FCC rules, devices operating under a watt of power can operate without a license. This technically forms a point-to-point radio wireless transport vehicle for PBXs and PCs, and point-to-multipoint PCN-type systems for multiple access to the base system.

PCS

Personal communications services can be deployed using the easier CT-2/CT-3 or the more complex PCN (personal communications network) technologies. The objective of PCS is to enable everyone to have their own personal telephone number (PTN), allowing everyone to communicate to anyone, at any place, anytime, using low power (10 mW), lightweight, hand-held sets. This is indeed a different technology than cellular mobile units' high-powered (MW) digital technology defined by GSM. The CCIR study group has recommended 170 MHz for vehicular use and a separate 60 MHz for low-powered, tetherless PCS access. These separate systems will continue to compete for radio spectrum as cellular growth is anticipated to move from 5 million in 1991 to 30 million in 1995 to 100 million in the new century.

The United Kingdom allocated a spectrum in the 2-GHz range for PCS, but the United States has far more usage in this spectrum, such that the 1.7- to 2.3-GHz choice, though highly desirable, may not be feasible because of unacceptable interference with this spectrum, due to preexisting microwave services. Based upon the Radio Consultive Committee Interim Working Party for WARC 92, the 60 MHz translates into 120 MHz for a minimum of two service providers. The House and Senate version identifies needs at 175 to 200 MHz, therefore U.S. PCS spectrum allocation will be approximately 200 MHz of spectrum, perhaps between 500 and 3,000 MHz. This area would require sharing and reallocation of spectrum by both governmental and nongovernmental users.

The 1,700- to 2,300-MHz frequency range is divided into five bands:

- 1,700 to 1,850 MHz — Government
- 2,200 to 2,290 MHz — Government
- 1,850 to 1,990 MHz — Private fixed microwave
- 1,990 to 2,110 MHz — Aux broadcast, CATV
- 2,110 to 2,200 MHz — Public fixed microwave

PCN

PCN has some challenges to resolve besides spectrum utilization, such as circuit quality, voice coding schemes, system complexity, power consumption, and system economics. The access technology—TDMA, FDMA, or CDMA—must be addressed. The tie to intelligent networks for call screening, selected call forwarding, and caller identification are valuable additions to PCS, requiring compatibility between wire-line and wireless networks. The PCN requirements for 156-channel or 20-MHz capacity to support a typical office building must be compared to outside requirements for similar frequencies. Dr. Ross, president emeritus of Bell Labs, has noted that "some of the key technologies for PCN architectures are microcells, digital voice coding, and CCS7 based intelligent networks. The microcell architecture permits increased capacity through frequency reuse and low cost radiators, that are used as an extension of PCN base stations. The increase in channel capacity is approximately inversely proportional to the square of the micro cell radius, thus reducing the cell radius from 5 miles to 500 feet, increases network capacity by a factor of 2,500. Smaller cell sizes reduce power requirements and thus increase the life of the battery within the terminal. The number of cell sites, however, increase maintenance costs and also the number of hand offs between cells."

Hence, very small cells, 500 to 1,000 feet, with tightly packed reusable frequencies, using complex voice compression techniques, with access to the intelligent network for wireless services and access to "the big database in the sky" to monitor where everyone is all the time is no trifling undertaking, especially as multiple carriers will be involved to ensure competition. Thus, PCS is indeed a formidable challenge for the 2000s, and is, as some in the standards groups have called it, "ISDN with wings."

Standards

Interconnectability

New organizations such as committee T1 (Standard Development Committee for North American Networks) and T1A (Telecommunications Industry Association) have come into being to help resolve the networking and terminal interface standardization issues. They are supporting the efforts initiated by older organizations: X3, EIA, and IEEE, as well as ANSI-Common Natural Standards Institute and CCITT, the International Telegraph and Telephone Consultative Committee of the ITU (International Telecommunications Union). The key considerations that these standard-setting bodies must handle are: the rate of change of technology, the integration of C&C, the internationalization of STDS, the multinetworking of multimedia services, the integration of private and public networks, and the "openness" of interconnectability, as well as open network architectures, open systems interconnection, portability of software, rapid service delivery by third-party software, and global networking. Those are but a few objectives.

Communication problems and issues are becoming more and more complex, as noted by one standards committee member: "There is consensus in large segments of the world that the standard's future challenges are: Broadband ISDN, advanced audio visual systems, intelligent networks, the telecommunications management network, and universal personal telecommunications (UPT). The first four pale in comparison with the fifth. The concept of UPT has been described as 'ISDN with Wings.' This is a complex networking arrangement to give each individual access to every conceivable service from any point on the globe. Each aspect of the design, from human engineering to protocol optimization through tariffing, security, and international frequency spectrum allocation, presents major challenges for the standards developers in any way concerned with communication in the broadest sense."

Complexity increases due to: delays in public offerings, supplier options for using framing and control bits, packetizing network address and routing information from multiple networks, enveloping overheads for SONET, and networking within networks as virtual private networks operate within value added networks (VANs). Finally, the global challenge of internetworking, interprocessing, and providing interservices for all aspects of information creation, storage, access, merger, manipulation, transport, utilization, and display requires a formidable group of committees. (See Fig. B-12.) However, these committees require both the providers and suppliers to commit full resources to addressing these standards on all fronts. Then the people on these committees, the French, the English, the Germans, the Americans, the Spanish, the Belgians, the Italians, etc., all with different backgrounds and languages, must give and take to form understandings and agreements. That is the information standards "challenge of the New Millenium." (See Fig. B-13.)

Technology

It's "MIPS versus bits" as processors approach mainframe computer power, mainframes become supercomputers, and supers become distributed, where many processors work together to solve problems using local or remote communication facilities that have advanced from transporting bits to kilobits to megabits to terabits (trillions of bits per second); memory sizes have also moved to greater and greater limits as consortiums strive to produce 60 million bit chips and hologram memories. Similarly, computer logic becomes more and more logical, in fact smart and then fussy as "thinking" machines enter the scene, functioning at "light" speed to become "light computers." Display-system pixels contain more and more information going from 4 bits to 24 bits to not only display the many gray levels, but the full color spectrum in greater depth and resolution; and as new programming languages become more and more conversational and database

Organizations Information Flow
Figure 2-12

AFNOR	Association francaise de normalisation
ANSI	American National Standards Institute*
AOW	Asian-Oceania Workshop
ARC	Administrative Radio Conference
BCS	British Computer Society
BSI	British Standards Institute
CCIR	International Radio Consultative Committee*
CCITT	International Telegraph and Telephone Consultative Committee
CEN/Cenelec	Comite Europeene de Normalisation Electronique
CEPT	European Conference of Postal and Telecommunication Administrations
COS	Corporation for Open Systems International
COSINE	Cooperation for Open Systems Interconnection Networking in Europe
DIN	Deutsches Institut fur Normung
DOD-ADA	U.S. Department of Defense--ADA Joint Program Office
ECMA	European Computer Manufacturers Association
ECSA	Exchange Carriers Standards Association
EDIFACT	Western European Electronic Data Interchange for Administration, Commerce, and Transportation
EMUG	MAP/TOP Users Group
ETSI	European Telecommunication Standards Institute*
EWOS	European Open Systems Workshop
GOST	USSR State Committee for Standards
IEC	International Electrotechnical Commission
IEEE	Institute of Electrical and Electronic Engineers*
IAB/IETF	Internet Activities Board/Internet Engineering Task Force
ISA	Integrated Systems Architectures
ISO	International Organization for Standardization
ITRC	Information Technology Requirements Council
JISC	Japan Industrial Standards Association
JSA	Japan Standards Association
JTC1	Joint Technical Committee 1--Information Technology*
NIST	National Institute for Standards and Technology
NNI	Nederlands Normalisatie-instituut
OSF	Open Software Foundation
POSI	Pacific OSI Group
SAA	Standards Association of Australia
SCC	Standards Council of Canada
SIGMA	[Unix Open Applications Group--Japan]
SIS	Standardiseringskommissionen i Sverige
SMPTE	Society of Motion Picture and Television Engineers
SNV	Swiss Association for Standardization
SPAG	European Standards Promotion and Applications Group
T1	Standards Committee T1--Telecommunications
TTA	Telecommunications Technology Association of Korea
TTC	Telecommunications Technology Council*
UI	Unix International
UAOS	Users Association for Open Systems
URSI	Union radioscientifique internationale
VESA	Video Equipment Standards Association
X/OPEN	[Unix Open Applications Group]

Figure B-12. Global standards organizations.

management systems become more and more constructive and accessible, we indeed are experiencing significant shifts in technology. As this has taken place over the 1990s, we have seen an ever-increasing trend to interconnect systems as control and feedback mechanisms go online, as problems are shared, as graphic systems display information obtained from various remote databases, and as computers design new computers.

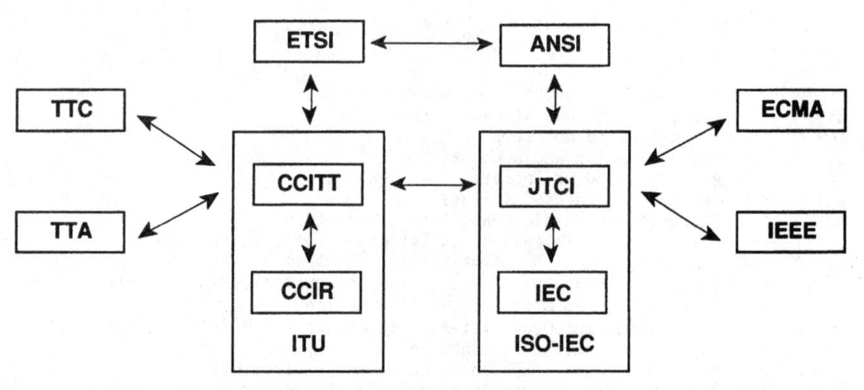

Figure B-13. Organizations' information flow.

We cannot ignore the traditional battle between C&C, nor should we say computers versus communications as "MIPS versus bits." Depending on the cost of transport and the availability of the needed capacity of transport, computers will interconnect the easy way—electronically—or the hard way—manually. Many a supercomputer center has been visited by a young scientist who flew halfway around the world to run his or her program. So the key to the "MIPS versus bits" issue is no longer capability but availability—economic availability. Compression techniques are attempting to reduce bandwidth requirements, as processor power makes up for lack of economic communication—throughput capacity, as noted in the new 64-Kbps, 128-Kbps and 1.544-Mbps viewphone/videophone/picturephone offerings. Similarly, due to compression, HDTV dropped from 1.2 Gbps to 600 Mbps to 155 Mbps, to what next? How near will it approach the 6.3 Mbps of commercial NTVC TV? But is it necessary to waste computer power if the communication capacity is really there, but artificial high transport (bit) pricing is inhibiting usage? On the other hand, high-capacity transport of bits may not be everywhere. The lower-speed copper-based distribution plant will be slowly charged out of the primary and secondary feeders over the 2000s. During this transition period there is a need for compression and a need for interoperable systems of different compression rates. For example, two picturephones over T3 or two over T1, or two over 128K (2B) need to be able to interexchange information at different rates from the same terminal. They need a terminal that can dynamically adapt to the different media. Hence, flexibility, versatility, and changeability will be requirements for ensuring connectability over the transition period of the Information Age (1990-2010).

Multimedia interoperability will also be essential as communications shift to wireless. This is a reasonable alternative for direct broadcast reception in the home in the more rural areas as technology for 6-MHz analog baseband video and 30-MHz FM-modulated video shift to digital video compression (digital QPSK) requiring 1.5-10 Mbps bandwidth for entertainment-quality video. Also, cellular, personal communication networks (PCN), wireless 10-Mbps digital radio data links, and PC-to-PC wireless 64-Kbps file transfer networks will need to overlay and interface with terrestrial fiber network.

Technology Challenges

Therefore, technology is changing. Complex software technology is advancing in leaps and bounds to object-oriented languages, enabling decision table inputs, versatile target-machine compilers, application-based programs, and object code outputs, as well as sophisticated database management systems. New software interfaces called application program interfaces (APIs) are needed to enable software portability. However, a major problem, besides the wireless issues noted above, will be managing the layers of interconnection that the OSI and the IEEE models provide, from physical to application.

Hence, the 1990s have been the decade where the walls between marketing and technology have been torn down similar to the walls between East and West Berlin. Technology is available to all who will not fear it, but will embrace it and learn to understand it and use it. Therefore, technology is only limited by *"die Mauer im Kopf"*—"the walls of the mind." This may be the biggest problem of all to solve. It is indeed technology's challenge of the new century.

Fiber Distribution Plans

Much has been written concerning the deployment of fiber. An acceptable topology must be finalized in order to obtain a reasonably "ubiquitous" broadband infrastructure to urban homes and businesses by 2010. Many aspects of fiber deployment have become understood, with various agreements specified over the 1990s timeframe. Bellcore has played a leading role in these endeavors, working with major suppliers such as Corning, AT&T, Siemens, etc.

Much is happening to help us better understand fiber's capabilities and how to use it properly. We are learning that fiber must be designed to perform successfully in harsh cyclic environments of –60°C to 85°C and salt solutions up to pH 12. Here, continuous testing is performed to determine the strip force, cooling abrasion resistance, and fatigue resistance for titan fiber and silica-clad fibers. Advances in expensive LASER (light amplification by simulated emission of radiation) and cheaper light emitting diodes (LEDs) are assessed and compared for application performance versus costs. Trade offs have been made for single mode versus multimode systems in terms of capacity, watts, attenuation, splicing time, splicing attenuation, repeater spacing, sharp bends, etc. Deployment topologies are especially important as active versus passive techniques are being researched and further considered using various star, double star, and ring arrangements.

Finally it has become apparent that it's essential to define user needs. Having a better understanding of how much future capacity is needed for what type of future services will help establish the fiber distribution plans for the providers of these services. Hence, as the 1990s unfolded, the challenge was to resolve these issues early and fine tune them in order to have the appropriate fiber facility infrastructure in place when the broadband switches will be available to send information over them to provide the identified (and not yet identified) future services.

Therefore, it appears that we could use a starting point based on past research and test conclusions, though some may disagree with this or that opinion. Hence, the following observations may set the stage.

1. While an expensive, high-speed modulatable laser with very good characteristics typically puts out 1 to 2 milliwatts, erbium-deployed fiber optical amplifiers will be able to increase the output to 10 or 20 milliwatts or even higher. They do not require the optical/ electrical/optical (O/E/O) conversion and can substantially boost power for the original high-quality signals. They are also adjustable to changeable bit rates without requiring upgrades. In tests, this type of power from these amplifiers easily delivers many high-capacity channels to numerous customers. Hence, once fibers are deployed, the future is bright for increased power capabilities.
2. For massive deployment of fiber in the loop systems, the 1310-nm and 1550-nm regions will be used. They are the same regions used for long-haul transmission. Transmission at 850 nm and 780 nm are alternatives, but as the cost of 1300-nm lasers drops, they will become more attractive.
3. Single mode fiber will support more and more gigabits for longer distances without repeaters, just as 1.2 Gbps gave way to 2.4 Gbps, and now 4.8 Gbps per second. Experimental systems in England are achieving 20 Gbps on fiber rings. Hence, capacity will greatly increase, as use of 45-Mbps DS3 systems having the capacity to handle 64,500 voice channels shifts to include 1.2-Gbps systems that handle up to 1,700,000 voice channels, etc.
4. Fiber in the loop (FITL) can be developed in several configurations. The deployment of fiber between the central offices (CO) and the remote terminal (RT) has taken place over the 1990s. Since primary and secondary feeder plant is upgradable to fiber, the costs were easy to justify as high-volume, high-capacity, low-error-rate data traffic supported its usage. Bellcore has defined

the hubbing point of fiber as a host digital terminal (HDT) that could be in a CO or remote terminal (RT), depending on the economics of the application, while an optical network unit (ONU) is located out at the curb, on a pole, or underground.

There are several topologies for providing fiber to the house (FTTH) or fiber to the curb (FTTC), as noted in Fig. B-14A and B-14B. It can be deployed in the traditional star back to the CO, as a double star from a remote terminal, as a ring, or as a branch from the remote terminal. (Bellcore has a series of documents defining fiber deployment, networks, systems and interfaces (TA-NWT-000909, SR-TSY-001681 and TR-TSY-000303), and TR-944 and TR-008, which define the central-office-to-remote-terminal interfaces as the next-generation digital-loop carrier-interface standards for broadband services.)

5. By 1994, the majority of the plant between the CO and the RT's DLC (Digital Loop Carrier Systems) will be fiber. This is 63 percent of the BOCs' plant. The remaining 37 percent is copper pairs going to the residence. The cost of delivering broadband fiber to the home competes with the cost of wire. For example, for short distances, copper costs $365 per line, versus $700 per line for

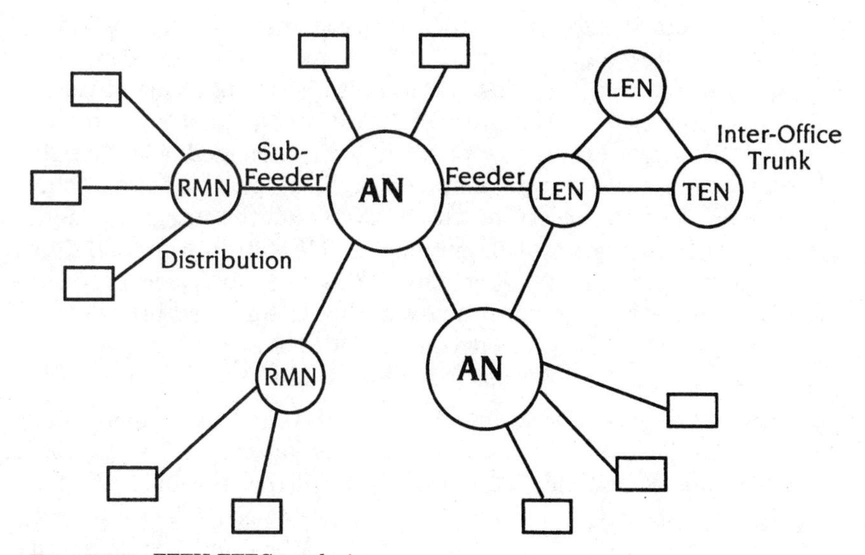

Figure B-14A. FTTH-FTTC topologies.

Logical : Double Star

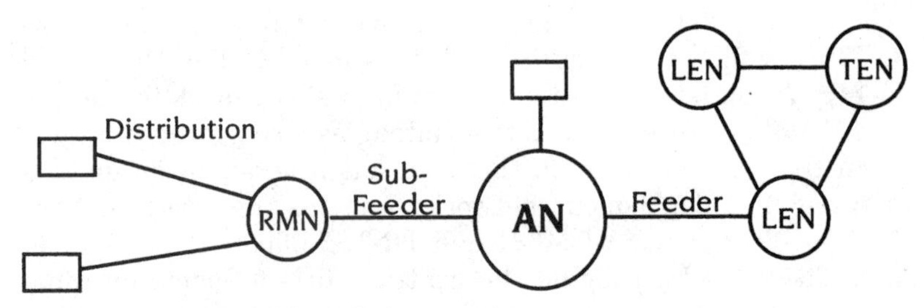

Figure B-14B. FTTH-FTTC topologies.

fiber; other figures show the full-deployment fiber costs dropping from $3,000 to $1,500, with copper going up from $1,300 to $2,000 for the 1,000- to 1,500-foot distribution plant from the RT to the home. Hence, the target appears to be approximately $1.00 per foot for economical fiber deployment.

6. Most believe there is a need for the home to have a user network interface (UNI) of 600-Mbps capacity in order to use four to six video channels, plus data links, and telephones, etc. The alternative broadband UNI is a 155-Mbps interface, while the narrowband UNI remains 2B + D (BRI-ISDN) at 144/192 Kbps with the wideband UNI for 23B + D (PRI-ISDN) at 1.544 Mbps.

7. Passive optical networks address the economical use of fiber in the loop (FITL) systems by sharing fiber and electronics serving a number of customers. This is accomplished using a time division multiple-access TDMA protocol that allows the payload of multiple ONUs (optical network units) to be transcended over a single pair of fibers. To separate combined payloads, passive optical splitters are used that conform with SONET and the integrated digital loop carrier general requirements (TR-303) to achieve 1:2 or 1:4 arrangements enabling four DSOs and six videoports per home; this arrangement anticipates that compressed high-definition TV can be nicely handled on 155 Mbps or even at the lower 45 Mbps, 19 Mbps or 6.3 Mbps for compressed LTV or NTVC.

8. The asynchronous DS3 rate of 44.736 Mbps and the Synchronous Transmission (SYNTRAN) DS3 rate, as well as other 90-Mbps, 139-Mbps, 405-Mbps and 565-Mbps rates will be overlaid onto the expanding Synchronous Optical Network (SONET) based on the STS-1 basic transmission rate of 51.84 Mbps and the standard European-U.S. agreed STS-3 rate of 155.520 Mbps. The SYNTRAN

format was established by ANSI T1.103-198, while the DS3 for national facilities is supported by add-drop multiplexing (TA-TSY-000010), digital cross-connect systems (TS-TSY-000233) and a direct digital switch interface (TA-TSY-000304).) The B-ISDN standards will extend these capabilities by use of DS1/DS3/STS-1/OC-1 wideband digital cross connect (W-DSC) and broadband (DS3/STS-1/STS-3/OC-N) cross connects (B-DCSs).

9. The extension of fiber from the CO to the carrier serving areas (CSAs) remote terminal (RT) nodes indicate the potential for a remote access/switch node where power and cooling can be provided. From this network hub, fiber can be extended directly to large businesses in the area, in a survivable ring structure. Alternatively, it can extend individually to a secondary hub where residences can obtain direct fiber to the home (FTTH) optical links, or it can proceed to a sub-subnode where a fiber-to-the curb arrangement can achieve a fiber-to-copper interface. The secondary node can function actively or as a passive optic splitter, depending on broadcast applications versus interactive conversations.

10. Both STM synchronous-transfer mode and ATM asynchronous-transfer mode switching fabrics and cross points will be extensively used by B-ISDN to switch information that is carried on the SONET format. As networks are implanted, new elements such as cell count, queue size, cell delay, cell loss, buffer overflows, and bandwidth utilization will require new traffic tables to better determine the correct fiber topologies for distributing these switching capabilities to the remote terminals as well as to new switch topologies in the distribution plant.

11. Network synchronization will be one of the most complex issues as remote terminals (RTs) have switching systems that can home on multiple central offices (COs) as well as other remote terminals (RTs) in a ring configuration. They will also be needed to access points of presence of alternate providers (POPs). They may become interface points for satellite (KU band and direct broadcast) networks, as well as interface points for local personal communication networks (PCNs). Hence, the selection and use of various levels of clocks such as the Stratum 1 clocks with minimum skip occurrences of $\pm 1 \times 10^{-11}$, Stratum 2 clocks $\pm 1.6 \times 10^{-8}$, Stratum 3 clocks $\pm 4.6 \times 10^{-6}$, and Stratum 4 clocks $\pm 32 \times 10^{-6}$, will play a major role as internetwork synchronization becomes more and more complex.

12. Rural communication networks can be overlaid with super-switch/front end switch configurations that bring wideband/broadband capabilities to the larger towns in the area, using areawide broadband deployment topologies based upon urban area technologies. In the interim, STM and ATM switching fabrics can be

added to existing digital host systems to provide local businesses broadband capabilities. The real challenge will be to ubiquitously provide videophone or HDTV to the rural community. This, of course, is tied to the ubiquitous deployment of videophone and HDTV in the urban environment.

A final observation: Distribution fiber topology deployment plans need to be finalized, as the BOCs and independents continue to drop 10 billion dollars or so in copper plant each year. Unfortunately, this delay is really a decision not to deploy 10 billion of new fiber plant, which means not to use the unlimited possibilities of the fiber for some time to come.

Private-Public Internetworking

We have traveled down the yellow brick road and seen some fairly complex and frightening technology. We cannot help but see the mounting pressure at every turn. If technology is ignored or blocked, it simply changes direction to take the path of least resistance and move on and on. There are three crossroads for the public network: narrowband, wideband, and broadband, as Fig. B-15 indicates. If technology is not used in the public network, it moves to the private network. This creates a mounting pressure to provide more and more private networks, first internally as LANs, then locally as MANs, then regionally as WANs, and finally globally as VANs, all the time shifting to VPN, where variable use of dynamically changing capacity and instantly reconfigurable destinations is the goal.

Alternative networks are provided by the VANs, both at the local and national level, as wireless systems invade the local arena, providing a viable option to traditional terrestrial evolution. Hence, time (delay) is not on the side of the traditional LECs and IXCs, as their price gaps open opportunities for new players to fill "the space between the stones."[2]

The pressure increases to internet private and public (P&P) services, noting the complexity of managing a thousand LANS, noting the complexity of routers that interface with everyone's local network, noting the new desire of computer mainframes to extend their capability by becoming more accessible to remote PCs. The power of a computer processor (CP) goes by a factor of the number of interconnected mainframes (MFs) and PCs.

$$\text{Total CP power} = \text{CP} + \frac{(n\text{MF} + m\text{PC})}{a}$$

[2]A knowledgeable marketing professor once said, "Put a bunch of large stones in a bag. See how they do not tightly fill all the space in the bag. There are pockets of air between the bigger stones. The more space between them, the more opportunity there is for smaller stones to fill the space between the larger—so it is with the marketplace."

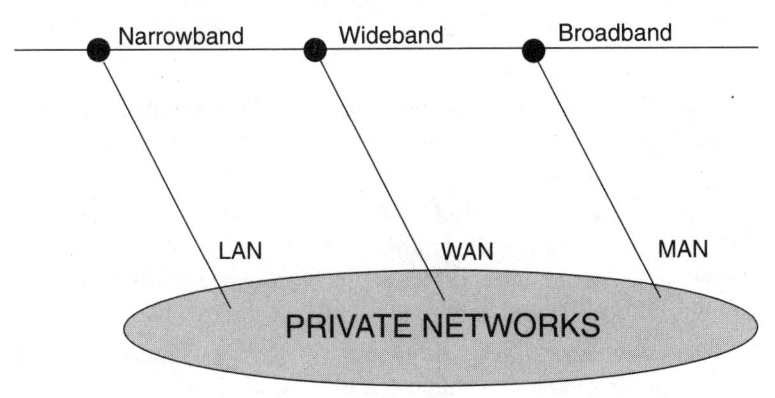

Figure B-15. Three crossroads.

As computer power becomes more and more dispersed in various applications, as mainframe power becomes more and more powerful, then the need to interconnect these dispersed workstations, PCs, and information systems brings a new directional drive to technology.

Integrated voice and data was the nebulous goal of the 1980s; C&C was thought by some to mean voice communications and computer data, where in reality it also meant integrating computer data and data communications as CD&DC or the computing and communications C&C of data. The 1990s have seen the less nebulous goal of the P&P (private and public) networking of data. Later, multimedia voice, data, and video will, of course, follow. One firm recently purchased ISDN narrowband Centrex, not for integrated voice-data option, but for wide area networking capabilities to move internal data from their mainframe computers to the firm's remote terminals that were served by the wide area Centrex.

This private-public networking objective (noted in the technology networking model) is therefore not only for voice, but also for data, image, and video. This is the basis for a layered networks' layered services architecture, where layers of private and public networking for both transport and services are clearly understood and achievable, separating the complexity so standards can be more easily identified and networks more quickly established. The P&P (private and public) information infrastructure is indeed C&C's "P&P challenge of the New Information Millenium."

Perspective: Information Technology

The private-public internetworking of various technologies will become increasingly complex and challenging as:

1. Voice returns to the public network, while separate and private data networking become the new emphasis for the early 1990s.
2. As LANs continue their growth, the need to perform more LAN-to-LAN networking leads to more intelligent bridges, routers, and gateways.
3. Switched ISDN public data networks enable remote personal computers to access mainframes for special-purpose processing, general shared processing, or access to databases.
4. PC networking over switched 64-Kbps or 128-Kbps narrowband ISDN data networks will enable higher-speed, less-error-rate transports for faster data distribution applications such as facsimile and E-mail systems.
5. Wide area networks using 64 Kbps and 128 Kbps can be established as permanent or switched virtual private networks. In the permanent mode, the paths are dedicated and "remain up" through the circuit switches for isochronous data streams. For switched VPN, the links are established at the time of call so that reconfiguration does not require the assistance of network operators, but can be performed under the control of the customer. In time, developments in automatic configuration control will blur the distinction as set-ups become online. For virtual packet switching, the virtual calls are connection-oriented. Here the route is selected by the initial packet, and then the packet switches simply wait for additional packets to be handled. In some transport schemes, selected time slots are preassigned or dedicated. In this manner, the virtual paths are set up in a manner that busy-hour or congested situations will still ensure availability. Switched virtual private networks can be easily established and changed to meet the more dynamic needs of day-to-day, hour-to-hour, or instantaneously changing capacity needs, as well as enabling route diversity or reconfigurations without requiring leased lines.
6. PBXs, cluster controls, and intelligent bridges, routers, and gateways will multiplex and concentrate higher data streams to distant users, as well as receive more numerous external data calls for access to internal databases. Hence, new internal broadband switches will be deployed on customer premises.
7. 56-Kbps data streams for larger businesses will give way to T1 rates once private broadband switching hubs are located closer to the customer, as shorter T1 leased-line distances become as economical as long 56-Kbps leased-line private networks. This private networking shift to T1 networking and then to T3 networking is taking place on customer premises where the private hubbing nodes are established at many firms' central locations. For large business enterprise networks, this trend may continue; however, it

may take another fork in the road where it returns to the public network to perform shared hubbing for T1 and T3 switching, thereby enabling permanent and switched virtual private networks, priced on a usage-based wideband and broadband DCS service.

8. Frame relay is an ISDN packet-mode bearer service for permanent and switched virtual circuits identified in CCITT I.122 based on CCITT Q.921 HDLC variable-length packets and LAPD multiple virtual circuits. It will be substantially faster than previous LAN-to-WAN systems where the full frame has less overhead, leaving error detection and retransmission to end-to-end CPE control. Frame relay has handled slower-speed interactions and less-volume calls up to 2 Mbps.
9. Fiber Distributed Data Interface (FDDI) (ANS1 X3T9 based on 802.5 100-Mbps counter rotating token rings), originally for multimode fibers for (limited) distances less than 2 kilometers, will be extended to single-mode fiber for up to 60 kilometers; FDDI-II will have isochronous video and voice capability of 16 multiples of 6-Mbps wideband channel capacity. It will be a private alternative to SMDS.
10. Switched Multimegabit Data Services (SMDS) will handle fast-speed, high-bandwidth, on-demand volume calls from 1.5 Mbps up to 45 Mbps initially, and then ingress/egress classes of 4, 10, 16, 25, and 34 Mbps. It is a segmented-oriented technology similar to cell relay. Its SMDS interface protocol (SIP) is based on 802.6 DQDB-Distributed Queue Dual Bus access mechanism. IEEE 802.6 metropolitan area networks use dual 45-Mbps buses. The DQDB access method for fixed-length slots is similar to ATM cells, for voice, video, and data and will serve as the interim technology. SMDS traffic will be later handled on future 155-Mbps buses. SMDS acts as a subnetwork between LANs. SMDS-to-SMDS traffic will be switched by B-ISDN ATM switches using SONET transport between ATM switches.
11. B-ISDN will use SONET and ATM as well as STM. Asynchronous transfer mode accepts synchronous time-division multiplexed channels and variable-length packets. Switched-fixed cells are based on a structure having a header of five octets. It carries a forty-eight-octet payload which is established by its ATM adaption layer (AAL) with a header (HDR) information and tail, as the incoming message is segmented into packets for transmission. The B-ISDN 5-byte header contains generic flow control, virtual path identity, virtual channel identity, payload type, cell loss priority and header error control information. All this header control information makes ATM a versatile vehicle for interfacing and switching with numerous types of input formats and protocols.

12. Multiple synchronous transfer mode fabrics for various isochronous circuit time division modulation (TDM) traffic will also enable the B-ISDN hybrid to handle groupings (64 Kbps, 1.5 Mbps, 45 Mbps, 155 Mbps, and 600 Mbps) of high-capacity, long-distance image-video traffic, rather than packetize it into segments. This will cause competition between ATM operations, depending on cost of ATM operations, depending on cost of ATM crosspoints and traffic mix and transport delays.
13. Traffic has continued to increase over the 1990s on these frame relay, SMDS, 802.6 MANs, and FDDI transports as global WANs, MANs, and VANs enable movement of LAN-to-LAN information, first locally, then regionally, nationally, and globally.
14. Private networking CPE systems will become more intelligent in order to perform external end-to-end transport control of concentrated high-volume messages, as well as internal broadband switching, as they control multiple internal rings, buses, and stars. Access and interface to the external networks will be through new broadband standard ISDN/IEEE/ANSI interfaces. They will interface to new points of presence of IXCs and VANS, as well as to the traditional public class structure of switching.
15. Public networking will need to interface to these private networks, providing new switched transport nodes closer to the private structure of internal networks. In introducing another level of access to the traditional switching, new B-ISDN broadband superswitches will have, in essence, a front end closer to the customer, providing narrowband, wideband, and broadband access and ring transport to move BRI-ISDN, PRI-ISDN, f-ISDN, T1, f-T1, T3, frame relay, FDDI, FDDI-II, 802.6 MAN, and SMDS data, voice, and video traffic using both ATM cell relay packet and STM circuit-switching capabilities to provide virtual private networks, both permanent and switched, connection-oriented and connectionless data handling, T1 and T3 hubbing, DS0/DS1/DS3/ STS-N/ OC-N DCS channel switching, and wide area networking of private LAN-to-LAN traffic.

The End of the Yellow Brick Road

X-rays, images, workstations, videophones, viewphones, picturephones, and HDTV will continue to require higher resolutions of 1K by 1K or 2K by 2K using color pixels having 24-bit codes for compressed or noncompressed data streams for multimedia (voice, data, video, graphic, image, and text), multivendor, multilocation, multiservice, multiuser applications. So, there is and will continue to be a growing need for multinetwork P&P (private and public) C&C (computer and communications) internetworking; and this indeed has been technology's challenge over the 1990s.

Therefore, we are at the edge of an ever-increasing array of new technical possibilities for meeting more and more functional needs of the information user. This is similar to the further advances in automobiles such as wishbone and double wishbone suspensions, under steer, over steer, integrated suspensions, various traction-type systems, overhead cams, double overhead cams, aluminum heads, sixteen valves, etc.

Technology is not standing still. The information marketplace will continue to expand. The more the drivers of the vehicles understand what their cars can deliver, the more selective they will be, the more in tune the cars will be for the applications. Similarly, the more the future users, providers, and suppliers understand what the information networks' products can deliver in terms of features and services, the more appropriately they will fit the specific applications. These expanding highways and pathways will then be used to bring us into the Information Age. There our tasks will become easier and our quality of life will become better as we substitute communication vehicles for transportation vehicles to enable new cities to develop into world communities, interconnected together, to form the future information society. (See Fig. B-16.)

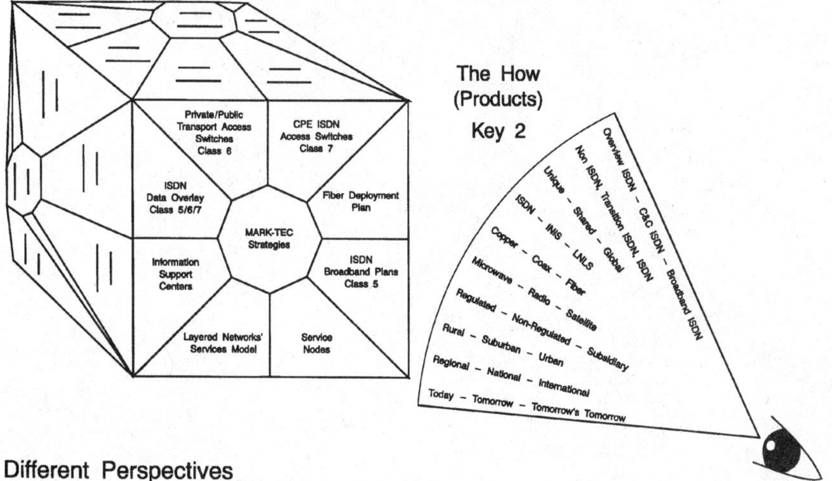

Figure B-16. The technology cubic.

Acknowledgements

The author is grateful to the following visionary leaders who contributed their views, definitions, opinions, and concerns as expressed throughout the 1990s: D. Auble, S. Ginn, R. Singer, D. Cox, J. Conlisk, P. Berteau, D. Hall, W. Anderson, J. McQuillan, T. Warren, G. Hawley, M. Pitchford, J. McConnell, C. Probst, M. Hurwicz, D. Raup, G. Mattathil, L. Huber, M. Tratner, L. Bender, J. Schatz, W. Kenupke, W. Ambrosch, A. Maher, B. Sasscher, Ming-Lei Liou, L. Mantelman, E. Roux, N. Lippis, J. Herman, D. Hoffman, D. Heath, A. Pearce, M. Aaron, M. Decina, H. Kasai, J. Yamagata, D. Powell, C. Peker, B. Lee, F. Bradley, J. Bush, M. McLoughlin, N. Miller, B. Pekarske, P. Maxwell, J. Steele, L. Campbell, C. Nelson, L. Righter-Mathews, J. Limb, T. Helmes, B. Paulson, A. Penzias.

Appendix

C

ISDN Reference

ISDN will be discussed in terms of its narrowband (n-ISDN) and broadband (B-ISDN) structures. ISDN may be viewed as having both a narrowband and broadband set of interfaces. In actuality, narrowband has both narrowband and wideband interfaces, where the primary rate access interface (PRI-ISDN) can be considered wideband. Hence, ISDN covers the full range of services, from narrowband through wideband to broadband. There are several key aspects to ISDN to keep in mind as various interfaces, standards, and services are considered. The first is a user-to-network physical interface model, which indicates the type of network and customer premise equipment for which various interfaces are required. The next aspect is the OSI seven-level reference model, which logically denotes layered separations for information exchange. The third is the services themselves, which ISDN provides on these transports facilities. With this in mind, let's take a brief look at ISDN to help determine what it is and what it is not in the grand scheme of things.

User-Network Interface (UNI) Model

There are several interfaces between user terminals and the network central office, depending on if one is in the United States or Europe. Due to the regulation for nontelco ownership of network terminating equipment on customer premises, there is a U interface required in America between the central office and the customer for terminating network signaling and line testing. The R, S, T, and U interfaces are so designated for the basic rate user network interface model, noted in Figs. C-1A and C-lB.

Here, NT1 (Network Terminating Unit One) provides the terminating of a twisted-pair line from a central office. It may be located up to a target of 18,000 feet from the C.O. Above this distance, normal-pair gain

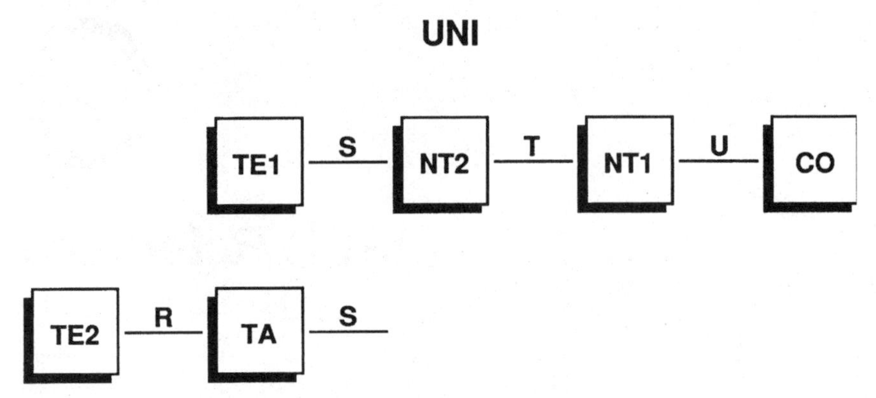

Figure C-1A. UNI.

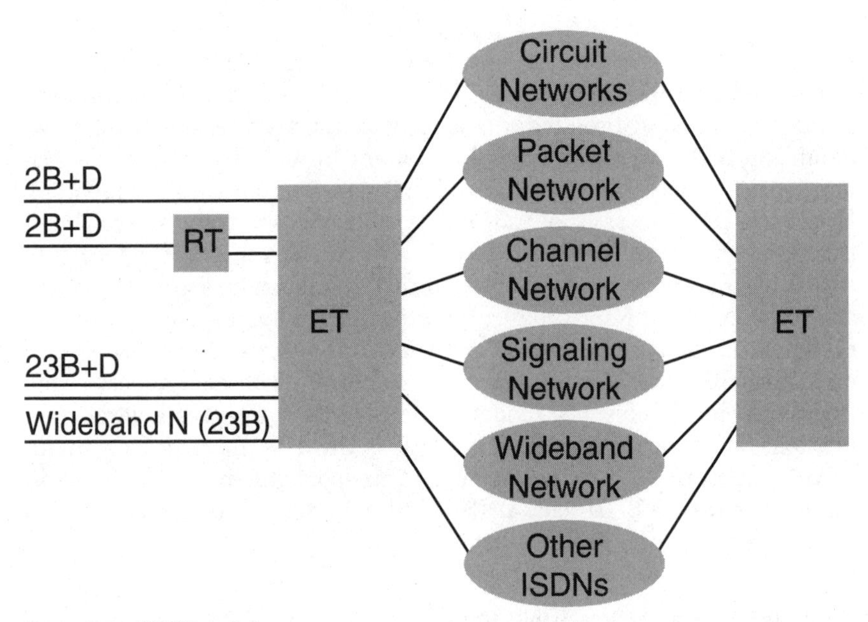

Figure C-1B. ISDN model.

devices need to be deployed, as loading coils can't be added due to their repairment of digital transmission. Information is exchanged over the U interface at 160 Kbps using the 2B1Q coding scheme (tests have adequately demonstrated distances of 13,000 to 15,000 feet as the 18,000 thousand feet target is approached). The NT2 interface unit on customer premises provides a four-wire interface to the ISDN terminals (TE1s). There may be as many as eight devices attached to this port using collision detection mechanisms to ensure orderly access. Non-ISDN terminals (TE2s) require a terminal adapter TA to interface to the NT2 or NT1 devices. Depending on the location, NT1 and NT2 devices can be combined in a single device. While the U interface is four wire, the T interface is two wire and can function up to 10 kilometers from the C.O.

Protocol functions for the Open Systems Interconnection Model (OSI), Figs. C-2 and C-3, are such that NT2 supports layer-two and layer-three protocols. It can function as a PBX, concentrator, multiplexor, or cluster control, where switching, concentrations, layer-two and layer-three multiplexing, maintenance, and terminal interfaces terminate. The T interface NT1 functions provided line transmission termination, layer-one maintenance, timing, power transfer, layer-one multiplexing and interface termination including multidrop termination employing layer-one contention resolution for passive bus, etc. This is a four-wire interface operating at 192 Kbps for 1,000 to 3,000 feet. It provides a basic rate of two 64-Kbps channels (B) and a signaling/channel of 16 Kbps containing a 9.6-Kbps packet network that uses an X.25 packet switching protocol operating over the first three layers of the OSI model. Since 2B + D = 2 × 64 Kbps + 16 Kbps = 128 Kbps + 16 Kbps = 144 Kbps, the 192 Kbps has a 48K overhead that is ample for additional control information for the eight-terminal shared-port set up. B channels are assigned by the C.O. exchange via signaling messages in the D channel. The D channel provides multiple access with collision detection (MA/CD) by comparing upstream D channel with echo D channel. R interfaces can be standard non-ISDN RS232C-type interfaces.

NT2 customer configuration can be in the star configuration, where direct interfaces exist using the S interface or where terminals (TE2) connect to it via TAs using two-pair 1KM standard interfaces. Or, with level-three protocol capability, the central office must support point-to-multipoint. Using the four-wire passive bus structure for distances up to 100 or 200 meters depending on the 75-ohm or 150-ohm cable, up to eight TEs can be connected to the bus at distances up to 10 meters from the bus. In this manner, we achieve basic rate access interface (BRI ISDN), which provides for a digital subscriber line (DSL) as a fully digital interface to the central office, enabling 2B (64 Kbps) full duplex circuit-switched with 1D 16-Kbps full duplex packet-switched or D channel signaling, 9.6-Kbps packet, and telenetworking data. This BRI interface is sometimes referred to as BRA, indicating the Basic Rate Access Standard. Primary rate interface ISDN (extended digital subscriber line) provides twenty-three 64-Kbps "B" channels and a 64-Kbps "D" channel connection to the central office operating from PBXs, multiplexes, and concentrators. This PRI interface is sometimes referred to as PRA, indicating it's the primary rate access (PRA) standard interface. PRI-ISDN operates at T1 rates, except, where T1 has twenty-four voice channels, PRI-ISDN has twenty-three 64-Kbps + a "D" 64K channel signaling information. Hence, PRI works consistently with T1 pair gain/repeater equipment.

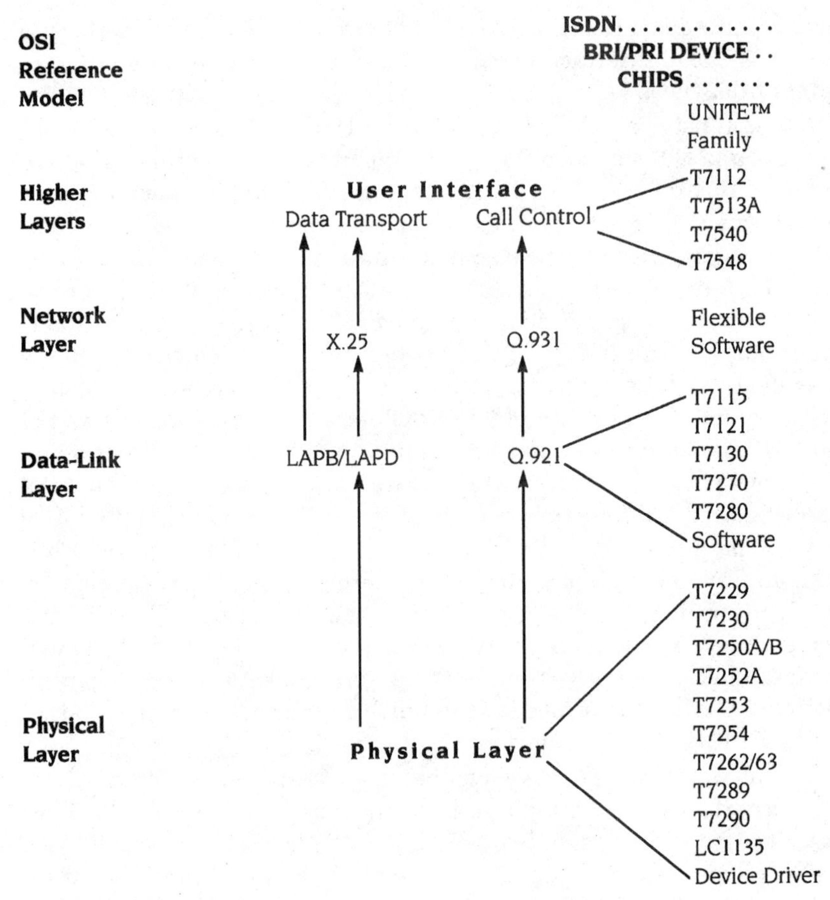

Figure C-2. ISDN chips.

In principle, ISDN offers two methods of communication that are complementary: basic transport/bearer services providing transport end-to-end digital channels. These services can be used by any proprietary application or standard teleservices, such as telephony and facsimile, which can be used independently of any terminal-based application. Pan-European ISDN services supported by new international ISDN switching systems, such as System 12, provide numerous new voice and data services. (See Pan-European services.)

As noted, ISDN will enable both 64-Kbps "B" channel circuit-switched or packet-switched services, as well as "D" channel packet-switched services. B channel connections can be established on demand or semipermanently according to CCITT X.31, Case A (minimum integration) or Case B (maximum integration) recommendations, where the access to an X.25 packet handler can be implemented in one of two ways (here the packet handler is or is not physically part of an ISDN exchange and may be part of U packet-switched public data network (PSPDN)). Q.921/Q.931 signaling for circuit-switched calls work together with LAPD (Link Access Protocol D Channel) lines and X.31 on demand or semipermanent service information, to use X.25 signaling (layers one, two, and three) to "out of band" inform the central office concerning what information is being provided over the "2B" basic rate interface.

The I-Series CCITT recommendation classifies the telecommunications services into bearer services (basic bearer service or basic bearer service + supplementary services) and teleservices (basic teleservice or basic teleservice + supplementary services). Here, the I-100 standard series defines the structure and terminology of ISDN, I-200 series the service aspects, I-300 series the network aspects, I-400 series the user network interface aspects, I-500 series internetwork aspects, and the I-600 series the maintenance principles and operations. They are updated at the CCITT Plenary Assembly every four years, with interim two-year outputs to help offset the long delays. U.S. contributions are provided by T1E1 carrier-to-customer interfaces, T1M1 internetwork operations, administration, maintenance, and provisioning, T1Q1 performance, T1S1 services architectures and signaling, T1X1 digital hierarchy and synchronization, and T1Y1 specialized areas. Their outputs go to ANSI for American National Standards, to the U.S. CCITT National Committee as contributors to CCITT, and to industry.

Due to the inconsistency of suppliers in choosing different options in X.25, Q.921, and Q.931 signaling, computer manufacturers' terminals do not necessarily work with different suppliers' ISDN network switches. Similarly, terminal adapters for OEMs (original equipment manufacturers) have difficulty designing chip sets for ISDN. This has delayed ISDN introduction so that pressure has been established to cause new user, computer, network consortiums to work out ISDN implementation inconsistencies with Bellcore to establish common interfaces. Efforts by groups such as the Corporation for Open Systems (COS), will greatly help to speed up agreements to resolve variations of

Level 7	Application Layer	X.500, FTAM, X.3 PAD
Level 6	Presentation Layer	ISO 8823
Level 5	Session Layer	ISO 8327
Level 4	Transport Layer	ISO 8073
Level 3	Network Layer	X.25, ISDN Q.931
Level 2	Data Link Layer	ISDN Q.921
Level 1	Physical Layer	Ethernet, Token Ring, FDDI, V.35, RS232

OSI Stack

Figure C-3. OSI stack.

the implementation of ISDN standards for CPE and network equipment supplies. Fortunately, manufacturers have now developed an array of ISDN chips, such as AT&T's full set of BRI and PRI chips, noted in Fig. C-2, where T7262/63 are the 2B1Q transceiver U interface chips that fully conform to the new ANSI standard. This chip family is used in C.O. switches, PBXs, key systems, DACs, ISDN terminals, channel banks, SLCs, MUXs, and packet switches.

New standards for bit rates adaption V.110 and V.120 are enabling various asynchronous 2.4-, 4.8-, 9.6-, and 56-Kbps data rates to fit within the 64-Kbps continuous "B" channel data stream. (V.120 is an error-detecting and correcting protocol for circuit-switching connections. It's reportedly greatly superior to the V.110 protocol for allowing bit rates under 64 Kbps to be sent over a B channel.)

ISDN extends point-to-point WANs (wide area networks), where dedicated circuits were used to link different business sites together. ISDN enables VPNs (virtual private networks) to provide a business with a network of circuits linking its various locations. In a VPN, each site is connected to the nearest local exchange (CO), where software establishes "on demand" connections. In physical private WANS, the connections between sites are permanent, whereas in VPNs connections are only set up when needed. The two variants of VPN are semi-permanent connections, which are set up without dialing as soon as one of the ends requires the circuit. These always link the same two end points. Or, dial-up VPNs connections require some form of user-to-network information exchange before setting up the connection. This provides greater flexibility, but requires the originating party to send signaling messages. However, because of the limited number of destinations, these messages can introduce only limited delay in establishing the calls. These permanent, on-demand, predetermined, flexible network options for both circuit and packet (connection-oriented and connectionless) basic rate, primary rate (wideband), and broadband ISDN has made the 1990s interesting, as more and more automation in changeable network set-ups occur via the customer or network operator.

ISDN Signaling

ISDN signaling is performed by the transmission of data frames continuing Q.931 call-control message packets over the D channel on the PRI or BRI line between the two switching points. Each Q.931 message packet contains information elements that can be further subdivided into octets, or groups of eight bits that specify call-control parameters. Since D channel information is independent of B channel information, data calls can potentially use all of the 64-Kbps bandwidth of the B channel for user data called "clear channel data." As various dialects by different ISDN vendors are resolved, this ensures better interoperability of systems to enable transmission of address information for call set by D channel Q.931 messages to multiple suppliers' central office equipment, thereby solving the need to "home" on multiple offices in the event of a system C.O. outage.

Signaling protocol

The signaling protocol uses the first three layers of the OSI reference model. The data link layer (layer two) protocol is based on the X.25 LAPB layer two protocol and is called LAPD. This protocol provides multiplexing, error detection and recovery, flow control, and sequencing. Layer-three protocol, called Q.931, provides for signaling for multiple calls between the user and the network, call establishment, maintenance, and call clearing. It also provides access to packet-mode services. Q.931 uses a sequence of messages to perform different functions for the user. The Q.931 messages and functions are as follows: the protocol discriminator, which identifies the packet type; the call reference, which identifies the call; and the message type, which identifies the message (e.g., disconnect, set-up).

Universal Digital Channel (UDC)

This equipment concept packages variable channels of information such as fractional T1, video, imagery, and variable data bit streams in common pipes to the home and small businesses. Here, 2B1Q, at the elevated rate of 80 Mbps per pair (160 Mbps), is applied to repeaterless T1 to enable ISDN narrowband and primary rate ISDN services to be active in the 2000s. In the 1990s, some equipment suppliers developed proprietary transmission schemes that were susceptible to crosstalk. The service based upon these vulnerable techniques such as data over voice (DOV), data voice multiplexor (DVM), and limited distance modem (LDM) were unprotected, while the 2B1Q (two bits in one quaternary or four-level signal) standard accurately transmits 16 Kbps full duplex on a single pair of wires up to 3 miles without repeaters. It' s guaranteed to run on any resistance-designated unloaded loop (99 percent of the dirtiest loops). This technique will be the key to achieving universality and connectivity over existing copper loop. The technique will also be essential to basic rate ISDN success, especially at targeted costs of $10 per chip and at 10 times the speed of traditional voice-grade modems.

Services Supported by Pan-European ISDN

Bearer services. Circuit-mode speech-bearer service; circuit-mode 64-Kbps unrestricted-bearer services; circuit-mode 3.1-kHz audio-bearer service; packet-mode bearer service: X.31 case A (B channel), X.31, case B (D channel), X.31, case B (B channel).

Teleservices. Telephony (3.1-kHz bandwidth), Teletex, facsimile group 4 (telefax), Telephony (7-kHz bandwidth), audiographic teleconferencing, videotex (alphageometric mode), videotex (photographic mode), teleaction, videotelephony, computerized communication service.

Supplementary services

Advice of charge (AOC) services. Charging information at call set up, charging information during the call, charging information at end of call.

Number identification services. Calling-line identification presentation; calling-line identification restriction; connected-line identification presentation; connected-line presentation restriction.

Conference services. Conference call: add on; meet-me conference.

Diversion services Call forwarding unconditional; call forwarding on busy; call forwarding on no reply; call redirection.

Closed user groups. Call waiting, completion of calls to busy subscriber, direct dialing in, freephone, malicious-call identification, multiple subscriber number, subaddressing, terminal portability, three-party service, user-to-user signaling.

ISDN Centrex

Areawide centrex, or wide area centrex, uses the 2B + D capabilities of basic rate ISDN to enable ISDN-LANs to compete with Ethernet by using 16-Kbps "D" channel with 64-Kbps "B" channel or 128-Kbps "2B" channel packet or circuit mode to internetwork terminals to host systems. Transfer for non-ISDN to ISDN can occur, as both stations can belong to the same functional groups; Customer Group, Call Pick-up Group, Hunt Group, Multiple Appearance Directory Number Group, Call Forwarding, Share Common Translations and Facilities (Automatic Route Selection, Attendant Access, Direct Dial, and Virtual Facility Groups). The ISDN Centrex users have access to eight terminals per loop, 64K data transport, call progress information with CCS7, ring again to recall busy numbers, name display, calling-party ID services, Group IV high-speed fax using B channel 64K circuit-switched data or dedicated channels for packet data, D channel X.25 and X.75/X.75 packet data services, 128K video teleconferencing, and virtual private networking (VPN).

ISDN-PBX

Dynamically assigned PRI-ISDN B channels can replace dedicated PBX-to-central-office trunks as ISDN integrated trunk access (ITA) services allow PRA B channel services to replace current non-ISDN trunking to a PBX, requiring the use of dedicated trunks for different call types such as direct inward dialing (DID), direct outward dialing

(DOD), wide area telephone services (WATS) and private office network trunks. To entice users to use ISDN PBX functionally, PBX attached automated call distribution (ACD) functions is one value-added capability for telemarketing and customer-service offerings, which, when integrated with BRA ISDN, enables behind-the-PBX connectability to computer databases. This allows data terminals access to integrated communication applications.

ISDN-CPE

ISDN CPE offerings include telephones, terminal adaptions, personal computer terminal adapters, ISDN PBXs, host computers, voice messaging systems, local area network terminal adaptors (19.2 Kbps), ISDN multiplexors, automatic call distributors, ISDN packet assembly/disassembly (PADS), ISDN premises controllers, Group III (9.6 Kbps) and Group IV (64 Kbps) facsimiles, telemarketing devices, ISDN integrated voice-data workstations, ISDN test and diagnostic products, ISDN C.O. emulations and simulations, alarm systems, meter reader, environmental control systems, audio centers, video phones, HDTV, and home communication centers.

PC terminal adaptors provide access to two independent bidirectional 64-Kbps B channels that can be either circuit switch or packet switch for data, and one 16-Kbps channel used for signaling. RS232C and V.35 (limit of 19.2 Kbps) and V.110 and V.120 rate adaption for 64 Kbps enable basic rate speed to contain subspeeds at digital error rates of one, even in 10^7 rates, instead of analog error rates of one error in 10^4 bits. This requires new throughput protocols to use the increased speed of communication and then take advantage of higher 64K and 128K rates. Hence, a multiplicity of TAs will be developed for non-ISDN-to-ISDN interfaces for cluster controls, multiplexors, workstations, and LAN services, as 64K rates give way to H0 and H1X and multiple PRI-ISDN trunks to achieve dynamic routing and bandwidth management capabilities, along with message services, calling-number display, message waiting, leave-word calling, video conferencing, reservation service, facsimile distribution, call management, compression, desk top conferencing, security and voice/data encryption, and the four types of ISDN telephones: A, B, C, and D, denoting various configurations for conference calling, etc.

ISDN-LAN

It has been said that ISDN is a standard without a product, while the LAN is a product without a standard. As data movement grows from 110 bits per second, 9,600 bits per second (VC), 19.2 Kbps, 64 Kbps (B), 1 Mbps (LAN), 1.544 Mbps (T1), 10 Mbps (Ethernet), 100 Mbps (FDDI),

155 Mbps (B-ISDN), 200 Mbps (HDTV), 565 Mbps, 600 Mbps (0C-12), 12 Gbps, 2.4 Gbps, and 4.8 Gbps, etc., ISDNs and LANs will merge. As front-end processors become more powerful, only message segments, analysis, conclusions, and data need to be interexchanged. (One test noted only a 1 second difference on delay between a 16-Kbps ISDN-PC network versus a 10-Mbps Ethernet network.)

PRI-ISDN will provide the wideband fractional-ISDN (*N* number of channels) for T1 hubbing and later multiple PRI-ISDN trunks for T3 hubbing over the copper plant at rates up to 6.3 Mbps with B-ISDN over fiber. Later, dedicated WANS will give way to VPNs of variable rates offered by VANS, with access through public switched nodes located closer and closer to the user, as LANS are first internetworked via ISDN and later augmented by future broadband internal CPE switches.

ISDN Network Management

IBM's Netview and Netview/PC provide a means to partially managing multiple networks, but they do not offer the long-term solution that the ISO network management standards offer. As the OSI network management standards for network management protocols and network management services cover the network configuration, fault management, network problem detection, network security issues, performance management and accounting management aspects, it becomes exceedingly clear that the real issue of the 1990s ISDN has been the management of the ISDN networks, as various ANSI committees, such as X3T5.4, .5, and .6 address the specific network management layers of the OSI model. This becomes increasingly complex as simultaneous voice/data (SVD) services are provided by numerous network interface nodes (NIN) interconnected to multiple-provider larger-scale networks or LANS. This complexity increases as we move to photonic logic gates and superconducting buses, cubicly arrayed multiprocesses requiring a layering of network management consistent with the ten-layered networks' layered services model.

OSI

Someone said that there are 64,000 permutations of how the standards under OSI can be implemented. With this flexibility comes complexity, as flexibility enables agreements by multiple parties to the standards, but also enables multiple parties to achieve multiple variations of their product's interfaces. The OSI model defines seven layers, from the physical layer one to the application layer seven. OSI protocols specify standard ways for systems to communicate at each of the various layers; in this way, different proprietary systems may work together at various levels of interoperability from simply passing data back and forth at the physical layer to exchanging formatted text, data and graphics at the application layer. Basically, CCITT Q.920 and Q.921 specify a means of pumping bits reliably from place to place; Q.931 covers call set up and supervision.

Some believe that while ISDN has grown out of the WAN public switched-network world, OSI has grown up in the LAN domain with ISDN oriented toward connection-oriented voice and OSI for connectionless data. There are some unresolved issues as the two worlds come together in interfacing WAN and LAN technologies to OSI, as seen by the following example.

The network layer three for WANs protocol ensures that everything is received correctly and in order at the other end for the connection-oriented WAN, while most LANS support a connectionless mode network service that relies on layer-four transport to make sure that data packets arrive at their destinations in order. This may cause WAN systems to support both connection-oriented and connectionless at OSI layer three in order to interface to ISDN. Similarly, at OSI layer four there are several classes of transport, where most OSI LANS support class 4 transport, while ISDN WANs support class 0. Class 4 is more robust than class 0, and does a great deal of checking for errors, whereas class 0 does not. LANS may have to support both class 0 and class 4 in order to talk to ISDN. (See Fig. C-3.)

GOSIP (Government Open Systems Interconnection Profile), Corporation for Open Systems (COS), the Bureau of Telecommunications and Information Technology Services (BTITS), and others are pushing for compliance to OSI as they wrestle with alternatives such as SNA's LU6.2. Similarly, EDI (electronic data interexchange) standards for businesses' electronic-forms handling can better streamline the exchange of all types of business documents, from purchase orders to invoices, using appropriate firm OSI standards. Hence, the need to more quickly resolve variations and obtain firm OSI standards is indeed the standards challenge of the new century.

OSI higher layers

To appreciate the higher layers of the OSI model, especially the formation and use of the application layer, let's review several views and observations expressed by Wallace, Lacovo, Young, Mantelman, and Roux during the critical years of the formation of the standards, realizing that future implementation changes will occur. But with this basic understanding comes appreciation of the objectives of the higher layers.

The OSI standards constitute a framework for defining the communications process between systems (computers and their users). The reference model, adopted by the ISO in 1984, includes seven layers that define the functions involved in communication between two systems, the services required to perform these functions, and the protocols associated with these services.

The implementation of these functions is achieved by software written to bridge the gap between the application process that starts the communication—for instance, a program in an automated teller machine that responds to a customer's balance inquiry—and the physical medium over which the communication travels, in this case the bank's private telephone lines.

The lowest five layers in a network that conforms to the OSI model ensure that the network provides a reliable connection, if it can. The sixth, the presentation layer, ensures that information is delivered in a form that the receiving system can understand and use. The top layer of the model, the application layer, reflects the behavior of an application process that is observable in its communication with other application processes.

For each architectural layer in the OSI mode, standards are being or have been defined that offer widely accepted means of meeting the layer's requirements. The standards for the four lowest layers (physical, data link, network, and transport) specify mechanisms for the transfer of information. Standards for the session and presentation layers specify, respectively, the mechanisms for checkpointing the data (for resynchronization purposes) and the syntactic representation of the semantics (that is, the encoding).

Of all the layers, the application layer contains the most functionality. Here, the decision is made whether to treat the communication as a file transfer, a virtual terminal session, or a computer-aided design session. After this choice, the complete stack of adequate protocols in the six lower layers is automatically selected.

For this reason, the widest variety of work is presently going on at the application layer. The protocols being studied in this layer include: file transfer, access, and management (FTAM); virtual terminal protocol (VTP); message handling system (MHS); transaction processing (TP); job transfer and manipulation (JTM); remote database access (RDA); and others. These protocols are at different levels of development.

The application layer is the only one that interfaces with the application process. But the application process itself is outside the scope of the OSI model; its form and function is the responsibility of the system user. (Here, "system user" refers to anything that will access the application layer. It can be a human being, a program, or a combination of the two.)

Structure

The structure of the application layer is currently being defined by ISO. This work will lead to international standards. A conceptual model of the application layer is being developed to interrelate all the standards at the layer.

The lower layers merely convey data. By contrast, OSI standards for the application layer are primarily concerned with communicating the semantics (the meaning) of information. Consider, for example, an important database that is split into several parts, each on a different computer. The computers are interconnected by a network. An application process might be responsible for some distributed information-processing task such as updating the database and keeping the components consistent. Using application-layer services, the process could perform or request actions related to the objectives of the update. That is, it might write data to or delete data from a server, or it might ask an applications process to take these actions.

In ISO terminology, an open system is a system that can interoperate with (run with) other computers, perhaps of a different manufacturer, through a network. In the ISO reference model, the interoperation of real (that is, actual hardware and software, as opposed to conceptual) open systems is modeled in terms of the interactions between applications processes in these systems.

An application process is an element within a real open system that takes part in the execution of one or more distributed information processing tasks. It can be either a user's program or a person deciding, for example, to answer a letter. The application process, which is part of the local operating environment, interfaces, via the user element (UE), with the application entity (AE), a program in the system that deals with the lower-level OSI protocols. Typically, application layer standards define conceptual schemes, which enable application processes to communicate successfully. Such schemes define the rules governing the transfer of data, the associated semantics, and the abstract syntax (or encoding conventions) to be used in data transfers. To meet the communications requirements of peer application processes, the upper layers of the OSI model provide services that: support the negotiation of semantics for the information being exchanged (for instance, whether to run file transfer with or without error recovery); support the negotiation of commonly understood representations (a common encoding or set of encodings) for the information being exchanged; enable the application processes to insert signals, uniquely distinguishable from each other, into the information stream, which allows the application processes to synchronize and resynchronize their activities; or allows the application processes to negotiate and manage the dialogue control needed to support the information-flow requirements (specifying, for example, whether they talk both at once or one at a time). The first of these requirements is met by services of the application layer, the second by the presentation services (that is, the services provided by the presentation layer), and the third and fourth by the session service.

Session layer

The purpose of the session layer is to transfer the application's data by means of data-transfer primitives. It uses other primitives to connect, release, abort, and resynchronize sessions.

Initially, the session layer's designers had assumed that data transfer would be used for the majority of user data. It therefore made perfect sense to limit the primitives, except for data transfer, to being no greater than 512 bytes long. Likewise, on an abort, no more than 9 bytes of user data could be carried.

In practice, however, the application-protocol writers decided that it would be more convenient to piggyback their protocol data units (PDUs) atop the sessions' PDUs. For example, it made sense for FTAM to do its connection negotiation at the same time as the session layer was negotiating its own connection (especially over long-haul networks, where negotiation can be expensive). Now the session protocol will handle such exchanges by allowing unlimited data transfer on all primitives. Early implementations of file transfer, access, and management (FTAM) were designed to work well with the session's 512-byte version 1, but later FTAM versions can transfer more than 512 bytes of user data on a connection. Products written to the newer FTAM specification would likely have problems with session version 1. For instance, when it sent out a long-winded connection request, such a product would not know if the other side could accommodate it. Work is under way at the National Bureau of Standards (NBS) to devise implementers' agreements, which overcome this mismatch.

Presentation layer

Just beneath the application layer, which is concerned with the meaning (or semantics) of transmitted data, is the presentation layer, which governs the way meaning is represented (the syntax). Like the transport layer, which ensures reliable delivery, presentation ensures reliable data content.

The presentation layer has two main functions. It changes the local abstract syntax (the representation native to the end node) into the commonly known transfer syntax. And it negotiates presentation syntaxes. In this negotiation, the local application gives the presentation layer a list of abstract syntaxes it wants to use. Presentation then determines which transfer syntaxes it needs to support those abstract syntaxes and attempts to agree with its peer on a set of abstract syntaxes that the two can use to communicate. The number of transfer syntaxes will grow, and so will the capabilities of the presentation protocol.

ASN.1

Right now, however, there is only one standard transfer syntax, called the basic encoding rule for Abstract Syntax Notation 1. ASN.1 is used to define protocols. It's analogous to the Backus-Naur Form, an abstract way of specifying programming constructs. ASN.1 has been in use for some time. The original X.400 and FTAM protocols both used it.

However, ASN.1 is a tool that has to be used in conjunction with something else to make it useful. Although it provides a standard way to define protocols, its drawback is complexity. Conformance to ASN.1 can be complicated. Moreover, it does not provide all the functions that a designer might want. There may be encrypting and/or compacting algorithms for transfer syntaxes without support.

ACSE

The Association Control Service Element was intended to be the single repository for all common functionality among application layer protocols. One such function is authentication. Since many application service elements might need an authentication mechanism, it makes sense to consolidate authentication in ACSE. Another such function is the application context, a shared understanding of what functions the applications may provide.

Application layer model

Within the OSI environment, an application process is represented by one or more application entities. That is, each application entity represents a different aspect of the communication behavior of the application process. An AE is accessed via the presentation address.

In each AE, the application process has to choose an *application service element (ASE)* to perform its task. A *service element* is a primitive defined at the interface between two adjacent layers. An *ASE* is a set of functionalities that supports a typical application. It represents different kinds of work that the user wishes performed, such as file transfer, mail service, or transaction processing, along with all the elements necessary to perform that kind of work. Each of the ASEs stands for a particular ASE implemented in the computer. For example, one ASE could stand for the X.400 standard for electronic mail and another ASE would be for FTAM.

An application entity contains one user element and a set of application service elements, The specific combination of these different elements determines the type of application entity. The user element represents, or acts on behalf of, the application processes. It allows the ASEs to communicate with other application entities to interoperate for a specific purpose. ASEs may be used independently or in combination to meet specific information processing goals. As mentioned earlier, the X.400 and FTAM ASEs are already defined.

Another type of ASE, the association control service element (ACSE) facilitates other ASEs working together. For example, FTAM can be used alone or in combination with another ASE, Commitment, Concurrency, and Recovery (CCR), for error-recovery purposes. (CCR will be described below.) The ACSE is also used to open and release the association.

ROSE

The remote operation service element is provided by a sublayer of the application layer. The ROSE users, which are ASEs, are located in the application layer. They use ROSE in the same manner, independent of the type of ASE and regardless of what the application is, provided it's built on top of ROSE. As is true in the client-server case, the initiating ROSE user establishes a ROSE association by a bind operation. If the ROSE association is established, operations may be invoked. When the initiating ROSE user wishes to release a ROSE association, an unbind operation is issued. It's important to understand that ROSE is only a tool for ASEs to transmit information. ROSE conveys some protocol information but does not know what type of information is conveyed.

RTS

Reliable transfer service was defined by the International Telegraph and Telephone Consultative Committee as part of the application layer. Its function is to interface electronic mail applications with the session layer through a presentation-layer entity. It does this by grouping several common primitives used by the electronic-mail application and one or more other applications. The RTS provides common services and is built on top of the presentation layer. RTS has minimal requirements from the presentation layer, but requires many of the functionalities of the session layer. Through RTS, the message transfer layer (MTL) passes messages from the user agent layer to the network. The user-agent layer prepares messages from the application-layer user to be sent.

X.400

If you took away their electronic-mail systems, most organizations would be hard-pressed to stay as productive and informed as they are today. E-mail provides a conversational interplay and a level of richness—graphics, spreadsheets, images, and words—that considerably enhance communication. And as electronic data interchange and other electronic links between organizations proliferate, it's likely they will use E-mail networks and standards as well. But because almost all E-mail systems are proprietary, getting different organizations' systems to talk together isn't easy. Gateways to support communication among E-mail systems must accommodate the protocols, conventions, and procedures of each product and platform. Fortunately, vendors of E-mail gateway software can use a nonproprietary, international standard for interconnecting E-mail systems—the International Telegraph and Telephone Consultative Committee's X.400 recommendations were first made in 1984 by the CCITT, a Geneva-based international standards body. They govern the interconnection of electronic-mail systems, providing the foundation for a reliable, robust, and international mail system. Most E-mail products conform to the standards of 1984; in addition, the U.S. Government Open Systems Interconnection Profile procurement specification incorporates the 1984 X.400 recommendations. GOSIP certification is overseen by the National Institute of Standards and Technology in Gaithersburg, Md., which oversees product testing. The Corporation for Open Systems International in McLean, Va., provides testing for organizations with no testing facilities.

In 1988, the CCITT issued revised X.400 E-mail recommendations, which were intended to be compatible with the 1984 standards. The 1988 recommendations are divided into a series of standards dealing with specific areas.

The 1988 recommendations include the X.400 standard itself, which is a high-level overview of the electronic message-handling system. The other standards include X.402, for the architecture of the X.400 system; X.403, for testing compliance with the 1984 recommendations; X.407, for a model of distributed information processing; X.408, for converting each type of electronic mail to any other type; X.411, for message-transfer system definition and procedures; X.413, for message-storage definitions and procedures; X.419, for definition of protocols for interaction of E-mail system elements; and X.420, for message-content specification and procedures for exchanging mail messages.

An important adjunct to the X.400 recommendations is the X.500 recommendations. X.500 describes and specifies directory services—essentially, it's a guide indicating who can send and get messages. Without X.500 or some alternative, interoperability among individual, proprietary electronic-mail systems would be difficult to manage.

Strategic protocols

Deployment of X.400 is half of the first phase of what may be considered OSI's grand scheme. An OSI planner noted that, "The number one goal of the OSI community (standardizers, product builders, and testers) must be to get the basic OSI platform, namely X.400 and FTAM, out there as working products. We can make all the standards in the world, but if you don't give people something they can use, they won't accept them. The strategy for getting OSI accepted is to provide platforms for electronic mail and file transfer, then virtual terminal capability, and finally transaction processing. A number of other standards, covering remote database access, network management, security, and other functions, will also sweeten the OSI pie."

Virtual terminal

Many network users face the problem of having to access applications from a terminal quite different from the type that the application was developed to support. Microcomputer users can bridge this gulf on a case-by-case basis with terminal emulation, but others may have to have two or more types of terminal devices on their desks. The OSI Virtual Terminal (VT) protocol is designed to provide remote, network-wide access to applications. This should allow vendors to write applications that run on a particular computer but that can also be accessed from any other processor vendor's attached terminals. For example, if an IBM 370 mainframe had an OSI VT link to a VAX, 3270 terminal users would be able to use the VAX's VT 100-oriented applications. VT Basic class represents the terminal screen in terms of a matrix of character cells, which precludes the support of bit-mapped or vector graphics. However, people will add forms, graphics, windows; these things will evolve, given the graphic, interactive nature of modern microcomputers.

Transaction processing

ISO's transaction processing (TP) standard lets applications process "atomic" units of work. In an atomic update, either a group of updates (such as to the same, replicated file in several different locations) is done in its entirety, or no update is done at all. There are no partial updates, which would lead to inconsistency in distributed databases. Atomic updates involve vendor verbs or primitives such as commit (which completes the transaction by telling everyone to proceed) and rollback (which aborts the transaction by telling everyone to undo). These primitives are akin to a distributed version of using semaphores for file-locking, which is a centralized means of maintaining consistency.

Remote database access

This ISO standard will appeal to micro-to-mainframe fans. Its purpose is to allow access to small portions of files. Today, users can only download large chunks, unless they buy idiosyncratic micro-to-mainframe link products or develop their own. ISO's RDA specification has a specialization that targets the ISO SQL standard, but it could also support other access languages. In principle, such support could facilitate the use of RDA in micro-to-micro applications. Remote database access and transaction processing compose the next big platform. It captures 90 percent of the business.

The idea of distributing pieces of a database, accessing those pieces in a uniform way, and creating secure, transparent links between them is not unique to OSI. It forms a key component of IBM's Systems Application Architecture (SAA) and Operating System/2's Database Manager.

Network management

Vendors can often sell the maturity of their proprietary network management schemes relative to that of standards. While the vendors' schemes are fairly well along, they are generally limited to their own environments. For example, to build large networks and sell a lot of modems, network modem vendors will need network management setups that can talk to computers from many vendors. Because most such modem makers do not sell computers themselves, they will need standards. IBM's umbrella Netview product, like SAA, will not cover all multivendor situations.

An ISO application layer protocol for passing information between managers and their resources is called common management information protocol (CMIP). Common management information services (CMIS) defines primitives that provide services to network management applications (either a manager or a manager's agent in a node being managed). The Defense Advanced Research Projects Agency's Internet Engineering Task Force, maker of the Transmission Control Protocol/Internet Protocol (TCP/IP) is considering the use of CMIS and CMIP.

Security

Another area likely to be of interest to corporations is security. We're not going to have open systems until we have secure systems! There are two basic building blocks of security: application authentication, which allows applications to be sure they are in contact with each other and not with an impostor; and encryption of the transport pipe, which ensures applications that the data passing between them is neither tampered with nor read.

FTAM

The file transfer, access, and management (FTAM) protocol may be doing as much to spur standards implementation in the United States as X.400 is in the rest of the world. The most critical layer-seven protocol to the industry right now is FTAM. Most of what we're doing is file transfer. FTAM has been called the pillar of MAP activity in the United States. FTAM has traveled a bumpy road on its way to becoming a standard, a road that illustrates the intricacy of standards development and deployment. MAP, TOP, and the government OSI procurement specification, or GOSIP, have gone through three stages: base standards put forth by ISO and the American National Standards Institute; a process of implementers' agreements, that narrow the standards down into practical, real-world specifications that can be implemented; and finally, further refinement into profiles, where the protocols are strung together vertically into stacks suited to the office (TOP), the factory (MAP), or government use (GOSIP). MAP 3.0 provides record-level access, in order to have a distributed file system that uses FTAM.

ODA/ODIF

Another upper-layer standard deserves mention. Here, information gets packaged in a universal or common format and then moved across a network so that it can be further processed and revised in the remote setting. This standard is known as the Office Document Architecture/Office Document Interchange Format (ODA/ODIF). The need for a document interchange standard is clear. Users who try to interchange documents between WordStar, WordPerfect, and MacWrite formats face the problem of how to send the files and have them still look the same. Most commercially available solutions use point-to-point, case-by-case translators. The problem is that, for *N* products, it's necessary to build *N*-squared translators. Each time a format is added, a translator must be written for every other existing format. With a standard such as ODIF, each new product only needs a translator to the common format.

Observation

The application layer is the one that the user—and the user's program—sees, so standards makers and implementers have to be more responsible about how it unfolds. Stability is needed for the programming interface to the upper layers; later generations of the protocols can then be viewed as extensions. At the inner layers, we can accommodate some volatility in the specifications, but the top of the stack has to be more solid so people will not be afraid to build on it. (See Table C-1.)

Application layers: a supplier's view

In the early 1990s, Lacovo and Young of Northern Telecom noted that today's multivendor market demands that any functional interface between the telecommunications and computer-processing environments be based on open industry standards. For this reason, their firm has proposed to standards-development bodies a computer-to-switch applications interface that is based on the Open Systems Interconnection (OSI) reference model, primarily on the application layer of this seven-layer architecture. Building on the flexibility of OSI, the concept incorporates the out-of-band signaling of CCS7 and ISDN, as well as the capabilities enabled through common ISDN interfaces. Although many of the architectural details-such as information flows and protocols-will be addressed in standards bodies, Fig. C-4 focuses solely on the application layer and on the way it can be used to support various applications in a telecommunications and computing environment.

The application layer can be supported by a variety of protocol stacks envisioned for voice and data environments. From the viewpoint of the user application, the underlying protocol stacks and associated

Table C-1. Application Layer

Protocol	Description
Abstract Syntax Notation 1 (ASN.1)	An abstract way of defining protocol specifications, currently supported by one set of basic encodong rules, which constitutes a transfer syntax used by the presentation layer.
Association Control Service Element (ASCE)	Provides basic facilities for the control of an association between two application entities.
Directory system (ISO); Also called X.500 (CCITT)	Provides information on names, addresses, routing, and other objects needed for global networking.
File Transfer, Access, and Management (FTAM)	Describes how to create, delete, read, and change file attributes as well as transfer and access (at file and record level) files stored at a remote site.
Job Transfer and Manipulation (JTM)	Supports distributed job processing, including job submission, file and resource access, and output redirection.
Manufacturing Message Service (MMS)	A command and control language used to communicate with manufacturing devices such as programmable and numeric controllers and robots.
Message Handling Systems (MHS)-CCITT X.400	
1984 version	Standard for the interchange of electronic mail among recipients served by diverse common carriers and computer vendors.
1988 version	A revision of X.400 to include changes in protocols and service elements, providing security, mailboxes, and physical delivery.
Network management:	
OSI management framework	Framework, service, and application layer protocol for passing information between managers and the resources they manage.
Common Management Information Service	
Common Management Information Protocol	
Office Document Architecture (ODA)	Specifies structures for the exchange of processible documents.
Office Document Interchange Format (ODIF)	An ASN.1 encoding that constitutes a particular representation for the office document architecture.

Table C-1. *(Continued)*

Protocol	Description
Presentation	Provides for transformation of data between local and common syntaxes and for negotiation of presentation syntaxes.
Remote Database Access	Facilitates access to databases from intelligent workstations and from other database environments via a client/server relationship.
Security	Security architecture and mechanisms that add application authentication and transport encryption to basic OSI reference model.
Session;	
Version 1	Lets users establish and release connections for data exchange as well as synchronize, interrupt, and resume services.
Version 2	Allows user data longer than 9 bytes on abort and primitives longer than 512 bytes.
Transaction processing	Provides services to process atomic units of work. A group of updates that must be done either in their entirety of not at all.
Virtual Terminal (VT)	Allows for the transmission across the network of terminal-oriented messages, including keyboard input, screen updates, and various types of terminal control information.

applications are largely transparent. This capability is key to ensuring interoperability between CCS7, basic rate access interface (BRI), primary rate access interface (PRI), ISDN signaling link (ISL), and X.25-based networks, and for promoting the development of intelligent, end-to-end, hybrid networks. In examining the application layer architecture (described in ISO Document DP9545), it's helpful to focus on the application that is to be enabled; for example, incoming call management. (In ISO terms, the application is called an application process, or AP.) To allow a specific result, such as a response to a query, there must be an unencumbered, open information flow—that is, common protocols—between the linked processing systems.

The OSI reference model and the Northern Telecom architecture use the application layer to create a common modeling structure. The application layer structure is initially segregated into specific components that support the AP. These components are called application entities. The application entity (AE) consists of subcomponents that perform specific services. There are two types of subcomponents: common, or base, service elements; and specific application service elements. Base service elements provide functions such as association

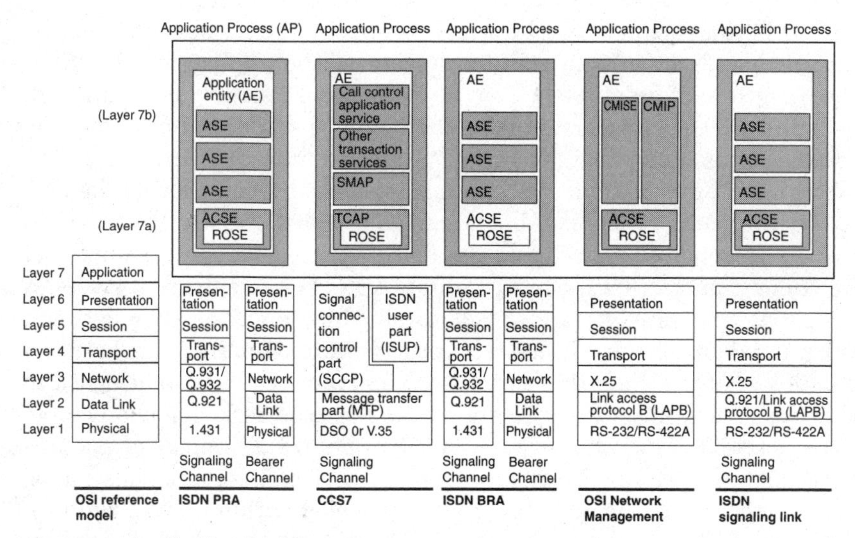

Figure C-4. Application stacks.

(connection) control and remote operations, and are common within each AE, regardless of the application process. Specific elements, such as voice/data association, can be constructed to support a distinct AP.

For the purpose of discussion, the application layer can be viewed as consisting of two separate sublayers within the AE to support a particular application. The first sublayer (7a) consists of base service elements—specifically the association control service element (ACSE), described in CCITT recommendations X.217 and X.227, and the remote operations service element (ROSE), described in CCITT recommendations X.219 and X.229. The ACSE provides the functions necessary to establish an association between two APs, and the ROSE provides the capability to perform remote operations between APs. In this architecture, the ACSE and ROSE are normally present. As noted, these base elements will be common within each AE, and thus form the foundation from which all applications can be structured.

The ACSE and ROSE from the first sublayer provide functions via a standard service interface (a set of primitives used to represent the information exchanged between layers or elements) to the next sublayer—layer 7b—which contains the application service elements (ASEs). The ASE is an OSI application layer term that applies to a set of specific functions that support an AP. The ASE functions—for example, telephony and data services—are then used with the ACSE and ROSE elements in an application entity to provide services to the AP via a service interface.

Although a specific ASE for each of the AEs has been shown in Fig. C-4, nothing prevents the addition of other ASEs to change the aggregate functionality of the AE. In other words, if a particular application will be better enabled by adding new services, then a new ASE can be included without affecting the provision of the application. Further, the definition of functions within an ASE may be strictly nonstandard, as defined by the implementor, or they may be standard ASEs, such as common management information service elements (CMISEs).

The main purpose of an ASE then is to structure and provide the necessary attributes, such as call hold and call transfer, to enable an application. Telephony services and database services are two examples of functions that can be modeled into an ASE. For telephony services, a telephony ASE provides a means of establishing an association between voice and data services. These services facilitate communication between a data terminal attached to a computer, and a voice terminal attached to a switch, to provide such applications as a message desk. Telephony functions between a switch and computer can enable a number of applications, including calling-line identification (CLID). These functions can be constructed within an ASE that, when modeled within an AE, will provide a standard service interface to an AP. The end result is an applications interface between a switch and a computer that enables peer-to-peer voice and data APs to communicate more effectively.

The second example of a function that can be modeled into an ASE is a database ASE. Database services are becoming a valued information resource that can be used in conjunction with call-related services or as general information to a user. In order to make the database information available in a ubiquitous manner to application users, a standard service interface to an AP needs to be defined to support functions for accessing, retrieving, and modifying database information in a secure manner. The functions to facilitate these services can be constructed within an ASE, from which a standard service interface to APs can be defined.

The computer-to-switch interface concept is based on a number of technologies currently developed, and is capable of covering a wide spectrum of networking requirements. Those requirements demand growing intelligence in network operations, management and, ultimately, user functionality. The evolving technologies of CCS7, ISDN, and OSI make possible new ways to exploit the respective attributes of different communications systems to develop an intelligent network.

The OSI model has been used to propose a standard mechanism for structuring an intelligent applications interface between open systems. Such an interface will serve two important functions. It will allow better association between computing and switching networks, and it will provide a base platform from which user applications can be more readily developed.

Broadband ISDN

Broadband ISDN will be based upon a new SONET transmission hierarchical structure. It will carry ATM (asynchronous transfer mode) fixed-cell (packet) and STM (synchronous transfer mode) circuit time division multiplex multifabric-type traffic for new broadband 155-Mbps and 600-Mbps rate user network interfaces (UNI) and network-to-network interfaces (NNI). It will use new wideband digital cross-connect systems (W-DCS) and broadband digital cross-connect systems (B-DCS) with new fiber deployment topologies that use new central office (CO), remote terminal (RT), and optical carrier nodes (OCN) and customer premises equipment (CPE) (Class 5, Class 6, sub 6, and Class 7) narrowband, wideband, broadband switching products.

Digital subscriber loop

To better understand B-ISDN requires a look at the North American Digital Network. It consists of sources and sinks of digital signal channel bank multiplexors, digital switches and digital cross-connect systems, which are interconnected by digital transmission facilities. The digital hierarchy is based upon a frame overhead and payload structure consisting of asynchronous DS1, having twenty-four 64-Kbps channels (DSO); DS1C, consisting of two DS1s at 3.152 Mbps; DS2, consisting of 4 DS1s at 6.312 Mbps; asynchronous and synchronous DS3, consisting of 28 DS1s or 7 DS2s at 44.736 Mbps. This could have continued through DS4, consisting of 139.264 Mbps, etc., but a new synchronous optical network called SONET was developed for the higher rates. (See Fig. C-5.)

During this process, 8-bit channels (octets) (bytes), representing a voice sample, were inserted into twenty-four-channel frames of 192-bit payloads with an overhead bit to form a 193 bit T1 frame. This was multiplexed into twelve-frame superframes of 2,316 bits and then into 4,632-bit extended superframes consisting of twenty-four frames, as payloads and overhead bits were thus integrated together to achieve higher bit streams. The extended superframe format "extends" the DS-1 superframe structure from twelve to twenty-four frames

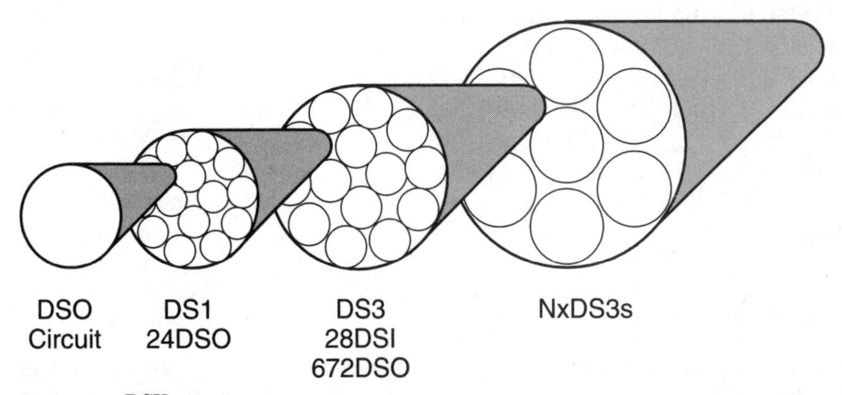

Figure C-5. DSX.

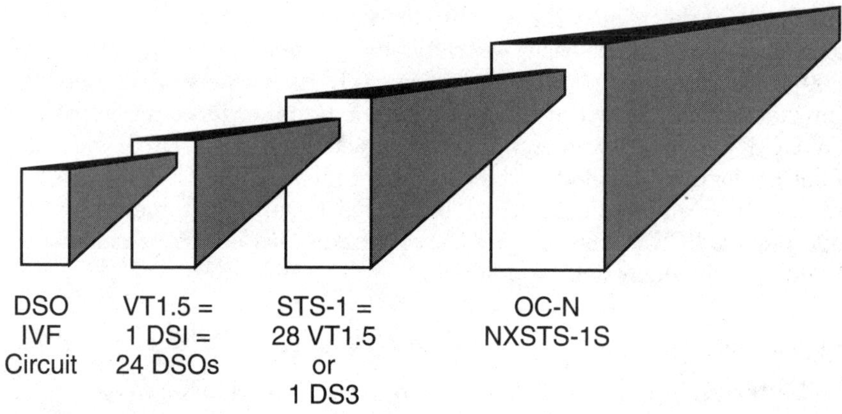

Figure C-6. STX-X, OC-N.

(4,632 bits) and redefines the 8 Kbps (framing bit position) previously used for robbed-bit signaling synchronization. The ESF format consists of a 2-Kbps channel for basic frame and signaling synchronization, 2 Kbps for cyclical redundancy check (CRC) code, and 4 Kbps for a maintenance data link. Complete specifications pertaining to the ESF requirements are contained in PUB 54016.

DS3 is 44.736 Mbps = 28 DS1s = 672 DS0s, using a master frame of 4,760 bits, as it's partitioned into seven M frames of 680 time slots. Each subframe is divided into eight blocks of 85 time slots, the first of which is a control bit. The remaining 84 are available for information bits. X, P, M, F, and C bits enable frame alignment and parity format (C bit). Various vendors use the C bit differently, causing internetworking problems. Timing is based on a hierarchy of stratum clocks, the highest of which has an accuracy of 1 in 1,011, which is equivalent to 72.338 days per slip.

The European digital hierarchy was as follows:

CEPT 0 64 Kbps (1 voice channel capacity).
CEPT 1 2.048 Mbps (30).
CEPT 2 8.448 Mbps (120).
CEPT 3 34.368 Mbps (480).
CEPT 4 139.268 Mbps (1,920).
CEPT 5 564.992 Mbps (7,680).

(See Fig. C-6.)

SYNTRAN

The next evolutionary step from asynchronous DS0, DS1, and DS3 was SYNTRAN (synchronous transmission) at the DS3 44.736-Mbps rate.

SYNTRAN is a standard for a new DS3 format that allows efficient synchronous electrical multiplexing, demultiplexing, and cross-connecting at the DS3, DS1, and DS0 levels. Because it's possible to identify individual DS1 signals from the DS3 bit stream, a device called an "add-drop" multiplexor or ADM has been introduced. With the SYNTRAN ADM, DS1 signals may be dropped from or added to the DS3 bit stream without the need for complete demultiplexing to the DS1 rate and remultiplexing back to the DS3 rate, as is required with today's asynchronous multiplexors. Up to twenty-eight DS1 signals can be multiplexed to form the synchronous DS3 signal. An additional advantage of the SYNTRAN concept is that it allows identification of individual DS0s (circuits) within the DS3. SYNTRAN ADMs can be designed to effectively cross-connect the DS0 signals, thus eliminating the need for conventional jumper work on a mainframe (MDF). The same cross-connect feature may be used to sort and cross-connect signals at the DS1 rate. In particular, any DS1 interface within the ADM can access up to 24 of the available 672 DS0s contained in the DS3 bit stream. All cross-connect features may be performed locally at the ADM or remotely via a system controller.

Many believed that the deployment of SYNTRAN would result in simplified network multiplex arrangements and reduced switch termination costs. It would also promote new and more flexible and efficient network architectures and offer many operational advantages such as dynamic provisioning, control, maintenance, and restoration. These features would be made available to the customer.

The counterpart of SYNTRAN that offsets its usage was found in the 1,300- and 1,550-nm wavelength fiber-optical world called SONET. This is a family of hierarchical synchronously interleaving optical signals. It is based upon the electrical building block called synchronous transport level 1 (STS-1), operating at 51.84 Mbps. As SONET became more and more ubiquitous, it diminished the role of SYNTRAN and asynchronous DS3. SYNTRAN's greatest application was in the transitional period, enabling 3/1/0 digital crossconnect channel switching and transport T1 and T3 hubbing. (See Fig. C-7.)

SONET

The SONET hierarchy, while compatible with the digital signal hierarchy in place today, differs from it in some fundamental ways due to overhead inherent in its signal format. The SONET format is a four-layered format. The lowest layer is the physical or photonic layer. This layer provides for the optical transmission of bits and is used by all SONET network elements such as repeaters. The remaining layers are the SONET section, line, and path layers. Each layer has its associated overhead.

STE (section terminating equipment). Network elements that perform section functions such as facility performance monitoring. The section is the portion of a transmission facility between a lightwave and a line repeater, or between two line repeaters.

LTE (line terminating equipment). Network elements that originate and/or terminate line (OC-N) signals.

PTE (path terminating equipment). Network elements that multiplex and demultiplex the payload and that process the path overhead that is necessary to transport the payload. (See Fig. C-8.)

The basic modular signal for SONET is the synchronous transport signal level 1 (STS-1), having a 51.84-Mbps rate. The optical counterpart of this signal is the optical carrier level 1 (OC-1), which is a result of direct electrical-to-optical conversion of the STS-1 signal after signaling. Higher rate signals are formed by byte interleaving the STS-1 signals. Thus, a STS-N signal at a rate of $N \times 51.84$ Mbps is formed by byte interleaving N STS-1 signals with no additional overhead. An STS-N signal could be carried on as OC-N (optical signal), having a similar $N \times 51.84$-Mbps rate. Allowable rates for N are initially 1, 3, 9, 12, 18, 24, 36, and 48. The STS-N signal is scrambled prior to transmission to an OC-N signal to balance 0s and 1s to protect loss of signal occurrences. (See Fig. C-9.)

STS-1 frame structure

The STS-1 frame structure consists of nine rows and ninety columns of 8-bit bytes, for a total of 810 bytes. The first three columns are the

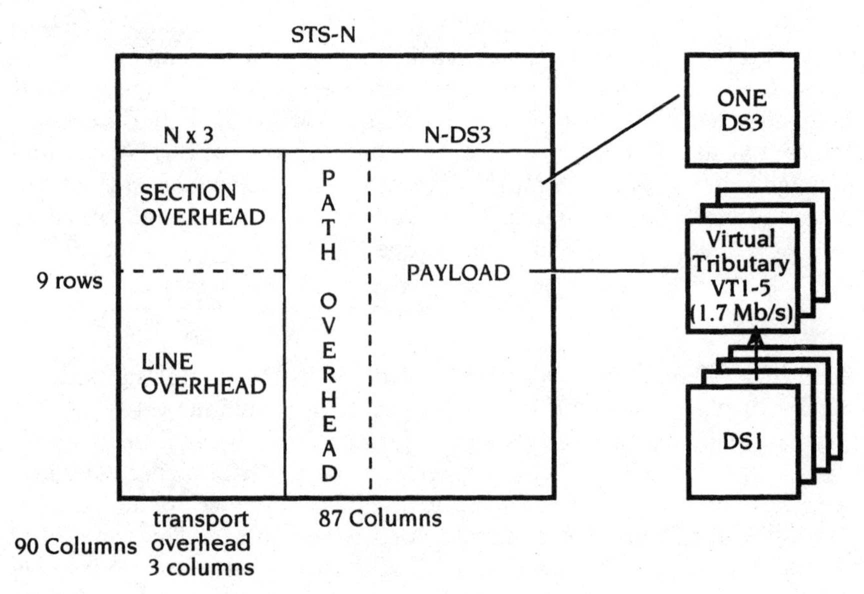

Figure C-7. SONET format.

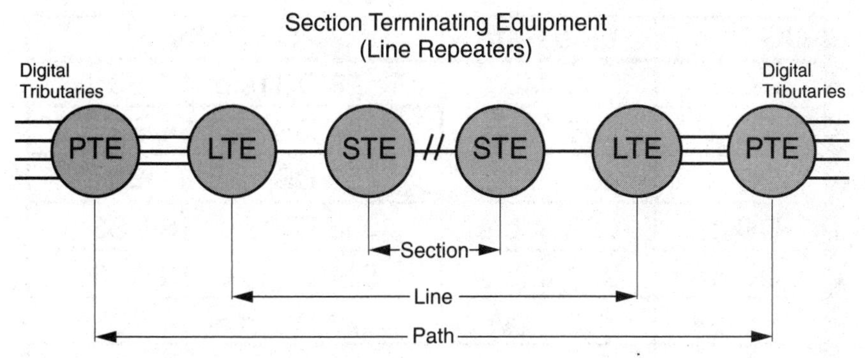

Figure C-8. SONET layers.

transport overhead, which contains overhead bytes for the section and line layers. The section overhead is the first three rows of the transport overhead, and the line overhead is the last six rows. Some transport overhead is specified only for the first STS-1 tributary of an STS-N signal; this STS-1 is referred to as STS-1 #1 of the STS-N signal. The remaining 9 × 87 bytes are available for payload, and are referred to as the STS synchronous payload envelope (SPE).

The STS path overhead uses 9 bytes of the STS SPE. The STS SPE floats within the frame structure; the STS pointer is used to locate the beginning of the payload within the SPE. SONET STS-1 frame size = 9 rows × 90 col or 810 bytes/frame × 1 frame/125 micro sec × 9 bits/byte = 810 bites or 51.86 Mbps.

Transport overhead = 9 rows × 3 col = 27 bytes
Payload capacity = 783 bytes (50.112 Mbps)
Payload overhead = 27 bytes
Total payload = 756 bytes or 48.384 Mbps

Several SONET STS-1 payloads can be concatenated together using byte interleaving to reach, for example, STS-3.

- The STS-(3)c format has 9 rows × (3 × 90) columns or 2,430 bytes × 64 Kbps = 155.520 Mbps.
- Transport overhead has 9 rows × (3 × 3) columns or 81 bytes.
- Payload with path overhead of 2,349 bytes—path overhead = 2,340 bytes at 149.760 Mbps.

The SONET STS-1/OC-1, STS-3/OC-3, STS-N/OC-N hierarchy will contain sufficient overhead to transport any asynchronous DS1, DS3, synchronous DS3 information using wideband W-DCS1/0 or broadband B-DCS3/1 digital cross-connects. (See Fig. C-10.)

OC Level	Line Rate	Capacity
OC-1	52 Mb/s	28 DS1s or 1 DS3
OC-3	155 Mb/s	84 DS1s or 3 DS3s
OC-9	466 Mb/s	252 DS1s or 9 DS3s
OC-12	622 Mb/s	336 DS1s or 18 DS3s
OC-18	933 Mb/s	504 DS1s or 18 DS3s
OC-24	1.2 Gb/s	672 DS1s or 24 DS3s
OC-36	1.9 Gb/s	1008 DS1s or 36 DS3s
OC-48	2.5 Gb/s	1344 DS1s or 48 DS3s

Figure C-9. SONET rates.

Virtual tributary

An STS SPE may contain a DS3 payload, or may contain DS1, DS1C, DS2, or the 2.048-Mbps European standard. In the latter cases, each low-speed signal is packaged into a virtual tributary (VT) within a VT group (VTG). A VT can be a VT1.5, which can carry a DS1 signal, a VT2, for carrying the 2.048-Mbps European standard signal, a VT3 for carrying a DS1C, or a VT6, which can carry a DS2 signal. VTs have two modes, floating and locked. In the floating mode they contain their own VT path overhead. A VTG contains all of one type of VT: either

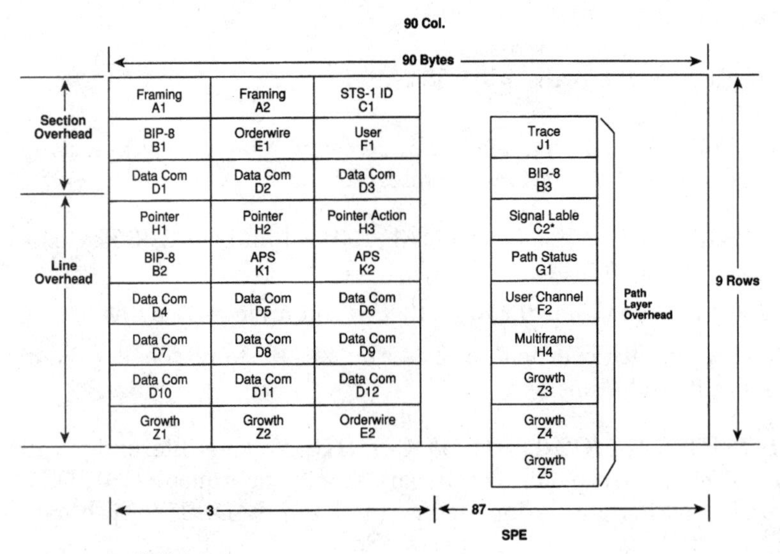

Figure C-10. STS-1 frame.

four VT1.5s, three VT2s, two VT3s, or one VT6. Seven VTGs compose an STS-1; thus, using an STS-1 to carry DS1s via the VT structure yields 28 VT1.5s (DS1s) in the STS-1. This is the expected primary use of VTs in the initial implementation of SONET.

Broadband services can be carried in an STS-1, or in a concatenated STS-Nc (c means concatenate) structure. There is also a concatenated VT6-Nc for carrying services between DS2 and DS3 rate. Initially, the primary broadband rates will be DS3, which will be carried in one STS-1 and the higher rate STS-3c.

All aspects of SONET are being pursued in order to bring it to a reality such as:

- Timing. SONET was originally established so point-to-point systems can communicate. In time, switched SONET will require plesiochronous operation that allows two or more synchronous timing regions to be timed from separate primary interface sources.
- SONET plesiochronous operation. Plesiochronous operation allows two or more synchronous timing regions to be timed from separate primary reference sources. Pointer processing provides the mechanism to accommodate modest frequency offsets between regions.
  a. Integrates into existing synchronization network.
  b. Timing may be received from an external timing reference source or extracted from the received optical line signal.
  c. Transmission is maintained even under synchronization failure conditions.
- Data communications channel (DCC) for OAM&P messages, as well as network element (NE) to NE messages. These are covered in overhead channels. X.25 interfaces also provide OS-to-NE message support.
- Cross-connects such as B-DCS and W-DCS will enable broadband digital cross-connects of DS3-to-STS-3 multiplexing and switching, as well as switching incoming combinations of OC-1s and OC-3s to outgoing OC-1s and OC-3s. (See Fig. C-11.)

B-ISDN

Broadband ISDN will use SONET transport ATM (asynchonous transfer mode) cell relay switching and STM (synchronous transfer mode) time division multiplexing circuit switching for multiple H_1, H_2, and H_4 fixed bandwidth fabrics, as well as a new fiber plant based upon new distribution topology.

Services for broadband range from 1 to 155 Mbps, using various compression techniques; but the return to fully integrated voice, data, image, and video multimedia workstations, office terminals, research

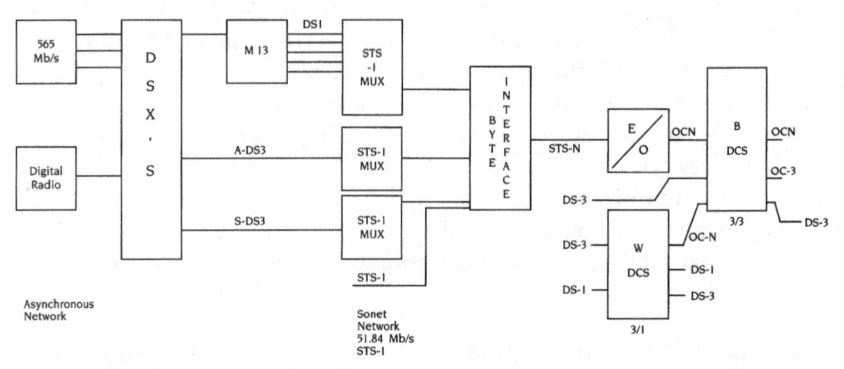

Figure C-11. SONET network.

facilities, and home terminals requiring better resolution, less delay, and more and more graphic/image information may expand the broadband range to 600 Mbps. Hence, two user network interfaces (UNI) have been selected, a one-channel 155-Mbps and a four-channel 600-Mbps, to address the new shifts in technology usage, as noted in Table C-2. Broadband ISDN will use SONET, ATM, and STM and SMDS in such a manner. (See Fig. C-12.) ATM cells can fit their header and payloads into the 9×90 (for STS-1) or $9 \times N \times 90$ payloads (for OCN); STM can be especially featured as concatenations of STS-1s, so they can be nicely handled.

Table C-2. Technology Shifts

Processor size	10-40 MIPS
Memory size	4-20 megabytes
Display	1-2 megabytes
Resolution	620/640-2000 x 2000
Encoding	8 bits/pixel-24 bits/pixel
Bit.s/frame	2.5M bits/F-96M bits/F

Distribution topologies

This then leads to the issue of new topologies providing unlimited bandwidth on demand, priced on usage or service bases with several new network nodes closer to the user. Besides the Bellcore terminology noted earlier, the T1X1 has recommended that the CPE-to-network entrance location for distribution and drop loop plant be called an RW (remote multiplexor node), which uses the subfeeder plant to an AN (access node). From there, primary feeders connect to the local circuit office as the local exchange network (LEN), and from there to other local central

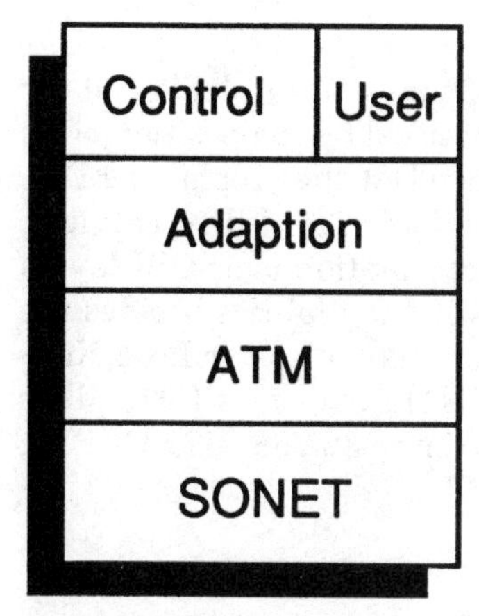

Figure C-12. B-ISDN format.

offices on to the world via IXC. (See Fig. C-13.) The access node (AN) provides STS-3N(c) circuit switching, ATM concentration, CO switching, and SONET multiplexing. The remote multiplexor node (RMN) includes SONET multiplexing and up and down stream-flow control.

For example, using the CO/RT terminology noted earlier, fiber goes from the CO to the RT (remote feeder terminal) to SAPs (service access ports), where the RMN can form a double star. The backbone runs at STS-24 (1,244.160 Mbps) or STS-48 (2,488.320 Mbps) data rate. The distribution and drop fibers from the RT can operate at either STS-3 (155.520 Mbps) or STS-12 (602.080 Mbps) to handle NTVC video, high-resolution video, SVP (selective video program), data packets, and telephone. Since the distance from CO to RT is proposed to be 10 to 15 km, lasers will be used in the backbone, where for distribution/drop fiber from the RT to the home at 1-3 km, the low-cost LEDs (light emitting diodes) or cheaper laser diodes can be used. (Multilongitudinal mode lasers have speeds up to 622 Mbps, while single-longitudinal mode lasers are up to 2.5 Gbps.)

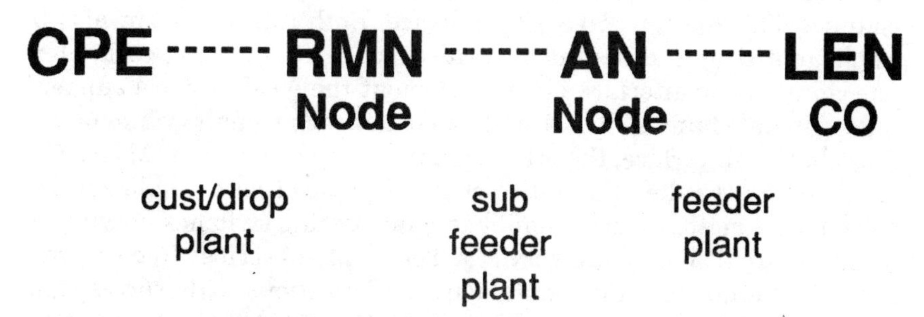

Figure C-13. Distribution plant reference model.

STM

STM allows time slots within receiving frames to be dedicated for the duration of the call. The channel is identified by the position of its time slots. STM is based on the traditional TDM that replaced FDM. CCITT has defined bearer-channel rates for STM ISDN services, where bearer channels carry end-to-end information over OSI layers one to three. STM will provide multiple switching fabrics besides the DSO digital switch to handle higher-bearer channels over B 64 Kbps such as: HO 384 Kbps, H1 as 1,536 Mbps (H11(NA)) or as 1,920 Mbps (H12(CEPT)); H2 as 43-45 Mbps (H21(NA)) or as 32.768 (H22(CEPT)); and H4 as 132.032-138.240 Mbps NA/CEPT.

ATM

ATM uses the following cell format:

- Generic flow control (GFC) for the user network interface (UNI) provides MAC functions and CPE assistance (4 bits).
- Virtual path identifier (VPI) provides path identification for channels. The VPI is 8-12 bits for the UNI interface and 16 bits for the network-to-network interface (NNI).
- Virtual channel identifier (VCI) identifies a virtual channel in a physical transmission facility, (12-16 bits UNI, 16 bits NNI).
- Payload type (PT) distinguishes user information from network (2 bits).
- Header error checks (HPLC) checks header bits.

User-network interface (UNI)

In the technical realization of broadband networks and services, two main interface categories must be considered: user-network interfaces (UNI), and network-node interfaces (NNI). The ATM-capable user network interfaces are outlined in CCITT recommendation I.121. UNI has two interfaces, the one-channel 155.52-Mbps (STS-3) or the four-channel 622.08-Mbps (STS-12) interface. Both interfaces can also be implemented with different bit rates to and from the network. Here one channel per interface must implement the ATM protocol supporting a virtual channel dedicated to user-network signals, while on the four-channel interface, the other channels may be either ATM or STM. Here the STM takes the entire channel payload as a single service, such as full motion video, similar to how existing switches provide for 64-Kbps circuit bases and services. The ATM subscriber line can connect to different terminal equipment configurations with: corresponding broadband network terminations (NT1 or NT2); a single terminal;

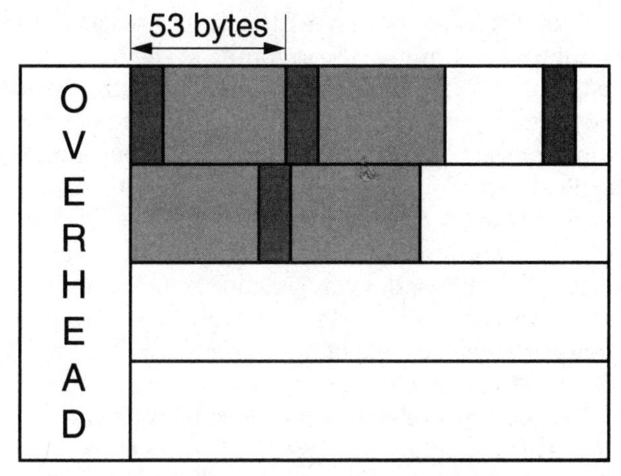

Figure C-14. ATM-SONET format. ATM in a SDH synchronous digital hierarchy.

several terminals without internal communication; a LAN or PABX (i.e. many terminals with internal communication); or several terminals with intercommunication via centrex functions in the exchange, or via short paths in the network termination, controlled by the exchange.

The trunks' interfaces between the ATM exchanges or cross-connectors (NNI) will probably differ in the various national networks, where CCITT recommendations G.707-G.709, with the new synchronous digital hierarchy (SDH) corresponding to the synchronous optical network SONET, are to be implemented among other countries in Japan, North America, Australia, Great Britain, and the Federal Republic of Germany.

Mapping ATM into SONET STS-3c is performed by dividing the STS-3c into fifty-three octet cells using the path overhead pointer (H1-H3) to locate the starting point of the ATM cells. Each STS-3c controls forty-four ATM cells. (See Fig. C-14.)

Global ISDN Standards Alignment

Donald E. Auble, Chairman of the ISDN Implementors' Workshop North American ISDN Users' Forum, made the following observations concerning the future of ISDN standards: Integrated Services Digital Network (ISDN) is defined in a group of international standards for a worldwide communications network for the exchange of all information (voice, data, and image) among all users, independent of any manufacturer, service provider, or implementation technology.

ISDN standards are being developed by the International Telephone and Telegraph Consultative Committee (CCITT) and for North

America in particular, by the Exchange Carrier Standards Associations' accredited standards committee, T1, under the umbrella of the American National Standards Institute (ANSI). The result of the ISDN standards process is one extensive standard with a tremendous variety of options and parameters. This is necessary to meet all the possible needs and applications for which the standards could be used. However, to ensure interoperability, interworking, and terminal portability for a timely, cost-effective, marketable offering, a uniform subset of options and parameters must be selected. Also, each application usually requires only a subset of functionality and in order for products to work together in a multivendor environment, common subsets of options must be selected.

To cope with this proliferation of choices and to provide practical products and services that meet users' needs, the specification process must be extended to include application profiles, implementation agreements, and conformance tests to promote interoperability. In the United States, these issues are being addressed by Bellcore and their client companies (BCCs), including the Regional Bell Operating Companies, in conjunction with their ISDN switch suppliers and in cooperation with the North American ISDN Users' Forum (NIU-Forum) and the Corporation for Open Systems (COS).

Bellcore and its client companies developed and published a robust set of ISDN implementation agreements in 1988 and 1989 identified as technical requirements (TRs). In cooperation with the emerging subset of implementation agreements being developed by the NIU-Forum and through successful negotiations with their predominate switch suppliers, namely, AT&T, Northern Telecom, and Siemens Stromberg-Carlson, Bellcore and the BCCs finalized plans for a deployable 1992 multivendor interoperable ISDN. This plan, known as National ISDN One, was announced by COS at a February 1991 press conference in New York.

Both the NIU-Forum and COS are essential for completing the full spectrum of post standards work needed to realize the introduction of standardized ISDN products. Standards begins the process and COS contributes the final step of verifying that a particular implementation conforms to the standards-based, time-sensitive implementation of ISDN. COS specifically develops the downstream executable tests and tester specifications, as well as offering a COS Mark program for ISDN customer equipment certification.

The NIU-Forum addresses the user need for standards-based, complete applications and contributes publicly developed implementation agreements and conformance test specifications. The NIU-Forum's principal objectives are:

1. To promote an ISDN forum committed to providing users the opportunity to influence developing ISDN technology to reflect their needs.

2. To identify ISDN applications, develop implementation requirements, and to facilitate their timely, harmonized, and interoperable introduction.
3. To solicit user, product-provider, and service-provider participation in this process. Although the NIU-Forum focuses on the requirements of the ISDN users in North America, participation and membership is open to anyone.

The users and implementors in the NIU-Forum have been working since the first meeting in June 1988 to realize the user agenda of standardized interoperable application profiles operational over the ISDN platform. The NIU-Forum has created a user voice in the implementation of ISDN and ISDN applications and has helped to ensure that the emerging ISDN environment meets users' application needs. The NIU-Forum is sponsored by the National Institute of Standards and Technology (NIST). The actual work of the NIU-Forum is accomplished in two workshops; the ISDN Users' Workshop (IUW) and the ISDN Implementors' Workshop (IIW). These workshops, which consist of various working groups and special project teams, meet several times a year and develop various products. The IUW produces application requirements that describe potential applications of ISDN and the features that may be required. The IIW develops application requirements, application analyses, application profiles, implementation agreements, conformance criteria, and an applications software Interface.

The following are the key ISDN standards upon which the majority of standards and implementation harmonization efforts are taking place. Also included are the related Bell Communications Research, Inc. (Bellcore) and North American ISDN Users' Forum (NIU-Forum) implementation agreements that are aligned and form the 1992 multiswitch vendor ISDN implementation known to the industry as national ISDN.

- Physical layer one—basic rate
  a. CCITT I.420, I.430 (four wire)
  b. ANSI T1.601 (two-wire U.S. "U-interface")
  c. ANSI T1.605-1989 (U.S. S/T)
  d. Bellcore TR-TSY-000397
  e. NIU-101 (two-wire "U"), NIU-105 (four-wire S/T)

- Physical layer one—primary rate
  a. CCITT I.421, I.431
  b. ANSI T1.403 (U.S. "U-interface") and T1.408 ("U," S/T)
  c. Bellcore TR-TSY-000754
  d. NIU-103 and NIU-103R1

- Link layer two—"D" channel, basic and primary rate
  a. CCITT Q.920 (I.440) and Q.921 (I.441)
  b. ANSI T1.602-1989 (U.S.)
  c. Bellcore TR-TSY-000793
  d. NIU-210

- Network layer three—"D" channel, basic and primary rate
  a. CCITT Q.930 (I.450), Q.931 (I.451), Q.932 (I.452)
  b. ANSI T1.607-1989 (U.S.) (basic call control procedures)
  c. ANSI T1.608-1989 (U.S.) (packet mode bearer services control)
  d. ANSI T1.610-1990 (U.S.) (supplementary services control)
  e. Bellcore TR-TSY-000268
  f. NIU-301 (BRI/class I), NIU-302 (PRI/class II)

- Rate adaption
  a. CCITT V.110 (Red Book), V.120 (Blue Book)
  b. ANSI T1.612 (U.S.)
  c. NIU-91-0001

The NIU-Forum's ISDN conformance test group has coordinated with NIST and ANSI to contribute its conformance test specifications as U.S. positions into CCITT's test specification developments. Considerable international ISDN test standards harmonization has begun as a result of these efforts. Progress to data has been at the layer-two link access procedures. Layer-one harmonization has also begun. Harmonization at this level of downstream implementation detail will make great strides towards a truly international interoperable and conformant ISDN.

The process of setting ISDN standards and then getting from there to a readily available interoperable ISDN commercial introduction is a complex trip indeed. It's worth the trip and worth completing the job so that the end users of ISDN can reap the full benefits of the standards promise. Unchartered cooperation amongst standards and standards promotion bodies is necessary to achieve the promise of ISDN. That cooperation is happening in the industry today.

Broadband Networking

Let's consider the future of broadband frame relay and cell relay as described by John McQuillan: "Many of the switching solutions for fiber optic-based broadband networking fall under the general category of fast packet switching. Unfortunately, this term has been used so loosely that it has come to mean different things to almost everyone who invokes it. It's best to think of fast packet switching as a concept rather than a specific technology, one with two underlying assumptions. First, networks are now highly reliable and run at very high

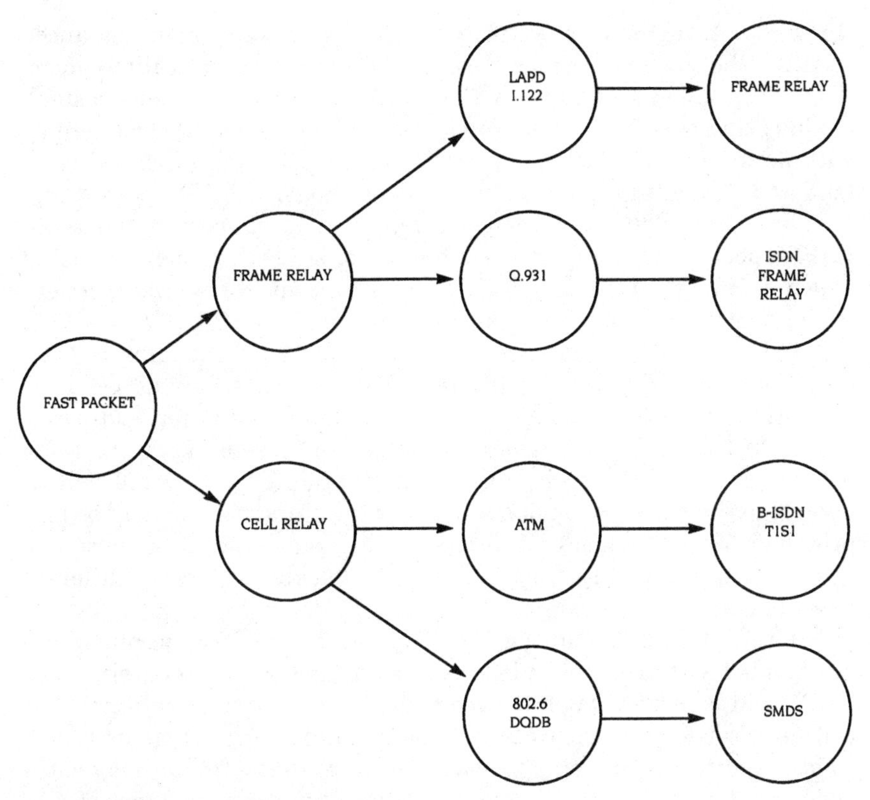

Figure C-15. Broadband technology flow.

speeds. Second, communications on such networks take place between intelligent senders and receivers capable of running high-level protocols that can deal with errors in transmission."

Fast packet switching is essentially a streamlined version of conventional packet switching that minimizes error-checking and flow-control overhead. Consequently, much less protocol processing is done on the network while traffic flows from switch to switch. In particular, the switches no longer set up connections or recover from errors because such errors are assumed to be infrequent and readily detectable by the end stations and their high-level software. (See Fig. C-15.)

The distinction between frame relay and cell relay is really quite simple. Frame relay operates with a variable-length frame (typically the same lengths as that generated by a LAN) and relays these frames at level two in the protocol hierarchy (the data link layer), similar to a MAC-level bridge. Frame relay can be thought of as the next generation of packet switching, the follow-up to X.25. Like that standard, the frame-relay standard link access procedure-D (LAPD, Q.921) has been established by the CCITT for use in both public and private networks around the world. LAPD is closely related to link access procedure-balanced (LAPB), the level-two protocol subset of high-level data link control (HDLC) specified for current X.25 links. Because LAPD is such a small modification to HDLC, many products can be upgraded to frame relay with no more than a change of software. In principal, X.25 switches, IBM and Digital Equipment Corp. network nodes, TCP/IP routers, and similar equipment could easily be modified to support frame relay. LAPD specifies a frame format made up of several elements, most of which is identical to LAPD. Flag bits delimit the variable-length frame, which has a 2-byte cyclic redundancy check (CRC). A new field in LAPD is the 2-byte header made up of an 11-bit data link connection identifier (DLCI) and bits that indicate congestion and echo the indication. The DLCIs denote connections, not destination addresses, and have local significance only, so they can be reused. Frame relay will be used as an interface standard, much like X.25. Internal switching inside the network is not specified by the standard and may be proprietary. Many vendors of bridges and routers will support the interface, so it should be possible for one backbone to carry different types of traffic.

The initial implementations will generally rely on permanently established virtual circuits between all bridges or routers, since the DLCI field is large enough to accommodate a fully interconnected logical network for most private networks. Thus, each bridge or router would require only a single physical interface to the frame-relay backbone and would specify the destination for each frame by appropriately setting the DLCI field. That makes the entire circuit bandwidth available to any subscriber on demand, while reducing the cost and complexity of configuring multiport bridges and routers.

Standardization is well under way. The basics of frame relay were specified in the CCITT Blue Book I.122 in 1988. Switched virtual circuits will use extensions to Q.931 for call setup. The access specification was completed by the mid-1990s. One of the key remaining issues is congestion control. So far, the standards describe the use of the DLCI in a locally significant context only. Work was completed to establish rules for a networkwide node-to-node standard by 1992. ANSI standardized frame relay in the T1S1 committee for usage throughout the 1990s.

From an implementation standpoint, the decision to go with frame relay, cell relay, or conventional TDM can be reduced to a choice between packet-oriented switching or a time-slot solution. TDM has proven very effective for voice and video. But it's less than optimal for data because it requires advanced knowledge of the traffic patterns and a semipermanent allocation of bandwidth to data traffic. As networks grow more data-intensive, alternatives to TDM become more attractive. The first to consider is frame relay, naturally suited to LAN interconnections because of its variable-length packets. But within the telecommunications industry it's generally felt that frame relay is inadequate for carrying voice because of the variable (and lengthy) delays associated with transmitting those frames. The second alternative is cell relay, which should, in principle, be able to provide nearly transparent communication for voice and video, as well as an underlying transport for data that may originate in frames. Thus, it's the only packet-oriented solution that represents a full substitute for TDM. Frame relay has been proposed as a part of the ISDN standard for DS1 transmission, and is feasible at higher speeds as well. The number of frames that would have to be processed each second (10,000 to 20,000) for it to support several key high-speed transmission rates are not unreasonable, given today's equipment.

Cell relay, though, will require much higher processing rates, both because of the small size of the cell and because of the high SONET transmission rates for which cell relay is intended. Clearly, the ability to process hundreds of thousands or even millions of cells per second will be a prerequisite, which demands a new generation of technology. Today's packet switches, bridges, and routers are limited to less than 100,000 packets per second and by a variety of constraints, including processor speed, bus bandwidth, and memory speed." (See *Data Communications*, June 1990.) (See also Fig. C-16.)

Future Broadband ISDN

Intensive studies concerning B-ISDN produced thirteen recommendations in the November 1990 MATSUYAMA meeting of SGXVIII. Here telecommunications services are categorized into: voice service that is sensitive to delay, data that is severely affected by transmission bit-rate errors, and video that is more sensitive to both delay and transmission errors. B-ISDN should provide various types of connection configurations, such as: point-to-multipoint and multipoint-to-multipoint, as well as point-to-point. The network node interface (NNI) of synchronous digital hierarchy (SDH) interface is now worldwide, uniquely standardized at the fully synchronized basic rate of 155.52 Mbps (STM-1). Higher levels are obtained as 2 to the *N*th power times 155.52 Mbps. This will support H4 (134 + Mpbs). B-ISDN will also need to satisfy variable-bit-rate services, continuous-bit-rate services, and respond to various quality of service (QOS) user demands. ATM is selected as the target transfer mode providing such features as universal transfer capability, dynamic bandwidth on demand, and simple network flexibility (using virtual path management). B-ISDN multimedia services provide for interactive services (conversational, messaging, and retrieval) as well as distribution services with and without user control.

The I Series B-ISDN recommendations provide for an overview of B-ISDN (1.121), service aspects (1.211), protocol reference model (1.321),

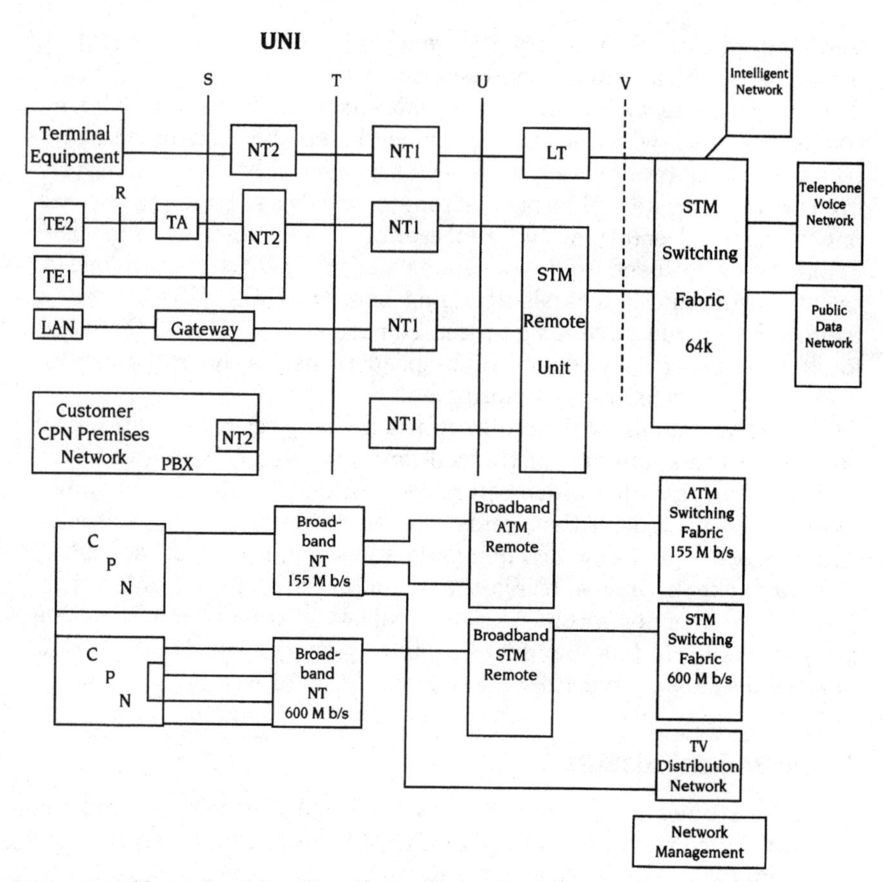

Figure C-16. Narrowband-broadband networks.

functional architecture model (1.327), OAM principles (1.610), network architecture and signaling (1.311), UNI interface (1.413 and 1.327), ATM layer (1.150 and 1.327), ATM adaptation layer (AAL) (1.362, 1.363), network node interface (NNI) (G.707, G.708, G.709), performance (1.358), and vocabulary (1.113). Similarly, 1.122 targets frame relay (which operates with a variable-length frame usually the length of those generated by the LAN) for applications ≤ 2.048 Mbps (CEPT).

As researchers study traffic patterns to cross match the best technology to handle both short-duration, bursty variable-bit-rate calls and long-duration continuous-bit-rate nonbursty message sets, synchronous transfer mode (STM) low-delay circuit switching may form a hybrid network solution with higher delay congestion sensitive ATM packet switching networks.

Conclusion

Nobel prize winner A. Penzia made the following visionary observations to help set the direction of the world beyond digital. "In the future, if we are to make advanced integrated communications universally available, we must provide a means of combining signaling and network control traffic with digital voice, wide and narrowband video, graphics and all the other kinds of data that must be moved, switched, multiplexed, etc., in an integrated system, which is more economically attractive than a series of individually optimized solutions. In the future, barriers to be bridged may well have to do with differences in logical, rather than physical, structure. Surmounting these barriers will require a synergistic combination of expertise in computing as well as telecommunications. This is the challenge which the designers of future network architectures must face."

"Tying it all together" still seems to be the name of the game!

Appendix

D

A Crossroad in Time

There is a tide in the affairs of man,
which taken at the flood,
leads on to fortune;
missed ne'er comes in again.

WILLIAM SHAKESPEARE

The excitement of the 1990s concerning the potential of new information technologies has given way to the disquieting realization that the full embracement of these opportunities has appeared to be just beyond our grasp. This has been somewhat similar to chasing toward an elusive oasis of water that seems just ahead of us as we race along an endless highway on a hot and steamy summer afternoon. After the many false starts, dead ends, and turmoil of the 1990s, we have abruptly come to a new juncture—a fork in the road—a crossroad in time—the new millennium.

We can continue along our current less productive route or venture down a less traveled path. Here the issues of the marketplace must be addressed. For not only must the what and how be considered, but the why, when, where and for whom must be resolved if the turn of the century is to see the dawn of a new age—the construction of the global information society in the new information millennium.

Or, we can wait for yet another day, but time continues, events change, and opportunity may or may not return.

Two roads diverged in a yellow wood,
And sorry I could not travel both
And be one traveler, long I stood
And looked down one as far as I could
To where it bent in the undergrowth;
. . .
I shall be telling this with a sigh
Somewhere ages and ages hence:
Two roads diverged in a wood, and I—
I took the one less traveled by,
And that has made all the difference.

ROBERT FROST

La Petite Histoire

Before beginning a major expedition, explorers have found it essential to understand previous endeavors in order to achieve future success. With this in mind, let's take a brief stroll through past accomplishments and disappointments.

Thirty years ago, young engineers in pursuit of the latest technology and the best-paying jobs joined the aerospace/military industries to build new data networks such as 465-L, 490-L, Autovon/Autodin and Minuteman, using stored program control and integrated circuit technology. By the mid-1960s a huge industry was flourishing, until the head of the Department of Defense made the now classic strategy statement, "Though I have built you, I will not sustain you." This triggered the transfer of technology from military to commercial endeavors. This technology, when applied to digital communications, fostered the new integrated digital networks (IDNs) of the 1970s. These were established to increase transport/switching effectiveness and automate operations. Since data networks were only 3 percent of the business, little was done to promote public data networks after the IBM-AT&T antitrust suit of the 1960s and the subsequent limited "transaction" network offering.

In the late 1970s, IDN was merged with services based upon integrating data with voice to form integrated services digital network (ISDN). However, few public data networks were actually established over the 1980s. This caused a shift by the large business players to private value-added networks (VANs) and local area networks (LANs). ISDN was marketed mainly at high prices to provide a few extended voice services such as a second voice line, CLASS-type offerings, and low-speed data packet switching.

On the private side, local area networks (LANs) needed to be connected together. Hence, the shift to MANs, WANs, and VPNs as private networking became the emphasis of the day. Computer manufacturers found the advantages of opening user groups to enable other systems to interconnect to their mainframes as the processing of information became more and more physically distributed.

This then leads to a rethinking of ISDN, as the private world looked for an integrated networks' integrated services (INIS) solution. This has shifted the focus to communications and computers (C&C). Their merger has mainly taken place in the private world, but by the mid-1990s, the benefits of public networks had become more and more apparent, hence the need for private/public internetworking (P&P) from both a transport and a service offering. This networking opportunity translates into a layered networks' layered services (LNLS) structure, which could also be called the "networking of information services" or "information networking." The time has come to foster and facilitate this transition of the information/communication public infrastructure from IDN to ISDN to INIS to LNLS, as well as to complete the extension of private networks from LANs to WANs and MANs to VPNs and VANs to LNLS. This resulting LNLS architecture will overlay and interconnect private and public networks' services to formulate the total solution to meeting the customer's needs and expectations. At this crossroad in time, this opportunity is indeed both the marketing and technical challenge of the late 1990s and new information millennium.

As we attempt to move from the industrial society to one based on an information revolution, we have come to realize that C&C means more than just the merging of communications and computers. The rapid introduction of revolutionary information networking technologies has caused intense change and complexity that result in difficult challenges and choices. During revolutions there is usually considerable misunderstanding and resistance to change. This affects the choice of paths; some paths lead to success, but others lead to failure and disillusionment as well as crisis and confusion.

Unfortunately, as we enter a new era, it is usually extremely difficult to see the complete picture of what is happening around us. However, it is during this transitional period that it is extremely important for us to make the correct decisions in order to help channel major changes in a desired direction.

As technology expands at an ever-increasing rate, as regulatory boundaries soften, as users become more and more aware of what new communication and computer features and services are needed to make their tasks easier and more effective, we have an opportunity to participate in a new game, the information game, with its new rules, new goals, and new equipment. Here, issues are becoming more global, more complex, more interrelated, and more dependent on providers and suppliers having similar understandings, perspectives, and expectations of what is needed for the new users. (See Fig. D-1.)

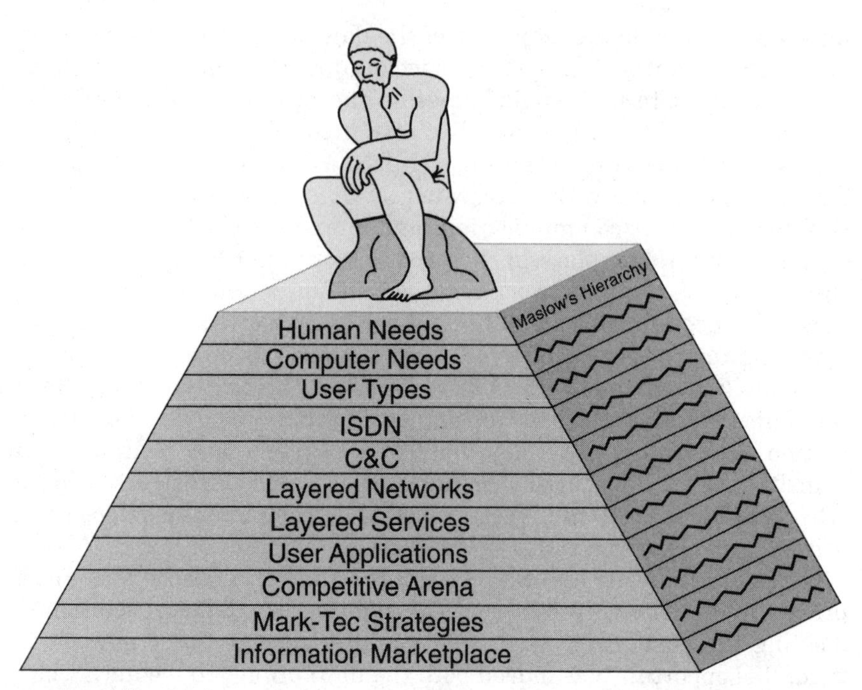

Figure D-1. The thinker.

Over the past thirty years, change has been somewhat evolutionary as new technology was utilized in three-year to five-year increments. During this period, traditional leaders embraced integrated circuit technology to obtain more powerful computer processors that mainly provided stored program control for automating financial or manufacturing operations. However, today, management decisions are no longer made simply on a technical basis; they are now tied to a dynamically changing marketplace. As the computer leaves the world of accounting to enter every facet of our daily lives, applications will become unbounded. Many of these applications will cause revolutionary changes to our daily lives, similar to those achieved by the introduction of the steam engine, electricity, the automobile, and the airplane.

As we use fiber-optic technology and its new broadband SONET-type transport standards, as well as new distributed STM-ATM switching technology, we will need to direct these new "engines" along the path we wish to take. Hence, there is a need to understand narrowband ISDN, wideband ISDN, and broadband ISDN in order to denote where one path leaves off and the other takes over. There is a need to understand user resistance or acceptance to various information presentation vehicles. We must obtain a better understanding of the potential of high-speed versus low-speed data movement in terms of pricing incentives and applications, as well as the standards that we should be pursuing now so they are available in the future when they are needed. We must learn how to price for growth, and we must make the necessary financial commitment to become active players, and not just watchers. (See Fig. D-2.)

As we consider the financial considerations of the new information marketplace, we need to become comfortable with its opportunities, its complexity, it risks, and it limits. The marketing challenge of the 2000s is to answer the why, when, where, and for whom questions in order to formulate a clear understanding of the opportunities. Both the novice and the specialist must become comfortable enough to venture off the beaten road and embark upon a new adventure. To do this we need to better understand our surroundings. We should never enter a strange town and purchase the first house we see. Some suggest that a helicopter ride over the area is desirable in order to help gain a better perspective of how the area fits into the landscape. Such rides often uncover other more desirable or less desirable surroundings such as a nearby landfill, chemical plant, or superhighway. We have also found from NASA's photographs taken from space that we are able to not only obtain a broad view, but once we have picked an area, we can then zoom down and completely analyze the selected location to better understand and appreciate it.

Because of these issues, an analysis was needed to provide a platform that fosters further thinking. Hopefully, this book will tickle our minds and enable us to better understand and formulate the broad

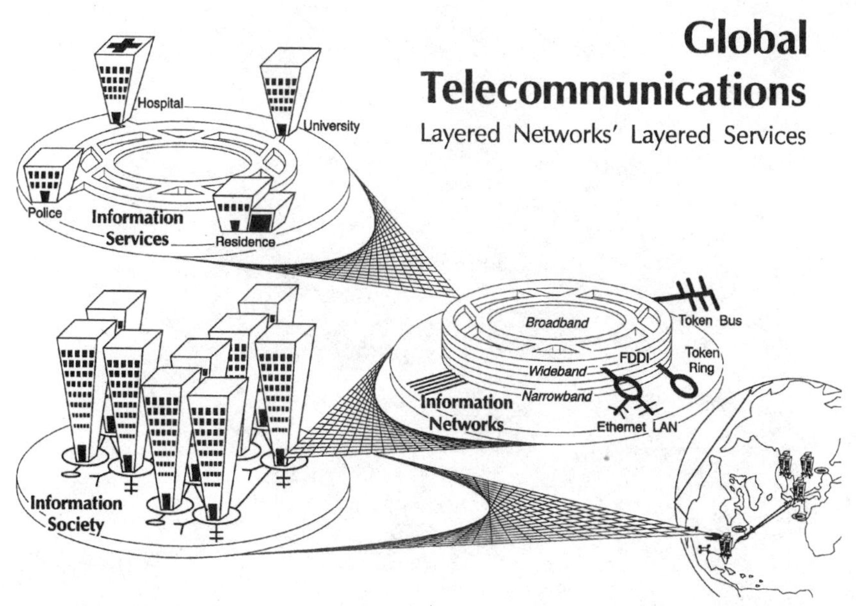

Figure D-2. Global pathways.

view, the global perspective of where we are going, to help identify where we might wish to go and determine how we can get there successfully as we proceed down the somewhat rocky and hazardous path to the information society.

To achieve this, we need to spend time formulating a comfortable vision of the future based upon specific MARK-TEC (market-technology) strategies that map nicely onto financially based directives such as RBOC's (Regional Bell Operating Company's) desire for "becoming a market-based firm, providing cost effective services that produce growing revenue from specific market applications, leveraging the existing network's capabilities to provide a satisfactory return on investment to its owners, and establishing the long-term revolutionary infrastructure to enable it to continue to be successful in the future." (See Fig. D-3.)

These MARK-TEC strategies should be structured around a layered networks' layered services model that is designed to enable competitive services to blossom in an interrelated private-public marketplace. Hence, one of the key outputs of this analysis should be to understand this structure and then overlay present and future services onto this model in order to appreciate how they must cross-relate and interrelate. Therefore, this book reviewed the many different types of current

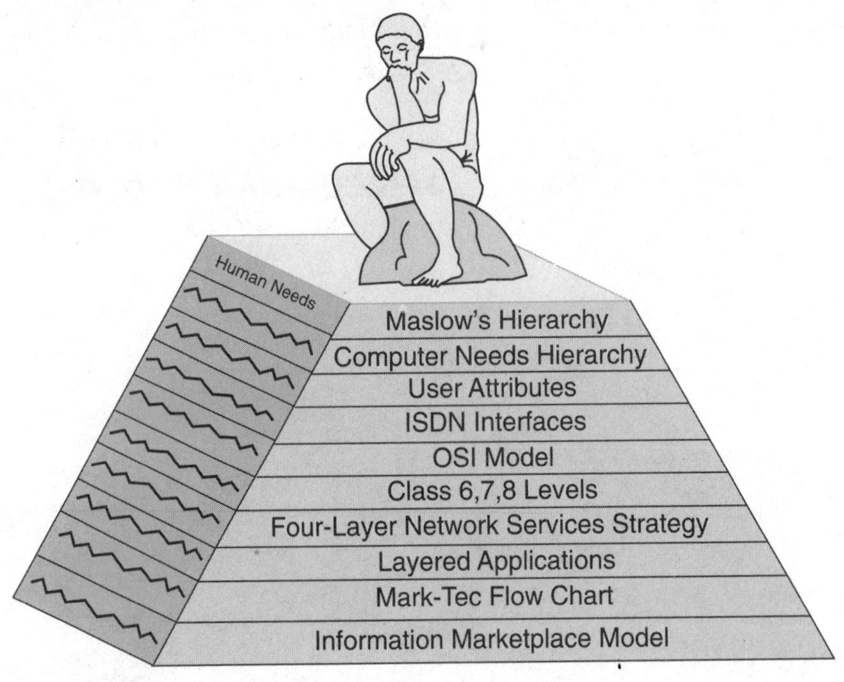

Figure D-3. The thinkers' models.

and potential information network services being discussed and considered today.

When players and planners talk about information movement and management (IM&M), they are talking about organizing, accessing, and managing information, as well as maintaining it. Some talk about IM&M in terms of searching, filing, composing, transforming, editing, and interpreting information; others view it (IM&M) in terms of reviewing and showing information. Everyone is attempting to look at it from one viewpoint or another until every possible aspect has been considered.

In reviewing the ever-expanding list of potential network services, we can see that a blossoming field of new information services could be available by the turn of the century. One visionary noted the following possibilities:

- Reading and answering your office mail at home.
- Browsing through sales items at remote shops and stores and pushing buttons to initiate your orders.
- Bringing up any movie or entertainment information program for play over advanced, enhanced, or high-definition television.
- Facsimile, teletype, or teleprinter machines located in every building and /or vehicle.
- Electronic editing via telecommunication hookups to remote computers, thus reducing paperwork and increasing productivity.
- Nationwide video teleconferencing with voice, data, and facsimile, along with hardcopy printout capability.
- Inexpensive electronic preparation and delivery of mail and newspapers.
- A variety of audiotex services, such as voice messaging and voice mailbox.
- Access to health and financial autotex information.
- Banking and credit transaction verification services.
- Remote security and fire alarm services.
- Energy management and smart-house-type services.
- Information storage, movement, and management, such as file storage, file transfer, and computer database timeshared services, and many more.
- Multimedia integration of services at the office desktop, the factory workstation, and the home.

Others look at new services in terms of just bandwidth management.

So it is with the forthcoming information era. As computers and communications integrate together, enabling distributed processing to become physically located closer and closer to the user, we need to step back and determine what's happening by taking a series of broader views from a more global perspective, and then, after focusing on particular areas of interest, zoom in to cover them in greater depth. For this reason this book was periodically shifted from the more expansive conceptual view to the details of the basics, but always with an application in mind. To do this successfully, we have proceeded from customer needs to networks and services, to visions and plans to meet the customers needs. (See Fig. D-4.)

Several models have been provided throughout the text to enable a quick grasp of the essential concepts, showing how nicely their various aspects are interwoven. Notes and footnotes are dispersed throughout to insure that every term is clearly defined, with further definition given to specific areas of interest.

Next came a review of market planning processes designed to transform technical possibilities into market opportunities, and vice versa, by having management participation. Thus, we all become "manager-planners" of new network services for the information marketplace.

. . . . THE INFORMATION USERS . . . .
. User Needs . . .
. . Services to Meet User Needs . . .
. . . Technology Tools to Provide Services . . .
. . . . Structure to Use Technology Tools . . .
. . . . . Networks Supported by the Structure . . .
. . . . . . Products to Build the Networks . . .
. . . . Applications that Utilize the Products . . .
. . . . Customer Solutions for the Applications . . .
. . . . Challenges to Achieve the Solutions . . .
. . . . Strategies to Match the Challenges . . .
. . . . Plans of Action to Meet the Strategies . . .
. . . . Society Resulting from the Plans . . .

Figure D-4. The global information society.

Finally, based upon this foundation, we have spent time prioritizing activities, covering supporting structures such as standards, pricing, global offerings, and product phasing in order to formulate a broad view, a global perspective of the future and how can we work together to achieve it.

We have covered the applications, basics, and concepts of information networking services in the information society. This will enable us to become formidable leaders positioned to be major competitive players in the information game as we enter the next millennium.[1]

The woods are lovely, dark and deep,
But I have promises to keep,
And miles to go before I sleep,
And miles to go before I sleep.

ROBERT FROST

---

[1]Heldman, *Telecommunications Information Millenium*, McGraw-Hill, 1995.

Glossary

address 1. A location that can be specifically referred to in a software program. 2. The identification of a physically and/or logically distinct entity in a network.

ADM (add-drop multiplexor) A network element that can add and drop standard DSn or SONET signals from a line signal.

analog private-line service A dedicated circuit that transmits information between two or more points. It uses analog transmission signals and is engineered for 300 to 3,000 Hz with a net maximum loss of 16 dB.

ANSI American National Standards Institute. A group, affiliated with ISO, that establishes standards for transmission codes, protocols and high-level languages.

ASCII American Standard Code for Information Interchange. A coded character set of seven bit-coded characters and a parity bit used for data communications. ASCII defines 128 characters.

ASN.1 (Abstract Syntax Notation 1) A standard ISO layer six that provides a flexible method of describing types of data. ASN.1 describes the form of the data, while the meaning of the data is described in layer seven (application layer).

asynchronous Transmission that is not related to a specific frequency, or to timing, of the transmission facility; transmission characterized by individual characters, or bytes, encapsulated with start and stop bits, from which a receiver derives the necessary timing for sampling bits.

asynchronous transfer mode (ATM) A multiplexed information-transfer technique in which information is organized into fixed-length cells and transmitted according to each user's instantaneous need. A high-bandwidth, low-delay, packet-like switching and multiplexing technique. Usable capacity is segmented into fixed-size cells, consisting of header and information fields, allocated to services on demand.

B-DCS (broadband digital cross-connect system) B-DCS is a generic term for an electronic digital cross-connect system capable of cross-connecting signals at or above the DS3 rate.

bisynchronous A method of transmission based on synchronous system clocks and two defined SYNC characters used to synchronize the transmitter and receiver. The SYNC characters (for example, 32, 16, or 96 in hex) are transmitted prior to data. BSC stands for binary synchronous communication and is a character-oriented protocol in half-duplex designed by IBM in 1964. These terms are often used interchangeably.

bit A binary digit; the representation of a signal, wave, or state as either a binary zero or a one.

bit interleaved parity N (BIP-N) A method of error monitoring. With even parity, an *N* bit code is generated by the transmitting equipment over a specified portion of the signal in such a manner that the first bit of the code provides even parity over the first bit of all *N*-bit sequences in the covered portion of the signal; the second bit provides even parity over the second bits of all *N*-bit sequences within the specified portion, etc. Even parity is generated by setting the BIP-N bits so that there are an even number of 1s in each of all *N*-bit sequences, including the BIP-N.

bit-oriented 1. A method of data communication (for example, ADCCP) defined by ANSI standards. 2. A protocol in which each bit may have independent significance and be without octet alignment.

BOC Bell Operating Companies.

bridges A device used to connect LANs.

broadband Today, "broadband" is used more loosely to describe circuits with very fast transmission rates. It is not a precise term because it is not always used consistently to refer to a specific transmission range. Here, terms are used as shown in Fig. G-1.

business voice messaging service (BVMS) Offers subscribers the capability to utilize voice mail functions without purchasing their own hardware and software. Calls are forwarded (when busy, or unanswered) to a central office messaging platform that responds to the caller with a personalized customer-recorded greeting and takes a digitally recorded voice message. Additional features include: group list distribution, future delivery of messages transfer, priority and private message delivery, outcall notification to pagers or telephone numbers and guest mailboxes.

byte Eight bits make a byte. A byte is typically the smallest addressable unit of information in a database or memory.

byte-interleaved multiplex structure A structure in which time division multiplexing is used to send a byte as a unit from one subchannel, interfacing it in successive time slots with bytes from other subchannels.

CCITT Consultative Committee on International Telephone and Telegraph. An international standards organization under the auspices of the United Nations.

central office automatic call distribution (CO-ACD) Offers small to very large customers the ability to have large numbers of incoming calls answered and distributed to available agents on a first-in and first-out basis. In addition, it provides sophisticated management information and reporting capabilities.

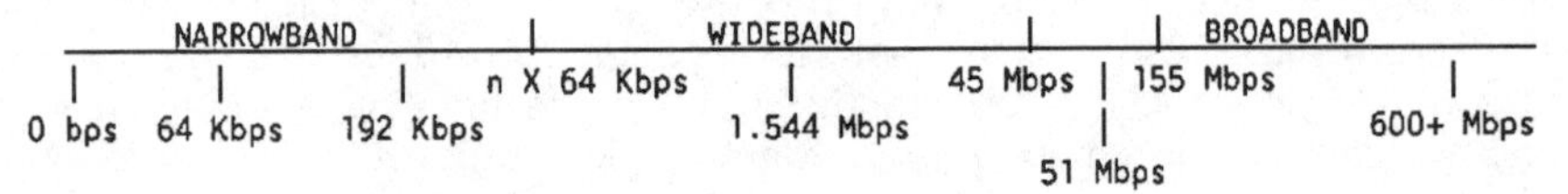

Figure G-1.

Centrex A central office-based service with a flexible combination of features that enhance the capability of the subscriber's telephone. Basic features include call transfer, conference calling, and call hold. Subscribers can tailor their service, choosing from a variety of optional features that can help them manage their communications better. Designed primarily to meet the needs of medium-sized businesses and branch offices.

channel time slots A time slot occupying a specified position in a frame and allocated to a particular time-derived channel.

circuit switching A continuous circuit that is set up on demand, such as when a telephone goes off the hook and is dialed. When the circuit is no longer required, as when a telephone is hung up, the circuit is broken. In contrast, a dedicated circuit is a continuous circuit that permanently connects two terminals, whether it is in use or not. Dial-up services are available over circuit-switched networks.

clock An oscillator-generated signal that provides a timing reference for a transmission link; used to control the timing of functions such as sampling interval, signaling rate, and duration of signal elements.

concatenated synchronous transport signal level N (STS-Nc) An STS-N Line layer signal in which the STS envelope capacities from the N STS-1s have been combined to carry an STS-Nc synchronous payload envelope (SPE), which must not be transported as several separate signals but as a single entity.

concatenated virtual tributary (VTx-NC) A set of virtual tributaries (VTs) in which the VT envelope capacities from N VTx's have been combined to carry a VTx-Nc, which must be transported not as several separate signals but as a single entity. Currently applies only to VT6-1.

CRC Cyclic redundancy check. A multiple bit check-sum calculated by a polynomial used for error detection. It is sent with the transmission and checked after being recalculated by the receiver.

custom calling services Features that are associated with an individual business line. These features are: call waiting, conference calling, speed calling, intracall, and various call forwarding features.

DACS (digital access and cross-connect system) An AT&T electronic digital cross-connect system that allows for centralized management of network functions.

DCC (data communications channel) A channel that carries information between network elements within the SONET overhead. Channels contained within section and line overhead used as embedded operations channels to communicate to each network element.

DCS (digital cross-connect system) A system that is optimized to provide connections of DS1 interfaces. A generic term for an electronic digital cross-connect system.

digital data service (DDS) A dedicated circuit that transmits information between two or more points. It uses digital transmission at speeds of 2.4, 4.8, 9.6, and 56 Kbps (19.2 Kbps—limited availability in 1991). This service provides a higher degree of accuracy and reliability than analog circuits.

direct inward dialing (DID) A special trunking arrangement that permits incoming calls from the exchange network to reach a specific PBX station or other type of device without attendant assistance.

DSO Digital signal, level 0.

DS1 frame structure A DS1 frame consists of a framing bit (S-bit) and 24 octets (bytes), for a total of 193 bits. Each octet represents 1/8,000 second of transmission in one DSO channel. Classical Transparency Scheme: For DSO speech channels, 1 bit in octet is flipped to "1" if all 8 bits are 0. For DSO data channels, 1 bit in each octet is permanently set to "1" at installation time.

DS1 service A dedicated circuit that transmits voice, data, and video between two points. It is capable of accommodating two-way transmission of digital signals at a speed of 1.544 Mbps. The subscriber may multiplex the signal via their own premises equipment.

EBCDIC Extended Binary-Coded Decimal Interchange Code. A set of 8 bit-coded characters, defined by IBM.

Ethernet One of several technologies used to allow workstations on a LAN to communicate at 1 to 10 Mbps. It uses the linear bus topology.

extended superframe structure (ESF) For each channel in which in-slot signaling is used, bit #8 is robbed and used for signaling in every sixth frame. Four consecutive robbed bits can be combined to represent up to 16 possible signal states in that channel. An ESF consists of twenty-four 193-bit frames, or a total of 4,632 bits. Every fourth S bit (called a C bit) is used to pass CRC error checking bits to the receiver. This CRC is based on the payload bits of all 24 frames in the previous ESF. Every fourth S bit (called an F bit) follows a specific sequence. The remaining S bits (called M bits) constitute a 4,000-bps "ESF data link," which may be used for maintenance or may be split between ZBTSI and maintenance. ESF prevents false synchronization due to aliasing. ESF provides a means for testing for logic errors (i.e., data errors) without interrupting data flow. ESF allows the ZBTSI transparency algorithm to be used, if desired. ESF provides a special 2,000- or 4,000-bps channel for system testing and control, apart from the infor payload channels. *2,000 bps if ZBTSI is used.

FDDI (Fiber-Distributed Data Interface) A standard that evolved out of the datacom industry to extend token-ring LANs. FDDI is a 100-Mbps token-ring LAN technology based on multimode fiber media. FDDI is limited to a range of about 120 miles. An American National Standards Institute–specified standard for fiber-optic links with data rates up to 100 Mbps. The standard specifies multimode fiber; 50/125, 62.5/125, or 85/125 core-cladding specification; LED or laser light sources; and 2 kilometers for unrepeated data transmission at 40 Mbps.

fiber-optic, DS1, DS3 service A dedicated facility that transmits voice, data, and video between two points. Fiber-optic technology transmits information in the form of pulses of light through fibers of ultrapure glass. Each application will be delivered to the subscriber at an electrical interface at the speeds of 1.544 Mbps (DS1) or 44.736 Mbps (DS3). Subscribers may multiplex the signal via equipment on their own premises.

foreign exchange service (FX) Uses a voice grade private line to provide a seven-digit telephone number from a central office located in an exchange other than the exchange from which the subscriber would normally be served. (FX is a telephone number extended to a location outside an exchange; FCO is a number extended to a location within an exchange.)

frame A group of bits sent serially over a communications channel; generally, a logical transmission unit sent between data-link-layer entities. A frame contains its own control information for addressing and error checking; the basic data transmission unit employed with bit-oriented protocols, similar to blocks. A logical grouping of data and control information usually used to define a bit-oriented sequence. A cyclic set of consecutive time slots in which the relative position of each time slot can be identified.

frame alignment time slot A time slot occupying the same relative position in every frame and used to transmit the frame alignment signal.

frame level The second or link level access procedure defined by ISO in its seven-layer OSI (Open Systems Interconnection) model.

frame overhead Bits (digits) that are added at regular intervals to a digital signal at the sending end of a digital link and used to provide network functions such as framing, operations, administration, and maintenance.

full-duplex Describes communications in which transmission occurs simultaneously in two directions.

gateway A hardware/software device that permits communications between two similar local or remote area networks. A device used to connect unlike systems. Usually used to connect mainframes or minicomputers to LANs.

half-duplex Describes communications in which transmissions occurs in two directions, but not simultaneously.

HDLC High-level data link control. A bit-oriented ISO protocol.

information payload Interface rate minus all overhead.

input-output (I/O) device A system component used to transfer data between the main storage and other devices.

interface Defines the physical specifications of the link between two devices.

interface rate The gross bit rate of the interface at the boundary between the physical layer and the physical medium.

interface signal A method of physical signaling defined by various specifications such as V.24, V.35, V.36, RS232, and RS449.

ISDN Integrated Services Digital Network. A digital network standard consisting of "bearer" (B) channels to carry voice, data, and image information and a "data" (D) packet-switched signaling channel. ***basic rate interface (BRI)*** An ISDN standard interface for customer station equipment. BRI is specified as two 65-Kbps B channels and one 16-Kbps D channel. Hence, it is referred to as **2B + D.** BRI enables a normal twisted-pair telephone circuit to carry two high-speed information channels and a feature-rich signaling channel to route information requests. ***primary rate interface (PRI)*** An

ISDN standard specifying twenty-three 64-Kbps B channels and one 64-Kbps D channel, or **23B + D.** PRI is compatible with existing T1 lines used extensively throughout North American telecommunications networks.

ISO (International Standards Organization) A worldwide organization standardizing the Open Systems Interconnection (OSI) reference model for computer and data communications.

jitter The slight movement of a transmission signal in time or phase, which can introduce errors and loss of synchronization in high-speed synchronous communication. Phase jitter, in telephony, is the measurement in degrees out of phase that an analog signal deviates from the referenced phase of the main data-carrying signal, often caused by alternating current components in a telecommunications network.

LAN (local area network) A method of reducing communications-related and computer-related expenses by connecting workstations (personal computers) together (a network) that are located near one another (local area) for the purpose of sharing resources (software and hardware) from a centralized computer (server) utilizing a specific topology for connectivity. A general-purpose local communications network that can serve a wide variety of devices, such as personal computers, minicomputers, mainframes, terminals, and other peripherals. LANs may be buses, trees, or rings, and may use twisted pair, coax cable, or fiber optics, with a wide range of data rates.

lightwave multiplexors High-speed multiplexors that support single-mode fiber applications at 565-Mbps and 1.13-Gbps transmission rates.

line A transmission medium, together with the associated equipment, required to provide the means of transporting information between two consecutive network elements (NEs), one that originates the line signal and one that terminates the line signal.

linear bus topology So named because it was one long (linear) cable (bus/ communications trunk) terminated at each end with a terminating resistor (terminator). Workstations (nodes) connect onto the bus via a "T" connector. Nodes send/receive information (data) up/down (both directions) the length of the cable. Each workstation (node) "listens" to the bus for other users (data) before sending data so as to prevent conflict (collision) for the usage of the bus.

LTE (line terminating equipment) Network elements that originate and/or terminate line (OC-N) signals. LTEs originate, access, modify, and/or terminate the transport overhead.

M13 A high-speed digital multiplexor, used with DS3 Service, which provides M13 (multiplex/demultiplex) functionality for DS1, DS1C, and 2.048-Mbps interfaces.

mapping In network operations, the logical association of one set of values such as addresses on one network with quantities or values of another set such as devices on another network, which include name-address mapping, internetwork-route mapping, and protocol-to-protocol mapping.

market expansion line (MEL) Provides subscribers a local identity without having a physical location in an exchange. It does this by enabling subscribers to use a local telephone number and have calls automatically routed to another local or long-distance location.

MD (mediation device) A device that performs mediation functions between network elements and OSs. Potential mediation functions include protocol conversion, concentration of NE-to-OS links, conversion of languages, and message processing.

message Any information-containing data unit, in an ordered format, sent by means of a communications process to a named network entity or interface.

mnemonic codes Operations translated to a symbolic notation to facilitate human recognition and memory. They must be converted back to their original binary bit patterns by a compiler or other device before a computer can interpret them. The symbolic representations—for example, SABM—are commonly called mnemonics.

mnemonic table A table showing the names and mnemonics, their binary equivalents and representation in other codes.

modem A communications device that modulates digital signals at the transmitting end and demodulates them at the receiving end. The name is a neologism that is formed from the words modulate and demodulate.

multiplexing The combining of multiple data channels onto a single transmission medium; any process through which a circuit normally dedicated to a single user can be shared by multiple users; typically, user data streams are interleaved on a bit or byte basis (time division) or separated by different carrier frequencies (frequency division).

multipoint Refers to a line with more than two stations connected.

NE (network element) A single piece of telecommunications equipment used to perform a function or service integral to the underlying bearer network.

network A system of connected devices used for communications.

network layer Layer three in the OSI model; the logical network entity that services the transport layer, responsible for ensuring that data passed to it from the transport layer is routed and delivered through the network.

network topology The physical and logical relationship of nodes in a network; the schematic arrangement of the links and nodes of a network; networks are typically a star, ring, tree, or bus topology, or some hybrid combination.

OC-N (optical carrier-level N) The optical signal that results from an optical conversion of an STS-N signal. N = 1, 3, 9, 12, 18, 24, 36, or 48.

optical carrier The SONET hierarchy is defined by optical carrier line rates, beginning with OC-1 (51.84 Mbps) and going up to OC-48 (2488.32 Mbps). Other OC rates include OC-3 (155.52 Mbps), OC-12 (622.08 Mbps), and OC-24 (1244.16 Mbps).

OS Operation system.

OSI Open Systems Interconnection is an internationally agreed upon model defining layers of communications functions to enable any OSI-compliant computer or terminal to communicate with any other OSI-compliant unit.

overhead In communications, all information (such as control, routing, and error-checking characters) that is in addition to user-transmitted data; includes information that carries network-status or operational instructions, network routing information, and retransmission of user-data messages that are received in error.

packet Short division of a message used in packet switching operations.

packet switching operations Long messages in the network are broken into packets and transmitted rapidly. Using short blocks facilitates error recovery and sending high-priority data.

packet switching A system in which information from one or more sources—all addressed to a common destination—is combined in a "packet." When the packet is "full," a circuit to that address is seized only for the time it takes to send that packet: then the circuit is disengaged until another packet, similarly addressed, has filled up. It is an economical, high-speed method of sending data, such as telecommunications signaling, in bursts. Packet switching is used, for instance, for the D channel of ISDN communications.

path A path at a given rate is a logical connection between the point at which the standard frame format for the signal is disassembled.

path overhead (POH) Overhead assigned to and transported with the payload until the payload is demultiplexed. It is used for functions that are necessary to transport the payload. These functions include parity check and trace capability.

payload Interface rate minus frame overhead. This is the net capability for information transfer provided for the next lower level of the hierarchy.

payload overhead Bits that are assigned at source and remain with the information payload until the payload reaches sink, and are used for functions associated with transporting the payload.

plesiochronous boundaries Networks timed by separate Stratum 1 clocks.

plesiochronous network(s) Networks that derive timing from more than one primary reference source. Network elements accommodate minor frequency differences between nodes.

plesiochronous operation Allows two or more synchronous timing regions to be timed from separate primary reference sources. Pointer processing provides the mechanism to accommodate modest frequency offsets between regions. Integrates into existing synchronization network. Timing may be received from an external timing reference source or extracted from the received optical line signal. Transmission is maintained even under synchronization failure conditions.

point-to-point A circuit that interconnects two points directly, where there are generally no intermediate processing nodes, computers, or branched circuits, although there could be switching facilities; a type of connection such as a phone-line circuit that links two, and only two, logical entities. A line with exactly two stations.

protocol A formal set of rules governing the format, timing, sequencing, and error control of exchanged messages on a data network; may be oriented toward data transfer over an interface, between two logical units directly connected, or on an end-to-end basis between two users over a large and complex network. A set of rules and procedures for establishing and controling conversations on a line.

ring topology Ring topology connects the nodes, including the file server node, in a circular fashion to form a ring. Only one node at a time can pass data to another node; therefore each node must "listen" for its unique address before receiving information.

signaling system 7-SS7 (Also referred to as Common Channel Signaling 7-CCS7.) A packet-switched network that carries both the called and calling number on a path separate from the voice path. The signaling path can be used to route the call. Because it is separate, it can be used to access additional subscriber information at network control points or other databases that contain network intelligence. This makes it possible to create innovative information services beyond simply switching a call to connect two parties.

simultaneous voice data service (SVDS) An end-to-end dedicated digital data service that provides integrated voice and data transport over a single line from the serving central office to end-user premises. SVDS can work with existing business lines, has optional transmission rates of 4.8 Kbps, 9.6 Kbps, and 19.2 Kbps, and is available for point-to-point or multipoint service.

sink A location at which a specified multiplex structure is terminated through connected equipment, removing the frame alignment signal and disassembling the channel time slots.

SL-100*[1] DMS-100 central office-based PBX. (*Trademark of Northern Telecom, Inc.)

SMDS (Switched Multimegabit Digital Service) A 1.544-155 Mbps public data service with IEEE 802.6 standard user interface. It can support Ethernet, Token Ring, and FDDI (OC-3c) LAN-to-LAN connections.

SNA Systems Network Architecture, developed by IBM.

SONET (synchronous optical network) A standard that evolved out of the carrier industry to provide for the interconnection of high-speed networks via single-mode optical fiber. SONET transmission speed starts with a building block of 51.840 Mbps. This rate is called STS1 (synchronous transport signal-level 1). The corresponding optical interface is called an OC1. The STS and OC prefixes are often used interchangeably, as are T1 and DS1. An ANSI interface standard defining a hierarchy for synchronous transmission beginning at a base rate of 51.84 Mbps.

SONET documentation (BELLCORE) TA 253: "SONET Transport Systems: Common Generic Criteria," TR 233: "Wideband and Broadband Digital Cross-Connect Systems Generic Requirements and Objectives," TR 496: "SONET Add-Drop Multiplex Equipment Generic Requirements and Objectives," TA 755: "SONET Fiber Optic Transmission Systems Requirements and Objectives," TA 782: "SONET Digital Switch Interface Generic Requirements," TA 917: "SONET Regenerator Generic Criteria."

---

[1]See AT&T, Bellcore, Fujitsu, NEC, Northern Telecom, Siemens, US West, etc., for further definitions from which this sample was selected.

source A location at which a specified multiplex structure is originated through connected equipment assembling separate channel time slots and a frame-alignment signal.

SPE (synchronous payload envelope) A frame structure composed of path overhead and bandwidth for payload. Payloads are carried within the SPE through use of payload mappings.

star topology To allow communication between nodes, start topology utilizes a centralized device such as a file server, a hub, or wire center (usually a high-performance personal computer) attached to workstations (nodes) by direct cabling that takes on the shape of a star.

STE (section terminating equipment) Network elements that perform section functions such as facility performance monitoring. The section is the portion of a transmission facility between a lightwave terminal and a line repeater or between two line repeaters.

STS synchronous payload envelope (STS SPE) A 125-microsecond frame structure composed of STS path overhead and bandwidth for payload. This term generally refers to STS-1 SPEs and STS-Nc SPEs.

STS-N (synchronous transport signal-level N) A signal obtained by interleaving N STS-1 signals together.

STS-Nc Concatenated synchronous transport signal-level N.

Superframe structure For each channel in which in-slot signaling is used, bit #8 is robbed and used for signaling in every sixth frame. Two consecutive robbed bits (A&B) are combined to represent four possible signal states in that channel.

Switchnet 56® A circuit-switched digital transport service that allows customers—with appropriate equipment—to transmit data or video as 56 Kbps. Transmission can be achieved on an intra-LATA or inter-LATA basis via a carrier over the public switched network.

synchronous Data communications in which characters or bits are sent at a fixed rate, with the transmitting and receiving devices synchronized; eliminates the need for start and stop bits basic to asynchronous transmission, and significantly increases data throughout rates. The essential characteristics of time-scales or signals such that their corresponding significant instances occur at precisely the same average rate.

synchronous network The synchronization of synchronous transmission systems with synchronous payloads to a master (network) clock which can be traced to a reference clock.

synchronous transport signal level 1 (STS-1) The basic logical building block signal with a rate of 51.840 Mbps.

synchronous transport signal level N (STS-N) This signal is obtained by byte interleaving N STS-1 signals together. The rate of the STS-N is *N* times 51.840 Mbps.

syntran Synchronous transmission.

T1 The ANSI Standards Committee responsible for telecommunications.

T1M1 A technical subcommittee to T1 responsible for internetwork OAM&P standards.

T1S1 A technical subcommittee to T1 responsible for standards related to service, architectures, and signaling.

T1X1 A technical subcommittee to T1 responsible for standards pertaining to synchronous interfaces and hierarchical structures relevant to interconnection of network transport signals.

TA (technical advisory) These publications are documents describing Bellcore's preliminary view of proposed generic requirements for products, new technologies, services, or interface.

terminal An input or output device used to send data to or receive data from another device in the system. A device using the services of a network, such as an X.25 network.

terminology/functionality

NARROWBAND	WIDEBAND	BROADBAND
narrowband ISDN	DS1/DS3	broadband ISDN (B-ISDN)
n-ISDN BRI	"N" × 64 Kbps	SONET
basic rate interface	1.544 - 45 Mbps	DS3/OC-X
64 Kbps - 128 Kbps	wideband ISDN	45 Mbps and above
class V switching	primary rate interface (PRI)	asynchronous transfer mode (ATM)
SS7 interoperability	frame relay	cell relay
DSO	SMDS (1.5 - 45 Mbps)	SMDS (155 Mbps)
	Ethernet	FDDI-II (100 Mbps)
	Token Ring	FDDI
		802.6 Standards

time slot Any cyclic time interval that can be recognized and defined uniquely.

TL1 (transaction language 1) A machine-to-machine communications language that is a subset of CCITT's man machine language.

token A right to transmit data.

token ring One of several technologies used to allow workstations on a LAN to communicate. It was the Star Ring Topology developed by IBM. Its uniqueness is the passing of a "token" around the ring for the purpose of requesting the use of the ring (ring topology) while utilizing the physical layout of the star topology. Its advantage is connectivity.

topology The physical layout or shape used to connect a file server (usually a high-performance computer) and workstations containing a network interface card (NIC).

TR (technical requirement) These publications are the standard form of Bellcore-created technical documents representing Bellcore's view of proposed generic requirements for products, new technologies, services, or interfaces.

traffic data reports service (TDRS) Provides subscribers a printed summary of traffic data on certain network facilities such as individual access lines, multiline hunt groups, trunk groups, CENTRON® features, and network access registers, etc. These reports are available on a one-week, one-month, or ongoing basis.

transport Facilities associated with the carriage of OC-1 or higher signals.

transport overhead The overhead added to the STS SPE for transport purposes. Transport overhead consists of line and section overhead.

TSI Time slot interchange.

two-way WATS Combines the features of Outward WATS with 800 service capabilities over a single dedicated intrastate access line. Combining services provides better rates and reduces the number of access lines needed on a phone system. Incoming calls (800) are at no cost to the caller.

UIS (universal information services) A goal that, when realized, will provide customers everywhere with access to any kind of voice, data, or image service, in any place, at any time, in any combination, with maximum convenience and economy.

UOS (universal operations services) A software application that supports UIS by providing traditional network operations functions.

Video Codec Supports the connection of the North American video hierarchy to 45-Mbps and 90-Mbps transmission links. (Rockwell).

virtual tributary (VT) A container for carrying signals below 50 Mpbs. VTs allow the easy identification of a transported signal such as a DS1 from the SONET 50-Mpbs frame. There are two modes of virtual tributaries: floating and locked. Floating VTs have pointers that enable cross-connecting VTs within the SONET signal with minimal delay. Locked VTs contain no pointers. This method was optimized for DSO transport. The frequency is locked to the SONET 50-Mpbs frame to allow easy DSO identification because it requires no pointer processing. A difficulty with local VTs, though, is that a VT cross-connect can only be accomplished by adding delay to the signal. Because VT cross-connects are thought to be important in management of the network, most people have opted to use the DSO-optimized floating VT mode, called the *byte synchronous floating VT*, for efficient DSO transport, which also allows VT cross-connects with minimal delay. In fact, Bellcore has specified this mode to be mandatory in the generic digital loop carrier-to-switch interface specified in TR303. A structure designed for the transport and switching of sub-STS-1 payloads. There are currently four sizes of VTs: VT1.5, VT2, VT3, and VT6.

VT (virtual tributary) A structure designed for transport and switching of sub-DS3 payloads.

VT envelope capacity Bandwidth within and aligned to the VT superframe that is available for the VT synchronous payload envelope.

VT group A 9-row-by-12-column structure (108 bytes) that carries one or more VTs of the same size. Seven VT groups (756 bytes) are byte interleaved within the VT-organized SPE.

VT synchronous payload enveloped (VT SPE) A 500 microsecond superframe structure carried by the VT composed of VT path overhead (POH) and bandwidth for payload. The envelope is contained within, and can have alignment with respect to, the VT envelope capacity.

VTx A VT size "x" (currently x = 1.5, 2, 3, or 6).

W-DCS (wideband digital cross-connect system) An electronic digital cross-connect system capable of cross-connecting signals below the DS3 rate.

WAN (wide area network) A method of reducing communications-related and computer-related expenses by connecting workstations (personal computers) or LANs together (a network) that are located near one another (local area) for the purpose of sharing resources (software and hardware) from a centralized computer (server) utilizing a specific topology for connectivity.

WATS Outward WATS service is provided over a dedicated access line that enables the subscriber to make outgoing long distance calls at a price that is less expensive than ordinary long distance calling.

X.25 A set of packet-switching standards published by the CCITT in 1976 and updated in 1980 and 1984, specifying the physical interface and message protocol between a network element or OS and a packet-switched data.

800 service Provided over a dedicated access line that enables a subscriber to receive and pay for incoming long distance calls.

800 serviceline™ Enables intra-LATA 800 calls to terminate on a regular exchange line. Subscribers can use custom calling features to enhance the efficiency of the line. Calling party does not pay for the call, and message detail is offered at no additional charge.

Index

A

B

C

D

E

F

R

S

T

U

V

W

X

Z

ABOUT THE AUTHOR

Peter K. Heldman is president of Telecommunications Planning Institute, a Denver-based think tank and consulting firm that helps businesses and state governments establish a future direction in the emerging telecommunications marketplace As the architect of South Dakota's future telecommunications infrastructure, he developed a comprehensive Vision & Plan for the Future and created a realistic program to successfully achieve its objectives. He also was among a few chosen specialists to offer input to the FCC on the impact of the Telecommunications Act of 1996 on rural America and on rural education. Previously, Mr. Heldman was Senior Manager of Strategic Opportunities for AT&T Network Systems and Bell Laboratories, and was a leader in defining future wideband and broadband network requirements. In addition, he has provided research and input for six major telecommunications and management planning books, which have been personally endorsed by the presidents and CEOs of the world's leading telecommunications firms.